# Ionic Liquids IIIB: Fundamentals, Progress, Challenges, and Opportunities

ACS SYMPOSIUM SERIES **902**

# Ionic Liquids IIIB: Fundamentals, Progress, Challenges, and Opportunities

## Transformations and Processes

**Robin D. Rogers,** Editor
*The University of Alabama*

**Kenneth R. Seddon,** Editor
*The Queen's University of Belfast*

**Sponsored by the ACS Divisions**
**Industrial and Engineering Chemistry, Inc.**
**Environmental Chemistry, Inc.**
**Petroleum Chemistry, Inc.**

American Chemical Society, Washington, DC

**Library of Congress Cataloging-in-Publication Data**

Ionic Liquids III : fundamentals, progress, challenges, and opportunities / Robin D. Rogers, editor, Kenneth R. Seddon, editor ; sponsored by the ACS Divisions of Industrial and Engineering Chemistry, Inc., Environmental Chemistry, Inc., Petroleum Chemistry, Inc.

    p. cm.—(ACS symposium series ; 901–902)

    Includes bibliographical references and index.

    Contents: A. Properties and structure — B. Transformations and processes.

    ISBN 0–8412–3893–6 (v. A : alk. paper) — ISBN 0–8412–3894–4 (v. B : alk. paper)

    1. Ionic structure—Congresses. 2. Ionic solutions—Congresses. 3. Solution (Chemistry)—Congresses.

    I. Title: Ionic liquids 3. II. Rogers, Robin D.. III. Seddon, Kenneth R., 1950- IV. American Chemical Society. Division of Industrial and Engineering Chemistry, Inc. V. American Chemical Society. Division of Environmental Chemistry, Inc. VI. American Chemical Society. Division of Petroleum Chemistry, Inc. VII. Series.

QD540.I58    2005
541′.34—dc22                          2004063111

The paper used in this publication meets the minimum requirements of American National Standard for Information Sciences—Permanence of Paper for Printed Library Materials, ANSI Z39.48–1984.

PRINTED IN THE UNITED STATES OF AMERICA

# Foreword

The ACS Symposium Series was first published in 1974 to provide a mechanism for publishing symposia quickly in book form. The purpose of the series is to publish timely, comprehensive books developed from ACS sponsored symposia based on current scientific research. Occasionally, books are developed from symposia sponsored by other organizations when the topic is of keen interest to the chemistry audience.

Before agreeing to publish a book, the proposed table of contents is reviewed for appropriate and comprehensive coverage and for interest to the audience. Some papers may be excluded to better focus the book; others may be added to provide comprehensiveness. When appropriate, overview or introductory chapters are added. Drafts of chapters are peer-reviewed prior to final acceptance or rejection, and manuscripts are prepared in camera-ready format.

As a rule, only original research papers and original review papers are included in the volumes. Verbatim reproductions of previously published papers are not accepted.

**ACS Books Department**

## In Memoriam

This volume is dedicated, with deepest respect, to the memory of Professor Robert Osteryoung, the grandfather of ionic liquids, who died of amyotrophic lateral sclerosis at the age of 77, in August 2004. His insight and inspiration will be greatly missed by the entire ionic liquids community.

# Contents

## Separations

ix

# Applications

# Fundamentals

# Reactions

# Indices

# Preface

The chapters in these two books (*Ionic Liquids IIIA: Fundamentals, Progress, Challenges, and Opportunities: Properties and Structure* and *Ionic Liquids IIIB: Fundamentals, Progress, Challenges, and Opportunities: Transformations and Processes*) are based on papers that were presented at the symposium *Ionic Liquids: Fundamentals, Progress, Challenges, and Opportunities* at the 226th American Chemical Society (ACS) National Meeting held September 7–11, 2003 in New York. This was the third and last of a trilogy of meetings, the second of which was *Ionic Liquids as Green Solvents: Progress and Prospects* at the 224th ACS National Meeting held August 18–22, 2002 in Boston, Massachusetts, which followed, by eighteen months, the first successful ionic liquids symposium *Ionic Liquids: Industrial Applications for Green Chemistry* at the San Diego, California ACS meeting in April 2001.

The success of the New York meeting can be judged by the simple fact that we need to publish two books to include the key papers presented, in a year in which we anticipate 1000 papers concerning ionic liquids will be published. The talks showed the depth of research currently being undertaken; the broad and diverse base for activities; and the excitement and potential opportunities that exist, and are continuing to emerge, in the field. The new industrial applications raise an especial level of interest and excitement.

The New York meeting comprised ten half-day sessions that broadly reflected the areas of development and interest in ionic liquids. We are indebted to the session organizers who each planned and developed a half-day session, invited the speakers, and presided over the session. The featured topics and the presiding organizers for each session were Ionic Liquid Tutorials (J. F. Brennecke, J. D. Holbrey, R. D. Rogers, K. R. Seddon, and T. Welton), Fuels and Applications (J. H. Davis, Jr.) Physical and Thermodynamic Properties (L. P. N. Rebelo), Catalysis and Synthesis (P. Wasserscheid), Spectroscopy (S. Pandey), Separations (W. Tumas), Novel Applications (R. A. Mantz and P. C. Trulove), Catalytic Polymers and Gels (H. Ohno), Electrochemistry (D. R. McFarlane), Inorganic and Materials (M. Deetlefs and J. Holbrey), and General Contributions (R. D. Rogers and K. R. Seddon). Page restrictions prevented the publication of all the presentations (despite having the luxury of two volumes); however, we tried to select a representative subset of the papers.

Among the wonderful chemical contributions, there was one chastening note; a minute's silence was held at 8:46 on the morning of September 11th to remember those thousands of innocent people who had died exactly two years earlier very close to the Javits Conference Center in which we were holding the meeting. It was hard to escape the emotional impact of those events on that day.

The symposium was successful because of the invaluable support it received from industry, academia, government, and our professional society. Industrial support was received from Cytec Industries, Fluka, Merck KGaA/EMD Chemicals, SACHEM, Solvent Innovation, and Strem Chemicals. Academic contributions were received from The University of Alabama Center for Green Manufacturing and The Queen's University Ionic Liquid Laboratory (QUILL). The U.S. Environmental Protection Agency's Green Chemistry Program also supported the meeting. Of course, we are (as always) indebted to the ACS and its many programs for their help, encouragement, and support. We especially thank the ACS Division of Industrial and Engineering Chemistry, Inc., the I&EC Separations Science & Technology and Green Chemistry & Engineering Subdivisions, and the Green Chemistry Institute.

Another measure of success was the impressive strength of the student c ontributions, i n t he t utorial, o ral, and poster sessions; on this basis, our future is in safe hands. And, as is definitely true in a burgeoning field, the future is always more exciting than the past!!

**Robin D. Rogers**
Center for Green Manufacturing
Box 870336
The University of Alabama
Tuscaloosa, AL 35487
Telephone: +1 205–348–4323
Fax: +1 205–348–0823
Email: RDRogers@bama.ua.edu
URL: http://bama.ua.edu/~rdrogers/

**Kenneth R. Seddon**
QUILL Research Centre
The Queen's University of Belfast
Stranmillis Road
Belfast, Northern Ireland BT9 5AG
United Kingdom
Telephone: +44 28 90975420
Fax: +44 28 90665297
Email: k.seddon@qub.ac.uk
URL: http://quill.qub.ac.uk/

# Contents (SS 901)

## Structure and Spectroscopy

## Theory and Modeling

## Physical Properties

## Phase Equilibria

# Separations

# Chapter 1

# The Road to Partition

# Mechanisms of Metal Ion Transfer into Ionic Liquids and Their Implications for the Application of Ionic Liquids as Extraction Solvents

**Mark L. Dietz, Julie A. Dzielawa, Mark P. Jensen, James V. Beitz, and Marian Borkowski**

Chemistry Division, Argonne National Laboratory, Argonne, IL 60439

The rational design of ionic liquid-based metal ion separation systems requires either that the ion transfer properties of these systems be predictable from the known behavior of conventional organic solvents or that the mechanism(s) of ion transfer be understood at a fundamental level. With this in mind, we have examined the transfer of selected metal ions from acidic aqueous media into a series of *N,N'*-dialkylimidazolium-based room-temperature ionic liquids (RTILs) in the presence of crown ethers, neutral organophosphorus extractants, or β-diketones. The results obtained indicate that although certain aspects of metal ion extraction into RTILs parallel the behavior of conventional solvents, ionic liquids frequently play a more active role in the partitioning process, with ion exchange involving the cationic or anionic constituents of the ionic liquid comprising an important mode of ion transfer. The implications of this observation for the application of ILs as extraction solvents are described.

# Introduction

Growing attention has recently been directed at the application of ionic liquids (ILs) in chemical separations *(1-17)*, most commonly as replacements for the organic diluents employed in traditional liquid-liquid (L-L) or membrane-based separations of organic solutes or metal ions. In many cases, the use of ionic liquids has been regarded as a key element in the design of more environmentally benign methods for effecting these separations. These solvents exhibit several properties that make them attractive as a potential basis for "greener" separation processes, among them negligible vapor pressure, a wide liquid range, and good thermal stability *(18, 19)*. In addition, they display an extraordinary degree of tunability. That is, considerable structural variation is possible in both the cationic and anionic moieties comprising an ionic liquid, and even minor changes in either the cation or anion structure can produce significant changes in the physicochemical properties of the ionic liquid *(20)*. This tunability, while obviously offering vast opportunities for the design of IL-based separation systems, also poses a formidable challenge to the separation scientist. That is, in the absence of predictive capabilities to guide the choice of solvent appropriate to a given separation problem, the selection of the "right" ionic liquid becomes a trial and error process, a clearly undesirable situation when so many candidates are available for consideration.

Recent research in this laboratory has concerned the application of ionic liquids in the liquid-liquid extraction of metal ions. With the need for information to guide the rational design of IL-based separations in mind, we have sought to develop a fundamental understanding of the process(es) involved in the transfer of a metal ion from an aqueous phase into an ionic liquid in the presence of various types of extractants. In addition, we have sought to draw parallels with and to elucidate differences between the behavior of ionic liquids and that of conventional diluents as extraction solvents. In this way, the relationship between various structural features of ionic liquids and their performance and utility as extraction solvents can be defined, and the advantages, if any, of ionic liquids over conventional diluents in this application can be delineated. To date, we have examined three systems: the extraction of strontium ion from acidic nitrate media by dicyclohexano-18-crown-6 (DCH18C6) *(14, 21)*; the extraction of uranyl ion, also from acidic nitrate media, by octyl(phenyl)-$N,N$-diisobutylcarbamoylmethylphosphine oxide (CMPO, shown below) / tri-$n$-butylphosphate (TBP) mixtures *(22)*; and finally, the extraction of europium ion (a representative trivalent lanthanide) from perchlorate media by 2-thenoyltrifluoroacetone (HTTA) *(23)*. In each of these cases, various water-immiscible $N,N'$-dialkyl-imidazolium-based RTILs were employed as a substitute for a conventional organic diluent. In this report, an overview of the results obtained for the three systems is provided, along with a brief discussion of their implications for the use of RTILs as extraction solvents.

# Mechanisms of Metal Ion Partitioning

## The Strontium-DCH18C6 System

For a number of reasons, a study of the extraction of strontium ion by DCH18C6 provides a good starting point for efforts to elucidate the fundamental aspects of metal ion transfer into room-temperature ionic liquids. First, of the IL-based extraction systems considered to date *(3-5, 11-12, 24)*, cation extraction by crown ethers has probably received the most attention. In addition, the Sr-DCH18C6 system is chemically simple; strontium ion exists in only a single oxidation state and although DCH18C6 exists in five stereoisomeric forms, the individual isomers are readily isolated *(25)* or synthesized *(26-28)*. Finally, the mechanism of strontium extraction into conventional molecular diluents by DCH18C6 is well understood, thus providing a basis for comparison of the results obtained in an ionic liquid.

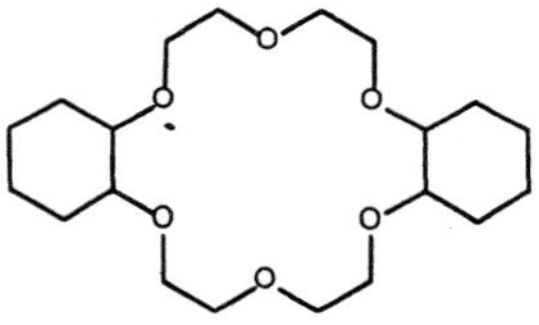

DCH18C6

Prior work with crown ethers has shown that among classical molecular solvents, chlorinated hydrocarbons typically yield the most efficient metal ion extraction *(29-31)*. These solvents are not acceptable for large-scale application, however, a result of their toxicity and potential for negative environmental impact. Paraffinic hydrocarbons, in contrast, have long been considered the solvent of choice for process-scale applications *(32)*, but solutions of crown ethers in them yield very poor metal ion extraction *(29)*. For this reason, efforts to devise improved crown ether-based extraction systems have focused on the use of oxygenated, aliphatic solvents such as 1-octanol *(33, 34)*. This solvent, unlike many other conventional diluents, possesses a number of desirable physicochemical properties, most notably, the ability to dissolve significant quantities of water. This high solvent water content facilitates transfer of incompletely dehydrated anions and cationic metal-crown ether ($M-CE^{n+}$) complexes, thus improving the efficiency of extraction *(35)*.

The extraction of strontium ion by DCH18C6 into 1-octanol from acidic, nitrate-containing aqueous phases has been shown to proceed via extraction of a strontium nitrate crown ether complex, as shown here:

$$Sr^{2+} + 2\,NO_3^- + DCH18C6_{org} \rightarrow Sr(NO_3)_2(DCH18C6)_{org}$$

As would be expected from this equation, increasing nitrate (*i.e.*, nitric acid) concentration is accompanied by an increase in strontium partitioning into 1-octanol. In contrast, for $N,N'$–dialkylimidazolium bis[(trifluoromethyl) sulfonyl]imides, such as the 1-ethyl-3-methyl compound ($C_2mim^+Tf_2N^-$), rising acidity is accompanied by a *decrease* in strontium partitioning. Also, despite considerable differences in the hydration energies of nitrate, chloride, and sulfate anion *(35)*, essentially the same acid dependency is observed when either hydrochloric or sulfuric acid solutions are employed as the aqueous phase, as shown in Figure 1 for $C_2mim+Tf_2N^-$. In contrast, in 1-octanol, the difference in $D_{Sr}$ values observed for acidic nitrate and chloride aqueous phases is nearly a factor of 100 *(35)*. Finally, unlike 1-octanol and other oxygenated aliphatic solvents, for the ILs, rising $[H_2O]_{org}$ is not necessarily accompanied by increasing strontium partitioning. These results indicate that the process involved in the partitioning of strontium ion between nitric acid and these ionic liquids is fundamentally different than for 1-octanol, and that this process does not involve extraction of a strontium nitrato-crown ether complex.

Subsequent extended X-ray absorption fine structure (EXAFS) measurements support this view *(21)*. Specifically, comparison of the EXAFS results obtained for solid $Sr(NO_3)_2(18C6)$, whose crystal structure has previously shown it to comprise a strontium cation coordinated by six ether oxygen atoms and by four oxygen atoms from axially coordinated nitrate ions *(36)*, to those obtained for the strontium-DCH18C6 complex extracted into 1-octanol show that the strontium coordination environment in the two systems is identical. This same coordination environment is observed when the solid complex is simply dissolved directly in water-saturated $C_5mim+Tf_2N^-$. Thus, nitrate forms only *inner-sphere* complexes with $Sr \cdot CE^{2+}$ cations in either a two-phase $H_2O/1-OAlc$ system or a single-phase water-saturated RTIL. If, however, the strontium complex is prepared by contacting an aqueous solution of $Sr(NO_3)_2$ with a solution of DCH18C6 in $C_5min^+Tf_2N^-$, the peak associated with the distal oxygen atoms nearly disappears. Also, the number of coordinated oxygen atoms drops by two, consistent with a structure in which a pair of water molecules now occupies the sites that had been occupied by nitrate ion.

In principle, it is possible that nitrate could still be present, but as an outer-sphere ligand this time. To determine if this is the case, measurements of the amount of nitrate transferred into the RTIL upon strontium extraction were made via $^{15}N$-NMR and ion chromatography. In both cases, only a few percent of the nitrate required to maintain charge balance is extracted. Thus, extraction of a strontium nitrate complex cannot be occurring to any appreciable extent. Instead, strontium partitioning into these RTILs, in contrast to conventional organic solvents, must occur *via* ion exchange, in particular a mechanism involving exchange of the cationic $Sr \cdot CE^{2+}$ complex for the cationic constituent

6

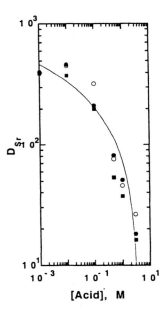

Figure 1. Dependency of $D_{Sr}$ on nitric (filled circles), hydrochloric (open circles), or sulfuric (filled squares) acid concentration for DCH18C6 (0.10 M) in $C_2mim^+$ $Tf_2N^-$. (T = 23 °C).

of the IL. This scheme is consistent with a couple of other observations. First, for this mode of extraction, it would be expected that as the alkyl chain on the IL cation becomes longer, strontium partitioning would decrease, since the transfer of an increasingly hydrophobic species into the aqueous phase is required. In fact, just such a decrease is observed (Figure 2). Also, it would be expected that the extraction of macro quantities of strontium would lead to increasing solubilization of the IL (in particular, the $C_n min^+$ cation) in the aqueous phase. This too is observed *(14)*.

These results raise a question whose answer has obvious and significant implications for the use of ILs as extraction solvents: Is ion-exchange a peculiarity of the Sr-DCH18C6 system, or is it a broad, defining characteristic of metal ion transfer into ionic liquids? In an effort to answer this, an analogous study of the extraction of uranyl ion by CMPO was carried out.

**The Uranyl-CMPO system**

CMPO

In this study *(22)*, two different ionic liquids, $C_4 min^+ PF_6^-$ and $C_8 min^+ Tf_2 N^-$, were employed and the extraction of uranyl ion into each compared to that observed using the classic TRUEX process diluent, dodecane modified by addition of tri-*n*-butylphosphate. (To keep the conditions in all systems as similar as possible, TBP was also added to the two RTILs.) The interaction of TBP with the uranium, it should be noted, is negligible, despite the not insignificant (1M) concentration of it present, as the phosphine oxide group of CMPO is a far stronger Lewis base than is the phosphate oxygen of TBP. Prior work on the TRUEX process has established, in fact, that the extraction of uranium from acidic nitrate media by CMPO proceeds as follows:

$$UO_2^{2+} + 2NO_3^- + 2CMPO_{org} \leftrightarrow UO_2(NO_3)_2(CMPO)_{2\ org}$$

As can be seen, the extracted complex is neutral and incorporates no TBP.

For the IL/aqueous systems, $UO_2^{2+}$ partitioning is consistently greater than for the conventional dodecane/aqueous system, suggesting the possibility of the

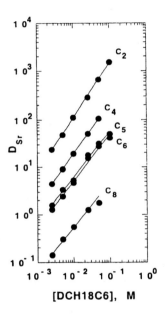

Figure 2.  Effect of alkyl chain length ($C_n$) on the extractant dependency of $D_{Sr}$ for DCH18C6 in 1-alkyl-3-methylimidazolium bis[(trifluoromethyl) sulfonyl] imides. ([HNO$_3$] = 0.10 M; T = 23 °C).

formation of a different complex in the ionic liquid. Several other experimental results support this conclusion. The UV-visible absorption spectrum of the complex in $C_4min^+PF_6^-$, for example, exhibits changes in the relative intensity of peaks in the $UO_2^{2+}$ fingerprint region and a red shift in the longer wavelength peaks *vs.* the spectrum in dodecane. Also, EXAFS studies of the complex in dodecane indicate that it is a hexagonal bipyramid comprising two bidentate nitrate anions and two monodentate CMPO molecules coordinated (through the P=O group) equatorially. For the IL systems, however, the number of equatorially coordinated oxygen atoms falls from 6 to 4. That the results are essentially the same for the two ILs indicates that neither anion is present in the inner sphere of $UO_2^{2+}$ in these systems. Because of the relatively high water content of the two ILs, it is difficult to determine unambiguously the denticity of either the nitrate anion or the CMPO molecule. Each may be either monodentate or bidentate, meaning that anywhere from 0 to 3 molecules of water may also be coordinated.

What is not ambiguous is the stoichiometry of the extracted complex, as determined by measurements of uranyl partitioning between the ILs and the aqueous phase as a function of extractant and acid concentrations. That is, log-log plots of $D_U$ *vs.* [CMPO] at constant acidity and $D_U$ *vs.* [HNO$_3$] at constant CMPO concentration yield straight lines of slope 1, indicating that the predominant extracted species is $UO_2(NO_3)(CMPO)^+$. (The complex extracted into dodecane is, as noted above, the neutral 1:2:2 species.) Accompanying this extraction is the appearance of the IL cation in the aqueous phase in a one-for-one ratio. Thus, as was the case in the extraction of strontium by DCH18C6, uranyl extraction by CMPO into the ILs involves the exchange of a cationic metal-ligand complex for the cationic constituent of the ionic liquid. This result suggests that the ion-exchange mechanism may indeed be a general feature of metal ion-neutral extractant systems in ionic liquids.

## The Europium(III)-HTTA System

As a final test of the prevalence of the ion-exchange mechanism, we have recently undertaken a series of studies of the extraction of trivalent lanthanides by β-diketones in various RTILs *(23)*. In the first of these studies, the extraction of $Eu^{3+}$ from perchlorate media by HTTA has been examined. This system is an

HTTA

excellent test system for at least two reasons. First, the complexes formed are amenable to study by fluorescence and EXAFS techniques. More importantly, depending on the experimental conditions, a variety of europium-TTA complexes can be formed. In a conventional, non-coordinating solvent, for example, lanthanides are known *(37, 38)* to form neutral, hydrated (n=1-3) 1:3 complexes with HTTA, as shown here:

$$Ln(H_2O)_n^{3+} + 3\ Htta_{org} \longrightarrow Ln(tta)_3(H_2O)_{n-6,\ org} + 3\ H^+ + 6\ H_2O.$$

If the solvent has some coordinating ability, some or all of the coordinated waters may be replaced by solvent molecules *(39)*. The same thing can happen in the presence of a neutral synergist in a non-coordinating solvent *(40-42)* or even a high concentration of HTTA, which can generate self-adducts *(43)*:

$$Ln(H_2O)_n^{3+} + 4\ Htta_{org} \longrightarrow Ln(tta)_3(Htta)_{org} + 3\ H^+ + n\ H_2O.$$

In addition to neutral complexes, extraction of lanthanides by HTTA can yield cationic complexes under the appropriate conditions *(44)*. Extraction from perchlorate media into 1,2 DCE with HTTA/CE mixtures, for example, proceeds as shown here *(45)*, with the lanthanide forming a cationic complex, which ion-pairs with $ClO_4^-$.

$$Ln(H_2O)_n^{3+} + ClO_4^- + 2\ Htta_{org} + CE_{org} \longrightarrow$$
$$Ln(tta)_2(CE)(H_2O)_{n-m}^+ ClO_4^-{}_{org} + 2\ H^+ + m\ H_2O$$

Anionic lanthanide/TTA complexes are also known. Lanthanide extraction from any of a variety of aqueous media ($NO_3^-$, $Cl^-$, or $ClO_4^-$) in the presence of a quarternary amine ($Q^+$) or a protonated tertiary amine, for example, involves a reaction in which an anionic 1:4 Ln:TTA complex is extracted *(46, 47)*:

$$Ln(H_2O)_n^{3+} + 4\ Htta_{org} + Q^+X^-{}_{org} \longrightarrow Q^+Ln(tta)_4^-{}_{org} + 4\ H^+ + X^- + n\ H_2O.$$

Given the resemblance of certain ionic liquids to conventional liquid anion exchangers, noted in a prior report *(48)*, it is not unreasonable to expect that the IL system might favor extraction of an anionic Eu/TTA complex. Figure 3 depicts the acid dependency of europium extraction by HTTA in $C_4mim^+Tf_2N^-$ and the extractant dependency of $D_{Eu}$ for the same IL at constant [$HClO_4$]. The slopes of these plots indicate that 4 molecules of HTTA are participating in the extraction process and that 4 protons are released to the aqueous phase for each $Eu^{3+}$ extracted, a result consistent with extraction of an anionic 1:4 ($Eu(TTA)_4^-$) complex. This stoichiometry is supported by UV-visible spectroscopic studies of the extraction of macro neodymium under the same conditions, which

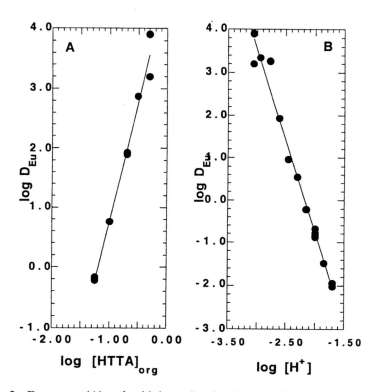

Figure 3. Extractant (A) and acid dependencies (B) of radiotracer Eu extraction. (A) Extraction from 1.0 M NaClO$_4$/0.001 M HClO$_4$ into C$_4$mim$^+$Tf$_2$N$^-$ solutions containing HTTA. (B) Extraction from 1.0 M (H,Na)ClO$_4$ into 0.50 M HTTA in C$_4$mim$^+$Tf$_2$N$^-$.

demonstrate that the organic phase lanthanide ions are present as $Ln(TTA)_4^-$. Specifically, comparison of the UV-visible spectrum of a 1:3 Nd-TTA complex in $o$-xylene, the Nd-TTA complex extracted into $C_4min^+Tf_2N^-$, and the 1:4 Nd-TTA complex formed when Nd is extracted into a solution of Aliquat 336 in $o$-xylene shows that the latter two spectra strongly resemble one another, both in terms of peak position and relative intensities, thus suggesting 1:4 complex extraction in the IL system. The observed absorbance is slightly lower in the IL system, raising the possibility that some of the Nd is present as another, less strongly absorbing complex. Fluorescence measurements, however, indicate the presence of only a single species in the Eu-HTTA-IL system. Curiously, these measurements show that this species (*i.e.*, the extracted complex) contains no coordinated water, despite the presence of significant concentrations of water in the IL phase. The absence of coordinated water is, however, consistent with the fully dehydrated 1:4 complexes observed in the solid state.

To gain additional insight into the structure of the extracted complex in the ionic liquid, EXAFS studies were performed, in this case, employing as a standard solid 1,4-dimethylpyridinium tetratris(thenoyltrifluoroacetonate) europate (III), whose crystal structure is known *(49)* to contain 8-coordinate $Eu(TTA)_4^-$. The results for the extracted metal complex are consistent with $Eu(TTA)_4^-$, both in terms of coordination number and bond distances. Taken together with the results of the partitioning, UV-visible, and fluorescence measurements, the EXAFS data clearly demonstrate that in the IL system, europium partitioning involves extraction of the anionic 1:4 complex. Since additional extraction studies have ruled out the possibility of extraction of either $Na^+$ and $H_3O^+$, this partitioning must involve anion-exchange, whereby $Eu(TTA)_4^-$ is exchanged for the anionic constituent of the IL, here $Tf_2N^-$.

### Influence of IL Cation Hydrophobicity on the Mechanism of Ion Transfer

In each of three quite different chemical systems ($Sr^{2+}/DCH18C6$, $UO_2^{2+}/CMPO-TBP$, and $Eu^{3+}/HTTA$), the ionic liquid, in contrast to conventional diluents, plays an active role in the extraction process, with exchange of either the cationic or anionic constituent of the IL representing the preferred mode of metal ion transfer. Although this is interesting from the perspective of fundamental chemistry, from a practical point of view, it is a troubling observation, as its implications for the "greenness" of ILs are not positive. Recent work in this laboratory, however, offers reason for hope *(50)*.

As noted above in the discussion of the extraction of strontium by DCH18C6 in various $N,N'$-dialkylimidazolium bis[(trifluoromethyl)sulfonyl] imides, a mechanism involving exchange of the cationic metal-CE complex for the cationic constituent of the ionic liquid requires that as the hydrophobicity of the IL cation is increased, cation exchange should become increasingly difficult, and that this increased difficulty should be reflected in a decrease in $D_{Sr}$ values under a given set of conditions. This is, in fact, observed (Figure 2). This raises a question: Can the IL cation be rendered so hydrophobic that ion exchange is no longer a viable pathway for strontium partitioning, thus favoring extraction of a strontium nitrato-crown complex, as is observed in conventional solvent systems? To explore this possibility, we have recently employed a combination of extraction, chromatographic, and EXAFS measurements to gain some insight into the influence of IL cation hydrophobicity on the partitioning process.

Table 1 summarizes the results of radiometric and ion chromatographic measurements of the partitioning of strontium and nitrate ions between water and a series of $C_n mim^+Tf_2N^-$ ionic liquids in the presence of DCH18C6. For the n-pentyl compound, the amount of $NO_3^-$ extracted is far less than the amount of $Sr^{2+}$ extracted, and is thus, not sufficient to produce a neutral strontium nitrato-crown ether complex. As the alkyl chain is lengthened, however, nitrate co-extraction becomes increasingly significant, suggesting a shift from ion exchange to extraction of the neutral complex. For $n = 10$, in fact, the amount of nitrate extracted is exactly that required if partitioning occurred solely by extraction of the neutral complex.

Additional evidence of a shift in the mode of partitioning is presented in Figure 4, which compares the dependency of $D_{Sr}$ on nitric acid concentration for solutions of DCH18C6 in 1-octanol, $C_5 mim^+Tf_2N^-$, and its $C_{10}$ analog. For $C_5 mim^+Tf_2N^-$, increasing acidity is accompanied by a decrease in strontium extraction, which as has been noted previously for $C_2 min^+Tf_2N^-$, is inconsistent with strontium nitrato-crown ether complex partitioning. In contrast, for the $C_{10}$ IL, increasing aqueous acidity (hence, nitrate concentration) yields an increase in $D_{Sr}$. This implies that, just as is the case for 1-octanol, the mechanism of strontium ion partitioning into $C_{10}mim^+Tf_2N^-$, must involve extraction of the neutral strontium nitrato-crown ether complex. Thus, just as relatively minor alterations in the structure of the cation or anion comprising an IL can lead to significant changes in the physicochemical properties of the IL, so too can such alterations have a significant (and in this case, beneficial) impact on the mechanism of ion transfer into these solvents.

## Conclusions

In contrast to many conventional solvents, ionic liquids frequently play an active role in the process of metal ion partitioning, with exchange of a metal complex for either the cationic or anionic constituent of the ionic liquid representing an important mode of ion transfer. Although this has obvious negative implications for the "greenness" of these solvents, this problem can, in certain instances, be reduced in significance simply by raising the molecular weight of the cationic portion of the IL. An analogous approach could

**Table 1.** Strontium[a] and nitrate[b] ion partitioning between water[c] and solutions of DCH18C6[d] in $C_n$mim$^+$Tf$_2$N$^-$

| Organic phase | %E$_{Sr}$ | %E$_{NO3-}$ observed | expected[e] | Partitioning mode indicated |
|---|---|---|---|---|
| $C_5$mim$^+$Tf$_2$N$^-$ | 96.5 | 9 ±7[f], 16[f,g] | 96.5 | cation exchange |
| $C_6$mim$^+$Tf$_2$N$^-$ | 82.6 | 16.1 ± 0.8 | 82.6 | cation exchange |
| $C_8$mim$^+$Tf$_2$N$^-$ | 39.0 | 20.9 ± 1.0 | 39.0 | mixed |
| $C_{10}$mim$^+$Tf$_2$N$^-$ | 20.2 | 20.0 ± 1.0 | 20.2 | extraction of neutral complex |

[a] Determined radiometrically.
[b] Determined by ion chromatography, unless otherwise noted.
[c] Containing 0.0310 M Sr(NO$_3$)$_2$.
[d] 0.202 M in the indicated ionic liquid.
[e] Assuming extraction of the neutral strontium nitrato-crown ether complex. For partitioning *via* cation exchange, no nitrate co-extraction (%E$_{NO3-}$=0) is expected.
[f] As reported in reference 21.
[g] Determined by N-15 NMR.

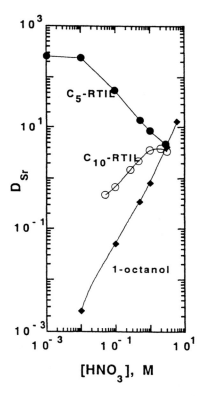

Figure 4. Nitric acid dependency of $D_{Sr}$ for DCH18C6 in 1-octanol (0.25 M), $C_5mim^+ Tf_2N^-$ (0.1 M), and $C_{10}mim^+Tf_2N^-$ (0.1 M). (T = 23 °C).

undoubtedly be used to suppress anion exchange in these solvents as well, and work is underway in our laboratory to demonstrate this. Although increasing the hydrophobicity of the IL cation (from $C_5mim^+$ to $C_{10}mim^+$) does markedly reduce the efficiency of metal ion extraction, the $C_{10}mim^+$ IL nonetheless yields greater extraction than the preferred conventional solvent, 1-OAlc, over a wide range of nitric acid concentrations, a likely result of the unique solvation environment provided by the ionic liquid. Taken together, these results offer renewed hope that improved, environmentally benign metal ion separation systems employing ionic liquids can be devised.

## Acknowledgements

This work was performed under the auspices of the Office of Basic Energy Sciences, Division of Chemical Sciences, United States Department of Energy, under contract No. W-31-109-ENG-38.

## References

1. Blanchard, L.A.; Gu, Z.; Brennecke, J.F. *J. Phys. Chem. B.*, **2001**, *105*, 2437-2444.
2. Huddleston, J.G.; Willauer, H.D.; Swatloski, R.P.; Visser, A.E.; Rogers, R.D. *Chem. Commun.*, **1998**, 1765-1766.
3. Dai, S.; Ju, Y.H.; Barnes, C.E. *J. Chem. Soc., Dalton Trans.*, **1999**, 1201-1202.
4. Visser, A.E.; Swatloski, R.P.; Reichert, W.M.; Griffin, S.T.; Rogers, R.D. *Ind. Eng. Chem. Res.*, **2000**, *39*, 3596-3604.
5. Visser, A.E.; Swatloski, R.P.; Reichert, W.M.; Mayton, R.; Sheff, S.; Weirzbicki, A.; Davis, Jr., J.H.; Rogers, R.D. *Chem. Commun.*, **2001**, 135-136.
6. Armstrong, D.W.; He, L.; Liu, Y.-S. *Anal. Chem.* **1999**, *71*, 3873-3876.
7. Yanes, E.G.; Gratz, S.R.; Baldwin, M.J.; Robison, S.E.; Stalcup, A.M. *Anal. Chem.* **2001**, *73*, 3838-3844.
8. Blanchard, L.A.; Brennecke, J.F. *Ind. Eng. Chem. Res.* **2001**, *40*, 287-292.
9. Branco, L.C.; Crespo, J.G.; Afonso, C.A.M. *Angew. Chem. Int. Ed.* **2002**, *41*, 2771-2773.
10. Branco, L.C.; Crespo, J.G.; Afonso, C.A.M. *Chem. Eur. J.* **2002**, *8*, 3865-3871.
11. Visser, A. E.; Swatloski, R. P.; Hartman, D. H.; Huddleston, J. G.; Rogers, R. D. In *Calixarenes for Separations*; Lumetta, G. J., Rogers, R. D., Gopalan, A S., Eds.; ACS Symposium Series 757; American Chemical Society, Washington, DC, 2000, pp. 223-236.

12. Visser, A. E.; Swatloski, R. P.; Griffin, S. T.; Hartman, D. H.; Rogers, R. D. *Sep. Sci. Technol.* **2001**, *36*, 785-804.
13. Jiang, T.-F.; Gu, Y.-L.; Liang, B.; Li, J.-B.; Shi, Y.-P.; Ou, Q.-Y. *Anal. Chim. Acta* **2003**, *479*, 249-254.
14. Dietz, M.L.; Dzielawa, J.A. *Chem. Comm.* **2001**, 2124-2125.
15. Bates, E.D.; Mayton, R.D.; Ntai, I.; Davis, J.H., Jr. *J. Am. Chem. Soc.* **2002**, *124*, 926-927.
16. Scurto, A.M.; Aki, S.N.V.K.; Brennecke, J.F. *J. Am. Chem. Soc.* **2002**, *124*, 10276-10277.
17. Scurto, A.M.; Aki, S.N.V.K.; Brennecke, J.F. *Chem. Comm.* **2003**, 572-573.
18. Seddon, K. R.: *Kinet. Catal.* **1996**, *37*, 693-697.
19. Brennecke, J. F., Maginn, E. J. *AIChE J.* **2001**, *47*, 2384-2389.
20. Olivier-Bourbigou, H.; Magna, L.: *J. Mol. Catal. A* **2002**, *182*, 419-437.
21. Jensen, M.P.; Dzielawa, J.A.; Rickert, P.; Dietz, M.L. *J. Am. Chem. Soc.* **2002**, *124*,10664-10665.
22. Visser, A.E.; Jensen, M.P.; Laszak, I.; Nash, K.L.; Choppin, G.R.; Rogers, R.D. *Inorg. Chem.* **2003**, *42*, 2197-2199.
23. Jensen, M.P.; Neuefeind, J.; Beitz, J.V.; Skanthakumar, S.; Soderholm, L. *J. Am. Chem. Soc.* **2003**, *125*, 15466-15473.
24. Chun, S., Dzyuba, S. V., Bartsch, R. A.: *Anal. Chem.* **2001**,*73*, 3737-3741.
25. Izatt, R.M.; Haymore, B.L.; Bradshaw, J.S.; Christensen, J.J. Inorg. Chem. **1975**, *14*, 3132-3133.
26. Yamoto, K.; Bartsch, R.A.; Broker, G.A.; Rogers, R.D.; Dietz, M.L. *Tet. Lett.* **2002**, *43*, 5805-5808.
27. Yamoto, K.; Fernandez, F.A.; Vogel, H.F.; Bartsch, R.A.; Dietz, M.L. *Tet. Lett.* **2002**, *43*, 5229-5232.
28. Yamoto, K.; Bartsch, R.A.; Dietz, M.L.; Rogers, R.D. *Tet. Lett.* **2002**, *43*, 2153-2156.
29. Blasius, E.; Klein, W.; Schon, U. *J. Radioanal. Nucl. Chem.* **1985**, *89*, 389-398.
30. Gloe, K.; Muehl, P.; Kholkin, A. I.; Meerbote, M.; Beger, J. *Isotopenpraxis* **1982**, *18*, 170-175.
31. Filippov, E. A.; Yakshin, V. V.; Abashkin, V. M.; Fomenkov, V. G.; Serebryakov, I. S. *Radiokhimiya* **1982**, *24*, 214-216.
32. Horwitz, E. P., Schulz, W. W. In *Metal Ion Separation and Preconcentration: Progress and Opportunities*; Bond, A. H., Dietz, M. L., Rogers, R. D., Eds.; American Chemical Society; Washington, DC, 1999, pp. 20-50.
33. Horwitz, E. P., Dietz, M. L., Fisher, D. E. *Solvent Extr. Ion Exch.*1990. *8*, 557-572.
34. Horwitz, E. P., Dietz, M. L., Fisher, D. E. *Solvent Extr. Ion Exch.*1991, *9*, 1-25.
35. Horwitz, E. P., Dietz, M. L., Fisher, D. E. *Solvent Extr. Ion Exch.* **1990**, *8*, 199-208.

18

36. Junk, P. C.; Steed, J. W. *J. Chem. Soc., Dalton Trans.* **1999**, 407-414.
37. Mathur, J. N.; Choppin, G. R. *Solvent Extr. Ion Exch.* **1993**, *11*, 1-18.
38. Hasegawa, Y.; Ishiwata, E.; Ohnishi, T.; Choppin, G. R. *Anal. Chem.* **1999**, *71*, 5060-5063.
39. Akiba, K.; Kanno, T. *J. Inorg. Nucl. Chem.* **1980**, *42*, 273-276.
40. Sekine, T.; Dyrssen, D. *J. Inorg. Nucl. Chem.* **1967**, *29*, 1481-1487.
41. Akiba, K.; Wada, M.; Kanno, T. *J. Inorg. Nucl. Chem.* **1981**, *43*, 1031-1034.
42. Kassierer, E. F.; Kertes, A. S. *J. Inorg. Nucl. Chem.* **1972**, *34*, 3221-3231.
43. Sekine, T.; Dyrssen, D. *J. Inorg. Nucl. Chem.* **1967**, *29*, 1457-1473.
44. Sekine, T.; Sakairi, M.; Shimada, F.; Hasegawa, Y. *Bull. Chem. Soc. Japan* **1965**, *38*, 847-849.
45. Kitatsuji, Y.; Meguro, Y.; Yoshida, Z.; Yamamoto, T.; Nishizawa, K. *Solvent Extr. Ion Exch.* **1995**, *13*, 289-300.
46. Kononenko, L. I.; Vitkun, R. A. *Russ. J. Inorg. Chem.* **1970**, *15*, 1345-1350.
47. Atanassova, M.; Jordanov, V. M.; Dukov, I. L. *Hydrometall.* **2002**, *63*, 41-47.
48. Dietz, M.L.; Dzielawa, J.A.; Jensen, M.P.; Firestone, M.A. In *Ionic Liquids as Green Solvents: Progress and Prospects*, Rogers, R.D., Seddon, K.R., Eds.; American Chemical Society, Washington, DC, 2003, pp. 526-543.
49. Chen, X. –F.; Liu, S. –H.; Duan, C. –Y.; Xu, Y. –H.; You, X. –Z.; Ma, J.; Min, N. –B. *Polyhedron* **1998**, *17*, 1883-1889.
50. Dietz, M. L.; Dzielawa, J. A; Laszak, I.; Young, B. A.; Jensen, M. P. *Green Chem.* **2003**, *5*, 682-685.

Chapter 2

# Potentialities of Room Temperature Ionic Liquids for the Nuclear Fuel Cycle: Electrodeposition and Extraction

Clotilde Gaillard[1], Gilles Moutiers[2], Clarisse Mariet[2],
Tarek Antoun[3], Benoît Gadenne[3], Peter Hesemann[3],
Joël J. E. Moreau[3], Ali Ouadi[1], Alexandre Labet[2],
and Isabelle Billard[1]

[1]Institut de Recherches Subatomiques, CNRS/IN2P3 and
Université L. Pasteur, B.P. 28, 67037, Strasbourg cedex 2, France
[2]CEA Saclay, INSTN/UECCC, 91191 Gif sur Yvette cedex, France
[3]CNRS UMR 5076, Hétérochimie Moléculaire et Macromoléculaire,
Laboratoire de Chimie Organométallique, 8 rue de l'Ecole Normale,
34296 Montpellier cedex 5, France
gilles.moutiers@cea.fr; hesemann@cit.enscm.fr;
clotilde.gaillard@ires.in2p3.fr

The potentialities of hydrophobic ionic liquids $BumimPF_6$ and $BumimTf_2N$ for their use in the nuclear fuel cycle were investigated, in particular for the electrodeposition and the liquid liquid extraction. The importance of water and impurities on the electrochemical behaviour of the RTILs was pointed out. We also demonstrated that the use of RTILs in replacement of the organic diluent for actinides partitioning is promising and we present a few examples of task-specific ionic liquids synthesis.

The introduction of room temperature ionic liquids (RTILs) in the nuclear fuel cycle can be envisioned through different approaches. In this respect, the RTILs' stability under alpha and gamma irradiation *(1)* and the enhanced safety they provide towards criticality *(2)* are additional interesting "green" properties of this new class of solvents. In a rather classical way, RTILs can just be considered as alternative solvents *(3)*, in replacement of the highly toxic and flammable kerosene mixtures that are in use nowadays *(4)*. In such a perspective, not all the potentials of RTILs are considered. Actually, hydrophobic RTIL's can be envisioned in biphasic systems (i.e. liquid liquid extraction) in replacement of the organic phase and can give access to unusual oxidation states *(5)*. Direct metal electrodeposition from RTIL solutions appears to be also an interesting route towards the separation/purification of nuclear wastes *(6, 7)*, since several recent electrochemical investigations have shown that some RTILs present a good ionic conductivity and a wide electrochemical window as compared to water or classical organic solvents *(8)*. Thus, in the absence of water, direct metal electrodeposition can be foreseen. Finally, another very promising objective is the use of functionalized RTILs, which would behave both as the organic phase and as the extracting agent, simply suppressing the problems encountered with extractant/solvent miscibility and the recovery of both the extracting species and the solvent. This last concept has already been the subject of some studies *(9, 10)*. Whatever the way RTILs are used in the nuclear fuel cycle, the question of the behaviour of water in those liquids is a key point for a deeper understanding of solvation effects, solute-solvent and solvent-solvent interactions.

Therefore, we present our first insights into the potential of RTILs in the nuclear fuel cycle, following the major routes summarized above. First, an electrochemical study has been performed to exhibit the influence of impurities and water in two hydrophobic ionic liquids (1-methyl-3-butylimidazolium hexafluorophosphate-BumimPF$_6$ and 1-methyl-3-butylimidazolium bis(triflyl)imide-BumimTf$_2$N in order to obtain information on their general organization, especially in terms of dissociation, solvation and ion migration. Second, extraction of uranium from an acidic aqueous phase toward a mixture of BumimPF$_6$ + 30% TBP has been performed. Finally, our first attempts to synthesize new functionalized RTILs are presented in the last section. The extracting abilities of these new liquids for americium are currently under investigation.

## Materials and Methods

The synthesis of BumimPF$_6$ has been previously reported *(5)*. The synthesis of BumimTf$_2$N is derived from this procedure and is classically as described: 200g of LiTf$_2$N is dissolved in 300 mL of water, 110 g of BumimCl is dissolved in 150 mL of water and the two parts are mixed under stirring at 70°C for 16 hours. Two phases are obtained and the upper one (aqueous phase) is decanted.

The bottom phase (BumimTf$_2$N) is washed several times with deionized water (10x350 mL). Typical yields of ≈80% are obtained.

Purification of the RTILs has been achieved (unless specified) as already described (5). The purification step allows lowering the water amount as compared to a water-saturated liquid stored under ambient air for several days. Nevertheless, although possible (11), no further drying of the RTIL was made, since we focused this study on liquid liquid extraction application. The so-called "purified water saturated" RTIL is obtained by stirring equivalent volume of deionized water and purified RTIL for 24 hours. The upper aqueous phase is then decanted. The water amount is determined by the Karl Fischer technique using standard procedures. The apparatus is composed of a Metrohm 703TI Stand titrator and a Metrohm 756KF coulometer.

Electrochemical measurements were carried out using an EGG/ PAR potentiostat-galvanostat (model 263A). The cell was a glass mini cell from EGG containing 4 mL of solvent equipped with EGG adapted electrodes. The working electrode was a Pt disc of 2 mm diameter and the counter electrode was a Pt wire. The ohmic resistance has been measured at a potential value where no redox processes occur following the classical procedure proposed by the M263A potentiostat, and positive feedback iR compensation was employed to minimise solution resistance. Each R value is an average of 12 experiments. In order to avoid any contribution of the semipermeable membrane of the reference electrode, a pseudo reference (Pt wire) directly immersed in the bulk solution was used. All measurements were carried out at 25°C. No particular controlled atmosphere was involved.

The solutions for the uranyl extraction experiments were prepared with reagent grade chemicals and deionised water. A standard 3 M HNO$_3$ aqueous solution of $^{233}$U (certified specific activity of $^{233}$U equal to 40 kBq/g, CERCA, France) was used as the parent solution and for all further dilutions and preparations. A solution of TBP in BumimPF$_6$ (30 % volume/volume) has been used as the organic phase. Extraction was performed by mixing 1 mL of the RTIL phase and 1 mL of the aqueous phase followed by vortexing (25 Hz, 5 min) and centrifugation (5000 rpm) to equilibrate the phases. The phases were separated and transferred into shell vials from which weighted uranyl aliquots of both phases were removed for radiometric analysis, following deposition onto stainless steel discs and evaporation to make alpha targets. The $^{233}$U activity has been measured by use of an ionization chamber (Numelec NU14B) coupled to a preamplifier, an amplifier and a multichannel analyser, taking into account back-scattering and self-absorption. At equilibrium, the distribution of uranium between the two phases (IL/water) is expressed in terms of a distribution coefficient $D_U$. It is defined as the ratio between the concentration of the extracted uranium and its concentration in the aqueous solution. Because equal volumes of both phases were chosen for analysis, the distribution ratio for the metal was determined as the activity ratio (IL/water).

## Electrochemical Study: Results and Discussion

Table I displays the water amount of the RTILs batches used in this work, together with the ohmic resistance. The values obtained for the water content are in agreement with the general trend indicating that BumimTf$_2$N is more hydrophobic than BumimPF$_6$ due to the effect of the counter anion $(SO_2CF_3)_2N^-$ which is not particularly able to involve hydrogen bonds with water. As can be seen, the purification procedure decreases the water amount roughly by a factor 3. However, these values also stress the high amount of water present in this ionic liquid family, even after purification. This is not negligible in terms of the physicochemical properties, and more precisely on the solvation effects. Moreover, this also indicates that the amount of water has to be well established before running experiments in order to obtain reliable and comparable values. Seddon *et al.* have shown recently *(12)* that water modifies dramatically the RTILs' properties and that depending on the water amount, a salt-rich region and a water-rich region have to be distinguished. Our samples are undoubtedly in the salt-rich region.

**Table I. Ohmic drop potentiostatically measured at 25°C**

| Ionic Liquid | Water Amount /ppm | R /Ohm |
|---|---|---|
| BumimPF$_6$ non purified | 20 000 | 700 |
| purified | 5 900 | 950 |
| purified, water saturated | 21 600 | 500 |
| BumimTf$_2$N non purified | 13 000 | 400 |
| purified | 1 800 | 470 |
| purified, water saturated | 14 000 | 300 |

R ± 8 % ; water amount ± 5 %

From our results, three major findings can be derived: (i) in all cases, the resistance values are high as compared to classical high temperature molten salts that are usually in the range of a few mΩ. (ii) The resistance values are in the range corresponding to classical molecular solvents (such as water and organic solvents) containing *supporting* electrolytes under usual electrochemical conditions. (iii) When the ionic liquid is purified and dried, the ohmic resistance increases dramatically, and the value is no more compatible with the general molten salts sense, which are considered to be composed solely of dissociated ions.

This leads to reconsider the structure of ionic liquids in their liquid form and their solvation action. For the purified batches, the high ohmic resistance values are characteristic of a poorly dissociated liquid and of a high global structure organisation. Moreover, it is now well-known that the Bumim⁺ cation is capable of hydrogen bonding *(13)* through the atom in position 2 (Figure 1) that will reinforce this organisation as compared to cations with a substituted position. Thus, it is suggested that an important ion pair formation occurs in the liquid between the cationic skeleton and the counter ion. This hampers the current circulation through ion migration. The decrease in the ohmic resistance for the water-saturated batches indicates that electromigration is strongly favoured by the presence of impurities and water molecules. Water molecules certainly play a role in the process by insertion in the ionic liquid lattice. This causes distortion of the organisation and favours ion dissociation. This is also emphasised by the fact that when interactions in the solvent are less pronounced (BumimTf₂N), the ohmic drop is less. These trends clearly indicate that hydrophobic ionic liquids composed of dialkylimidazolium cations exhibit a high local organisation when dried and purified which finally lowers their electrochemical properties. This has to be taken into account when preparative (instead of analytical) electrochemistry is involved because high current intensity will therefore result in high cell voltage and important ohmic drop.

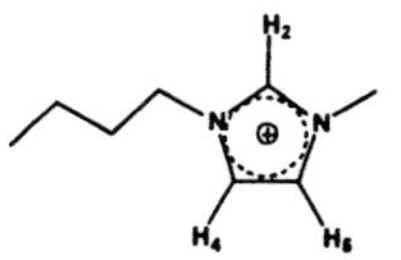

*Figure 1. Bumim⁺ cation.*

## Uranium Extraction with TBP/BumimPF₆ Mixtures

Our aim is to investigate the potential of hydrophobic BumimPF₆ in biphasic systems for uranyl ($UO_2^{2+}$) extraction from acidic water. In order to make comparisons, we used conditions as similar as possible as those in use in the PUREX process (Plutonium, URanium EXtraction).

It is well-known *(14)* that hard metal cations are preferentially solvated by water molecules, so that they remain in the aqueous phase and require in consequence the addition of an extractant to enhance their affinity for the non-polar extracting phase. In the PUREX process, TBP (tri-n-butyl phosphate) is used as the uranyl extracting agent and is incorporated at 30 % (in volume) in n-dodecane *(15)*. Indeed, TBP does not dissolve entirely and undiluted TBP does not generally behave very well in contactors. Therefore, TBP is almost always used in conjunction with a diluent, usually consisting of hydrocarbons, which are essentially inert in a metal extraction context. Under these conditions the diluent serves two main purposes: i) it reduces the extracting power of undiluted TBP, which would often be a too powerful extractant that will cause difficulties to strip the metal back into the aqueous phase. ii) It improves the physical properties of the organic mixture, shifting the density and the viscosity towards those of the diluent itself. As a rule, 20% to 30 % (in volume) of TBP in a hydrocarbon diluent provide a solvent of suitable strength. We extended the possibilities of IL-based separations to actinides by incorporating TBP as the extractant in the same conditions: TBP was dissolved in BumimPF$_6$ and the extraction of U(VI) from nitric aqueous solutions examined. Although it is indicated in the literature that biphasic systems containing PF$_6^-$ anions may become monophasic at high HNO$_3$ concentrations *(18)*, this has not been observed under our experimental conditions.

Figure 2 displays the amount of extracted uranyl in the BumimPF$_6$ phase as a function of the uranyl concentration in the aqueous phase.

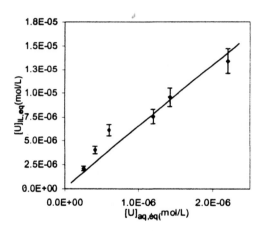

*Figure 2. Extraction isotherm of $UO_2^{2+}$ in water/HNO$_3$ 3 M by TBP (30 % V/V) in BumimPF$_6$.*

As the uranyl concentration in solution is very low (trace level), the distribution coefficient reaches a constant value $D_{U,limit}$. From the slope of the curve displayed in Figure 2, $D_{U,limit}$ can be estimated at 11 ± 3. The overall efficiency of 30 % TBP in BumimPF$_6$ is lower than that of 30% TBP in dodecane but this had to be expected because it is known that sometimes the diluent can play a role in the extraction process. In general, varying the diluent does not result in a great deal of control over distribution ratios or separation factors but there are exceptions to this rule. For example, it occurs when the TBP stoechiometry in the extractable species varies or for ionic salts like uranyl nitrate. Several authors have collected distribution data for TBP extraction of various metal salts *(10)*. These data clearly demonstrate that diluents can be conveniently classified into two broad classes – polar and non polar. For an ionic salt, as is the case for uranyl nitrate, the distribution ratio is about the same with all non polar diluents although it slightly increases following the diluent order: aromatic ≤ aliphatic < halogenoalkane. By contrast, for polar diluents, a decrease in the extracting capacity of a factor of 2 to 20 as compared to the case with non polar diluent is observed *(16)* and the order is alcohol < chloroform < dichloroethane. It was shown that RTILs and especially $BumimPF_6$ have polar properties similar to those of polar aprotic solvents or short chain alcohols *(17)*. Therefore, a low value of the $D_{U,limit}$ has to be expected and is experimentally confirmed. The values of $D_{u,limit}$ as a function of the $HNO_3$ concentration in the aqueous phase are displayed in Figure 3.

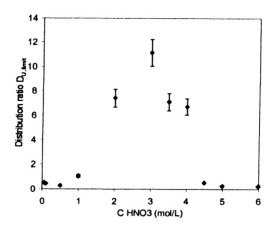

*Figure 3. Distribution ratios for $UO_2^{2+}$ with 30 % $TBP$-$BumimPF_6$ as a function of $HNO_3$ concentration in the aqueous phase.$[U(VI)]$ initial = $10^{-6}$ mol $L^{-1}$.*

All the distribution ratios shown in Figure 3 are at least one order of magnitude lower in $BumimPF_6$ as compared to similar extracting conditions in dodecane *(18)*. However, the usual effect of an increase of the acid concentration onto the $D_{U,limit}$ values could be observed and showed the characteristic behaviour of a neutral extractant. Indeed, increasing the acid concentration leads to an increase in the $D_{U,limit}$ values (due to increased amounts of the extractible $UO_2(NO_3)_2$ species in the aqueous phase), followed by a decrease in the $D_{U,limit}$ values (due to the formation of the $TBP$-$HNO_3$ adduct). This behaviour has already been observed in a similar study devoted to the extraction of $Sr^{2+}$ with RTILs *(19)*.

An important consideration lies in the fact that these values of distribution ratios were obtained with TBP 30 % v/v in $BumimPF_6$ in an aim of comparison with actual processes. However, for further development and a renew in such processes, the concentration of TBP in BumimPF$_6$ could be increased. In these conditions higher values of the uranium distribution coefficient could be expected.

## Functionalized Ionic Compounds: New agents for Liquid-Liquid Extraction

Owing to their immiscibility to numerous solvents such as water, diethyl ether, alkanes and supercritical CO$_2$, room-temperature ionic liquids (RTILs) are suitable solvents for liquid/liquid separation processes *(19-21)*. The usefulness of RTILs for the extraction of simple organic compounds such as naphthalene has already been reported *(20, 22, 23)*. However, the partitioning of metallic species is largely limited by the low complexation properties of the hydrophobic ionic liquids: in general, hydrophobic RTILs are non-coordinating, and the highly hydrated metal ions remain in the aqueous phase *(10)*. Several attempts have been made to enhance the affinity of metal ions towards the IL phase. For example, the addition of organic coordinating compounds increases significantly the distribution ratios of metal ions between the ionic liquid and the aqueous phase *(19, 24, 25)*.

Partitioning of metal ions can also be achieved by grafting complexing moieties on the organic cation of the IL. ILs bearing urea, thiourea and thioether *(10)* or ethyleneglycol groups *(26, 27)* enhance the partitioning of various metal ions to the IL phase from water. This approach may also minimize leaching of the extracting ionic compound to the aqueous phase. Furthermore, various functions can be grafted on the RTIL giving rise to task-specific ionic liquids (TSILs) which allow a selective separation of targeted metal ions.

Actinides also exhibit significant partitioning towards IL phases from aqueous solutions with the addition of an extractant. Functional RTILs can therefore be considered as new media for *f*-element chemistry *(28)*. We focused on the synthesis of functionalized ionic liquids for the partitioning of actinides and lanthanides.

Here, we describe the synthesis of ionic compounds bearing different types of metal complexing groups, in particular malonamides, β-diketones and 2-hydroxy-benzylamine entities. These coordinating groups have widely been investigated for the complexation of various metallic species. They should increase the affinity of metals to the IL phase and give rise to suitable media for the liquid-liquid extraction of metals.

### Ionic Compounds Bearing Malonamide Entities

Malonamides are efficient metal complexing compounds which can be used for the separation of actinides *(29, 30)*. The synthesis of ionic compounds containing malonamide groups started from the malonic acid esters **1a,b** *(31)* (Figure 4). The general three step synthesis of the *bis*-imidazolium hexafluorophosphates or *bis*-(trifluoromethanesulfonyl)imides from the dicarboxylic acid esters **1a,b** is represented in Figure 5.

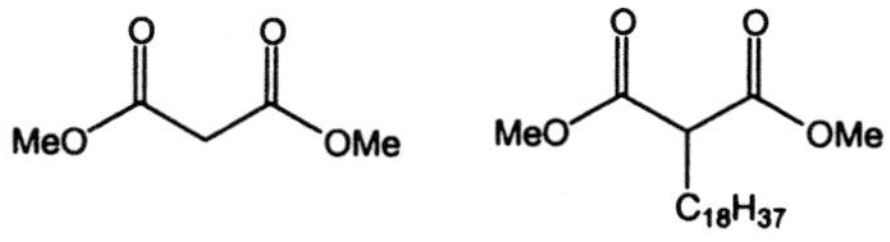

**1a**    **1b**

*Figure 4. Structures of the malonic esters **1a,b***

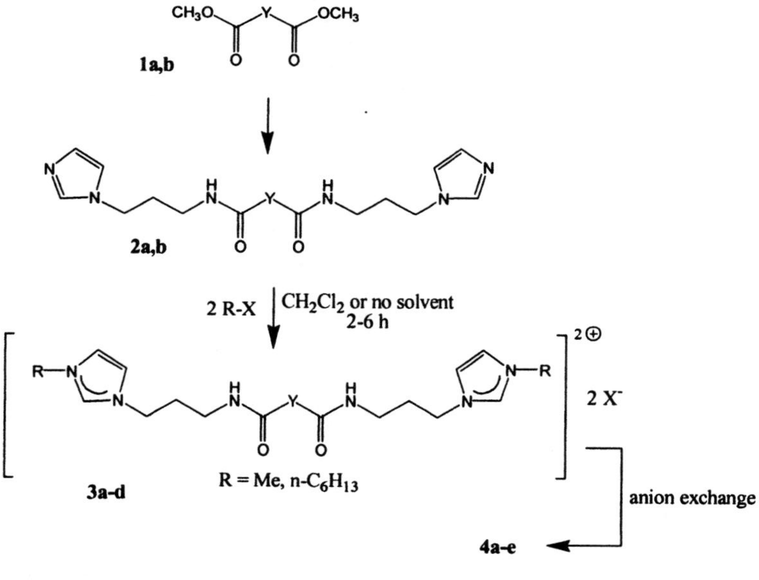

*Figure 5. Synthesis of the functionalized ionic liquids **4a-e**.*

Aminolysis of the diesters **1a,b** with N-(3-Aminopropyl)imidazole gave the corresponding malonamides **2a,b** with terminal imidazole groups. The *bis*-amides **2a,b** were alkylated either with iodomethane or 1-bromohexane affording the corresponding water soluble *bis*-imidazolium halogenides **3a-d**.

Finally, anion exchange with either potassium hexafluorophosphate or *N*-lithio-*bis*-trifluoromethanesulfonimide gave the *bis*-imidazolium salts **4a-e** in global yields of 85-90%. Table II gives a survey of the various functional ionic liquids prepared by this general synthetic route.

**Table II. Composition and structure of the ionic liquids 4a-e**

| Product | Y | R | Anion X | solubility $H_2O$ | solubility $CH_2Cl_2$ | Melting point /°C |
|---------|---|---|---------|----------|---------------|-------------------|
| **4a** | -CH$_2$- | -CH$_3$ | PF$_6^-$ | + | + | < r.t. |
| **4b** | -CH$_2$- | -CH$_3$ | NTf$_2^-$ | - | - | < r.t |
| **4c** | -CH$_2$- | -C$_6$H$_{13}$ | PF$_6^-$ | - | + | < r.t |
| **4d** | -CH(C$_{18}$H$_{37}$)- | -CH$_3$ | PF$_6^-$ | - | - | 106 |
| **4e** | -CH(C$_{18}$H$_{37}$)- | -C$_6$H$_{13}$ | PF$_6^-$ | - | + | 56 |

The properties (solubility, cristallinity) of the *bis*-imidazolium malonamides **4a-e** are highly dependent both of the structure of the organic cation and the nature of the counterion. The methylated hexafluorophosphate **4a** is well soluble in water and in numerous polar organic solvents such as alcohols and acetonitrile. In consequence, the anion exchange is difficult to perform, and even after repeated precipitation of potassium iodide in acetonitrile, **4a** contained considerable amounts of the halogenide. We therefore investigated different ways to increase the hydrophobicity of the ionic compounds in order to synthesize water-immiscible compounds.

The hydrophobicity of the ionic compounds can be increased by a judicious choice of the anion. By replacing the hexafluorophosphate with the less polar *bis*-trifluoromethanesulfonimide cation, the methylated bis-imidazolium malonamide **4b** becomes water immiscible. However, the compound is also immiscible to numerous organic solvents such as CH$_2$Cl$_2$.

Hydrophobicity of ionic compounds can also be achieved by grafting hydrophobic alkyl chains on the organic cation. We prepared imidazolium-*bis*-malonamides bearing alkyl chains on two distinct positions: *i)* on the terminal imidazole unit and *ii)* on the central carbon (C$_2$) on the malonamide unit.

The product **4c** with two terminal hexyl groups is immiscible in water but soluble in dichloromethane. In consequence, the anion exchange affording **4c** can easily be performed in a biphasic water-CH$_2$Cl$_2$ system and leads to a halogen-free product. The hydrophobicity of **4c** is clearly related to the influence of the terminal hexyl chains, and the processing of this product is considerably simplified.

Alkyl substitution on the central carbon of the malonate unit by $C_{18}$ chains also influences the miscibility properties of the ionic compounds. The hydrophobicity is increased and results the formation of the water-insoluble products **4d** and **4e**. Solubility in $CH_2Cl_2$ was achieved in the case of the hexylated compound **4e**, whereas the methylated product **4d** is immiscible in chlorinated solvents.

Alkylation on the central carbon ($C_2$) on the malonamide unit increases the hydrophobicity of the ionic compounds **4d-4e**, but do not enhance their solubility in organic solvents. Similar results were obtained for ionic compounds substituted with central octyl chains. It has to be pointed out that alkylation on the central carbon ($C_2$) on the malonamide unit also influences the phase behaviour of the ionic compounds: **4d** and **4e** are crystalline products with melting points of 106°C and 56°C, respectively.

In conclusion, the miscibility and phase transition behaviour of the ionic compounds **4a-4e** is strongly influenced both by the anion and substitution of the functionalized bis-imidazolium cation. The use of non-coordinating anions ($NTf_2^-$) and substitution of the organic cation with long alkyl chains increase the hydrophobicity. The position of the alkyl chain on the organic cation has considerable importance on the properties of the compounds. Substitution with terminal alkyl groups shows great influence on the solubility of the compounds: the hexylated compounds are generally well soluble in $CH_2Cl_2$. The substitution on the central carbon $C_2$ of the malonamide unit rather influences the phase behaviour, in particular the cristallinity of the compounds.

### 1-(N-alkylpyridin-3-yl)butane-1,3-dione hexafluorophosphates

The preparation of ionic compounds based on the 1-(N-alkylpyridin-3-yl)butane-1,3-dione-entity was accomplished in a multi-step synthesis starting from nicotinic acid methyl ester (Figure 6). Claisen condensation with acetone in basic medium afforded the 1-(N-alkyl-3-pyridyl)butane-1,3-dione, which was alkylated using iodomethane or bromohexane resulting in the formation of the halogenides **6a/b**. Anion exchange finally gave the corresponding hexafluorophosphates **7a/b**.

Both products present similar physical properties in terms of cristallinity and solubility. **7a** and **b** are crystalline compounds with melting points of 139-140°C and 72-73°C, respectively. Both products show considerable solubility in aqueous media and can be purified by recrystallisation from water. The 1-(N-alkylpyridin-3-yl)butane-1,3-dione hexafluorophosphates are light sensitive and should be stored and handled in a dark room.

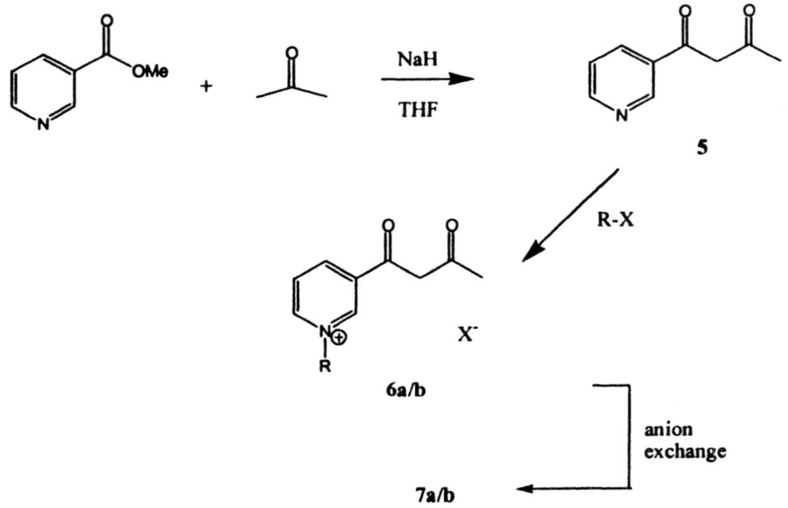

*Figure 6. Synthesis of the pyridinium-β-diketone salts 7a/b.*

## Ionic Liquids Bearing 2-Hydroxy Benzylamine Entities

2-Hydroxy benzylamines represent another class of efficient metal complexing compounds *(30)*. These groups have been grafted on IL structures by a three step synthesis starting from salicylaldehyde and 3-aminopropylimidazole (Figure 7). The formed imine was first alkylated using bromobutane, and the resulting ionic imine was finally reduced with sodium borohydride. The water soluble bromide **9** was finally transformed into the hexafluorophosphate with KPF$_6$. The water immiscible imidazolium hexafluorophosphate **10** was isolated as a highly viscous orange oil.

In conclusion, we synthesized three new types of ionic compounds with various complexing groups: malonamides, β-diketones and 2-hydroxy-benzylamines. The products are generally highly viscous oils or crystalline solids with melting points up to ~100°C. All compounds show excellent solubility in hydrophobic ionic liquids such as BumimPF$_6$. The use of these materials in liquid-liquid extraction is in progress.

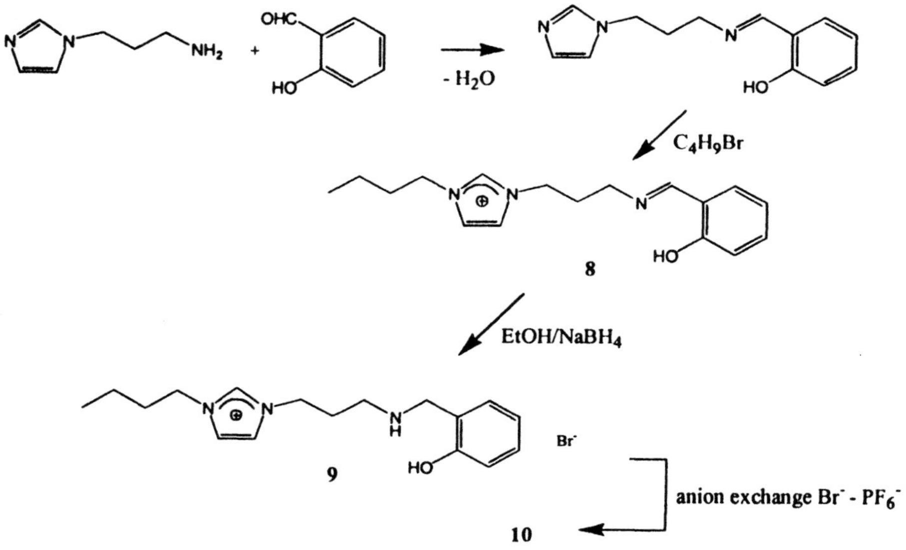

*Figure 7. Synthesis of the imidazolium-hydroxybenzylamine salt 10.*

# References

1.  Allen, D.; Baston, G.; Bradley, A. E.; Gorman, T.; Haile, A.; Hamblett, I.; Hatter, J. E.; Healey, M. J. F.; Hodgson, B.; Lewin, R.; Lovell, K. V.; Newton, B.; Pitner, W. R.; Rooney, D. W.; Sanders, D.; Seddon, K. R.; Sims, H. E. Thied, R. C., *Green Chem.* **2002**, *4*, 152.
2.  Harmon, C.; Smith, W. Costa, D., *Radiat. Phys. Chem.* **2001**, *60*, 157.
3.  Visser, A. E. Rogers, R. D., *J. Solid State Chem.* **2003**, *171*, 109.
4.  Mathur, J.; Murai, M. Nash, K., *Solv. Extrac. Ion Exch.* **2001**, *19*, 357.
5.  Billard, I.; Moutiers, G.; Labet, A.; El Azzi, A.; Gaillard, C.; Mariet, C. Lützenkirchen, K., *Inorg. Chem.* **2003**, *42*, 1726.
6.  Baston, G. M. N.; Bradley, A. E.; Gorman, T.; Hamblett, I.; Hardacre, C.; Hatter, J. E.; Healy, M. J. F.; Hodgson, B.; Lewin, R.; Lovell, D. W.; Newton, G. W. A.; Nieuwenhuyzen, M.; Pitner, W. R.; Rooney, D. W.; Sanders, D.; Seddon, K. R.; Simms, H. E. Thied, R. C.s, *in Ionic liquids, industrial applications of green chemistry*, Rogers, R. D.Seddon, K. R., Eds.; ACS Publishing Series, San Diego, California,2002; 162.
7.  Oldham, W. J.; Costa, D. A. Smith, W. H.s, *in Ionic liquids: Industrial applications for green chemistry*, Rogers, R. D.Seddon, K. R., Eds.; ACS, Washington, D. C.,2002; 188.
8.  Suarez, P. A. Z.; Einloft, S.; Dullius, J. E. L.; de Souza, R. F. Dupont, J., *J. Chim. Phys.* **1998**, *95*, 1626.

32

9.   Visser, A. E.; Swatloski, R. P.; Reichert, W. M.; Mayton, R.; Sheff, S.; Wierzbicki, A.; Davis, J. H. Rogers, R. D., *Envir. Sci. Technol.* **2002,** *36,* 2523.
10.  Visser, A. E.; Swatloski, R. P.; Reichert, W. M.; Mayton, R.; Sheff, S.; Wierzbicki, A.; Davis, J. H. Rogers, R. D., *Chem. Commun.* **2001,** 135.
11.  Billard, I.; Mekki, S.; Gaillard, C.; Hesemann, P.; Moutiers, G.; Mariet, C.; Labet, A. Bünzli, J. C. G., *Eur. J. Inorg. Chem.* **2004,** 1190.
12.  Seddon, K. R.; Stark, A. Torres, M. J., *Pure Appl. Chem.* **2000,** *72,* 2275.
13.  Hitchcock, P. B.; Seddon, K. R. Welton, T., *J. Chem. Soc. Dalton Trans.* **1993,** 2639.
14.  *Ion solvation,* Eds.; John Wiley and sons limited, 1985;
15.  *Science and Technology of Tributyl Phosphate,* Schulz, W. W.; Burger, L. L.Navratil, J. D., Eds.; CRC Press, 1990; Vol.III.
16.  *Science and Technology of Tributyl Phosphate,* Schultz, W. W.Navratil, J. D., Eds.; CRC Press, 1984; Vol.I.
17.  *Ionic Liquids in Synthesis,* Wasserscheid, P.Welton, T., Eds.; Wiley-VCH, 2003;
18.  *Reactor handbook: fuel reprocessing,* Stoller, S. M.Richards, R. B., Eds.; Interscience, 1961; Vol.II.
19.  Visser, A. E.; Swatloski, R. P.; Reichert, W. M.; Griffin, S. T. Rogers, R. D., *Ind. Eng. Chem. Res.* **2000,** *39,* 3596.
20.  Huddleston, J. G.; Willauer, H. D.; Swatloski, R. P.; Visser, A. E. Rogers, R. D., *Chem. Commun.* **1998,** 1765.
21.  Visser, A. E.; Swatloski, R. P.; Griffin, S. T.; Hatman, D. Rogers, R. D., *Sep. Sci. Technol.* **2001,** *36,* 785.
22.  Blanchard, L. A.; Hancu, D.; beckman, E. J. Brennecke, J. F., *Nature* **1999,** *399,* 28.
23.  Visser, A. E.; Swatloski, R. Rogers, R., *Green Chem.* **2000,** 1.
24.  Visser, A. E.; Jensen, M. P.; Laszak, I.; Nash, K. L.; Choppin, G. R. Rogers, R. D., *Inorg. Chem.* **2003,** *42,* 2197.
25.  Dai, S.; Ju, Y. H. Barnes, C. E., *J. Chem. Soc., Dalton Trans.* **1999,** 1201.
26.  Branco, L. C.; Rosa, J. N.; Moura Ramos, J. J. Afonso, C. A. M., *Chem. Eur. J.* **2002,** *8,* 3671.
27.  Fraga-Dubreuil, J.; Famelard, M. H. Bazureau, J. P., *Org. Proc. Res. Dev.* **2002,** *6,* 374.
28.  Spjuth, L.; Liljenzin, J. O.; Hudson, M. J.; drew, M. G. B.; Iveson, P. B. Madic, C., *Solv. Ext. Ion Exch.* **2000,** *18,* 1.
29.  Bourg, S.; Broudic, J. C.; Conocar, O.; Moreau, J. J. E.; Meyer, D. Wong Chi Man, M., *Chem. Mater.* **2001,** *13,* 491.
30.  Dey, M.; Rao, C. P.; Saarenketo, P. K. Rissanen, K., *Inorg. Chem. Commun.* **2002,** *5,* 924.
31.  Kates, M. J. Schauble, J. H., *J. Org. Chem.* **1996,** *61,* 4164.

Chapter 3

# Ionic Liquid Technologies for Utilization in Nuclear-Based Separations

Keith E. Gutowski[1], Nicholas J. Bridges[1], Violina A. Cocalia[1], Scott K. Spear[1], Ann E. Visser[1], John D. Holbrey[1], James H. Davis, Jr.[1,2,*], and Robin D. Rogers[1,*]

[1]Department of Chemistry and Center for Green Manufacturing, The University of Alabama, Tuscaloosa, AL 35487
[2]Department of Chemistry, The University of South Alabama, Mobile, AL 36688
*Corresponding author: email: rdrogers@bama.ua.edu
*Corresponding author: email: jdavis@jaguar1.usouthal.edu

The favorable properties of Ionic Liquids (ILs) containing organic cations, such as low melting points, lack of vapor pressure, wide liquidus ranges, and tunable physical properties, make this class of liquids particularly interesting to study as potential replacements for traditional volatile organic solvents (VOCs). While nuclear separations are mostly done using non-volatile organic phases such as octanol or dodecane, the potential to use ILs as alternatives is appealing because of the unique solvent environments and solvation properties that are presented for coordination and extraction mechanism studies. Here, we show results that illustrate the application of ILs to fundamental problems surrounding nuclear separations and waste remediation.

33

## Introduction

**The nuclear legacy.** The development of nuclear energy in the United States over the past 60 years has left nuclear facilities with stockpiles of high- and low-level nuclear waste approaching tens of millions of gallons. At many legacy sites, primarily those involved in defense-related activities (i.e., processing of irradiated fuel for recovery or uranium and plutonium), the aging and corrosion of immense tanks that store these wastes, as well as the severe environmental threat that wastes pose, has led to an increasing emphasis on nuclear waste remediation via volume reduction (by removing the most hazardous nuclides) and/or long-term disposal (i.e., storage). Remediation of these wastes involves a fundamental understanding of the origin of the waste (i.e., the type of processing that generated the waste) as well as the type of radioactive components present in the liquids (supernatant) and solids (sludge, salt cakes) that comprise the waste. Nearly all tank wastes are basic in nature due to post-reprocessing neutralization of the acid used in most extraction schemes (i.e., nitric acid in the PUREX process) to reduce steel tank corrosion. Additionally, most high-level nuclear wastes contain intensely radioactive fission products such as $^{137}Cs$ and $^{90}Sr$, as well as long-lived actinides including $^{235}U$ and $^{239}Pu$. Reducing the volume of high-level waste to yield low-level waste involves removing these dangerously radioactive components and requires a basic understanding of the chemistry associated with fission product and actinide elements (*1*).

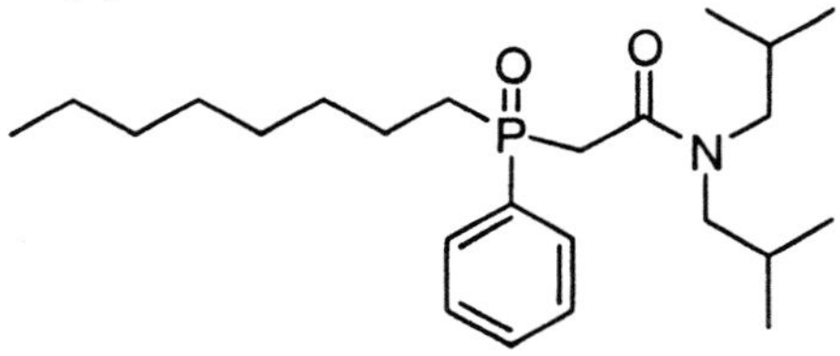

*Figure 1. CMPO*

The chemical complexity of actinides, evident in the variety of oxidation states possible, makes separations inherently difficult. Traditionally, separations schemes for actinide elements typically involved the use of dodecane, odorless kerosene, or select chlorinated solvents as an extracting phase, resulting in the thorough study and characterization of these solvent systems. In the PUREX process, *n*-tributyl phosphate (TBP) is used as an extractant for the removal of uranium and plutonium from irradiated fuel dissolved in nitric acid. The development of more specialized extractants, such as octyl(phenyl)-*N,N*-diisobutylcarbomylmethylphosphine oxide (CMPO, Figure 1), has further advanced the area of selective extraction. CMPO is used in combination with TBP in the TRUEX process for the partitioning of tri-, tetra-, and hexavalent actinides from nitric acid media into organic solvents. Despite many advances, the need for further progress in the area of actinide separations chemistry is needed due to the magnitude of the nuclear waste problem (*2*).

**Ionic Liquids (ILs).** ILs are, simply put, organic salts that melt at, or near, room temperature. They have been widely investigated recently as solvents for synthetic and catalytic chemical and biochemical processes, as liquid electrolytes for electrochemistry, and as solvent phases for separations processes (*3*) and a large variety of examples are known. The versatile nature and unique solvent properties which include wide liquidus temperature range, high thermal stability, electrical conductivity and tunable physical properties based on suitable alterations of either the cationic or anionic component(s) of ILs provide the basis for their interest. ILs are composed of an organic cation, typically imidazolium-, pyridinium-, ammonium-, or phosphonium-based, and organic or inorganic anions; the most widely studied cations for forming ILs are 1-alkyl-3-methylimidazoliums (Figure 2).

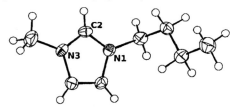

*Figure 2. Structure of the 1-butyl-3-methylimidazolium ([C₄mim]⁺) cation, representative of many ionic liquids. From the crystal structure of [C₄mim]Cl (4).*

Recently, ILs have been investigated in liquid-liquid separations applications (*5*). While ILs are currently specialized chemicals, and are relatively expensive to purchase or synthesize, especially compared to conventional commodity solvents, they do offer several advantages in separations systems. First, ILs have in general, no measurable vapor pressure and are not flammable. These are significant characteristics with regard to reduction of chemical and environmental hazards associated with exposure and the increasing regulation or restriction of volatile organic compounds (VOCs), particularly in industry. Second, in liquid form, ILs are composed completely of cations and anions, fully ionized and dissociated, and loosely arranged in the liquid state in a cation-anion-cation array in the liquid phase with no long-range ordering. These features make them unique systems for studying the coordination chemistry and extraction mechanisms of metal ions, particularly actinides (6).

In light of their extraordinary chemical and physical properties, we are studying ILs as alternative solvent media for actinide and fission product separations. While many traditional solvents are efficient and well-characterized media for such studies, ILs offer the potential to alleviate the hazards associated with many VOCs as well as provide unique coordination environments not observed in many molecular solvents. In addition, radiolytic studies with some imidazolium-based ILs have shown that they possess comparable or superior stability when compared to traditional organic solvents(7). At The University of Alabama, we are focusing on core actinide and fission product research in an effort to assess the degree to which novel chemistry in nuclear separations can be assisted by, or done uniquely with, ILs. The main thrust of our research, the highlights of which are presented here, includes utilizing ILs as extraction solvents combined with traditional extractants, with a focus on elucidating the joint roles of solvent extraction and ion exchange mechanisms, applications of Task-Specific Ionic Liquids (TSILs) in which extractant functionality is incorporated into either the cationic or anionic components of the IL forming salt, and novel IL/aqueous biphasic systems.

## Traditional Extractants in Ionic Liquids

**Methods.** Using ILs as direct replacements for organic solvents in organic/aqueous liquid-liquid biphasic systems is currently limited to hydrophobic ILs that form biphasic systems when contacted with aqueous solutions. These ILs usually contain perfluorinated anions such as in $[C_n mim][PF_6]$ or $[C_n mim][(CF_3SO_2)_2N]$ ILs, where $[C_n mim]^+$ is the common 1-alkyl-3-methylimidazolium cation (but can also include pyridinium, ammonium, and phosphonium salts). Fundamental studies of hydrophobic ILs typically involve $[PF_6]^-$ ILs as a standard, but its implementation in practical separations schemes is hindered by its susceptibility to hydrolysis and the formation of HF (8). As a result, the focus for hydrophobic ILs has recently largely shifted to the bis(trifluoromethanesulfonyl)imide anion, $[(CF_3SO_2)_2N]^-$ (Figure 3) that offers much more stability to hydrolysis and acidic conditions but is much more expensive. A later section will center on the use of more environmentally friendly cations that form the basis of hydrophilic ILs in novel IL-based aqueous biphasic systems.

We have investigated the use of hydrophobic ILs in IL/aqueous biphasic separations systems for the uptake of both actinides and fission products from acidic media using traditional extractants, with a principle focus on radiotracer studies to determine distribution behavior under infinite dilution conditions, as a function of extractant, acid, or salt concentrations in the systems. Liquid-liquid samples containing equal volumes of the extractant-containing IL and salt or acid solution (varying between $10^{-3}$ M and $10^{1}$ M) are contacted and spiked with a desired activity of a given radiotracer. Partitioning of actinides ($^{230}$Th, $^{233}$U, $^{239}$Pu, and $^{241}$Am) and fission products ($^{137}$Cs, $^{85}$Sr, $^{99}$Tc, and $^{125}$I) have been investigated. Partitioning of radiotracers between the two phases is analyzed either by gamma counting or scintillation counting (for alpha and beta emitters) and distribution ratios for a given radiotracer due to partitioning by the extractant can be calculated.

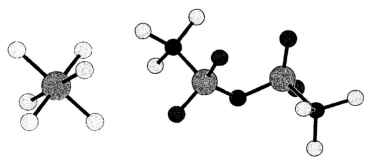

*Figure 3. Both hexafluorophosphate (left) and bis(trifluoromethanesulfonyl)imide anions (right) are highly perfluorinated and yield hydrophobic ILs. [PF$_6$]$^-$ is octahedral and charge diffuse whereas the charge carrying (S-N-S) core of [(CF$_3$SO$_2$)$_2$N]$^-$ is protected by trifluoromethyl and S=O groups.*

Analysis of distribution data from both nitric acid dependency and ligand dependency experiments allows for the elucidation of the nature of the extraction mechanism. Solvent extraction mechanisms involve the formation of adducts where metal cations bind with the extractant, along with certain anions to maintain charge neutrality, and are collectively transferred to the diluent phase as a neutral species. Ion exchange, on the other hand, is dependent upon the mutual exchange of cationic or anionic components between the phases to facilitate the transfer of a desired charged complex and maintain electroneutrality of each phase. Details surrounding the extraction mechanisms in the following sections will be discussed where appropriate and where supportive data is available (*9*).

**CMPO.** Octyl(phenyl)-*N,N*-diisobutylcarbomylmethylphosphine oxide (CMPO) is an effective extractant for tri-, tetra-, and hexavalent actinides from nitric acid media and has seen large-scale implementation in the TRUEX process with TBP as a phase modifier. The extraction of all actinides increases with increasing nitric acid concentration, with distribution ratios increasing in the order Pu(IV)>U(VI)>Am(III). Typical flowsheet studies use 0.2M CMPO along with 1.2M TBP in a dodecane solvent. Selective stripping is feasible at low acid concentration and with suitable choice of additional stripping agents.

We have studied the effect of actinide partitioning by substituting dodecane with hydrophobic ILs such as [C$_4$mim][PF$_6$]. Due to the limited solubility of CMPO in our ILs, studies were performed at concentrations of 0.1M CMPO and 1M TBP in the IL as well as in dodecane while varying nitric acid concentration. The maximum nitric acid concentration was limited to about 1M HNO$_3$ due to acid-induced hydrolysis of the anion. Figure 4 shows the distribution ratios of Am(III), Th(IV), U(VI), and Pu(VI) in the IL-based systems as compared to dodecane. While comparisons to previous studies are limited by CMPO solubility, it is clear that the distribution ratios at 0.1M CMPO extractant exceed those observed in dodecane by nearly an order of magnitude over the studied range. In all cases, addition of TBP as a phase modifier enhanced the extraction and gave higher distribution ratios. The distribution ratios for each species increase over the range of $10^{-2}$-$10^0$M HNO$_3$ (range limited due to hydrolysis of PF$_6^-$) with distribution ratios for $^{233}$UO$_2^{2+}$ - being greater than $10^3$ at the highest acid concentration (with 0.1M CMPO and 1M TBP). Stripping conditions have yet to be determined in the ILs and it appears that the high distribution ratios may hinder efficient removal of the actinides from the IL phase.

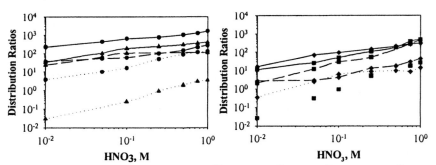

***Figure 4.*** *Distribution ratios for Am$^{3+}$ (▲), UO$_2^{2+}$ (●), Th$^{4+}$ (■), and Pu$^{4+}$ (♦) ions in [C$_4$mim][PF$_6$]/aqueous systems with CMPO as an extractant. Solid lines represent 0.1M CMPO/1M TBP/IL, dashed lines represent 0.1M CMPO/IL, and dotted lines represent 0.1M CMPO/1M TBP/dodecane.*

Preliminary studies using the more acid stable [C$_n$mim][(CF$_3$SO$_2$)$_2$N] ILs show the same trend as observed in [PF$_6$]$^-$ ILs with the distribution ratios for tri-, tetra-, and hexavalent actinides higher than the dodecane/aqueous systems over an even wider nitric acid concentration range. In $[C_6mim][(CF_3SO_2)_2N]$, distribution ratios follow the trend Am(III) > Pu(IV) > U(VI) in the $10^{-1}$-$10^1$ M $HNO_3$ range, with the values being as high as $10^3$ for Am(III) between 1 and 10M $HNO_3$ (with 0.1M CMPO). For IL systems using both $[PF_6]^-$ and $[(CF_3SO_2)_2N]^-$ anions, the nature of the extraction mechanism was analyzed and indicates an ion exchange mechanism rather than a solvent extraction mechanism as observed in many molecular solvents (*10*).

The coordination environment of $[UO_2]^{2+}$ extraction complexes from nitric acid into hydrophobic ILs has been investigated using Extended X-Ray Absorption Fine Structure (EXAFS) spectroscopy. In addition to yielding higher distribution ratios than in traditional dodecane, ILs provide a unique coordination environment not observed in molecular solvents. In dodecane systems, a $[UO_2]^{2+}$:$[NO_3]^-$:CMPO stoichiometry of 1:2:2 has been observed, while in $[C_4mim][PF_6]$ and $[C_8mim][(CF_3SO_2)_2N]$ systems, EXAFS suggests a 1:1:1 stoichiometry (*11*). This indicates that An coordination and extraction into IL solvents may be more efficient in terms of extractant required than into conventional solvents.

**Crown ethers.** The ability of crown ethers to selectively bind alkali (*12*) and alkaline earth metals (*13*) is well established. Their selectivity arises from a variety of factors including ring size and rigidity and donor atom types. For example, exchanging oxygen for sulfur may allow one to take advantage of the soft nature of the sulfur atoms in improving selectivity say, for example, of hard metal ions over soft metal ions (*14*). $Cs^+$ and $Sr^{2+}$ partitioning from nitrate media using a series of crown ether molecules dissolved in $[C_nmim][PF_6]$ (n = 4, 6, 8) ILs has been studied. The distribution ratios for these metal ions to the IL phase was observed to depend on both the hydrophobicity of the crown ether complexant and on the composition of the aqueous phase. The best distribution ratios were obtained for the most hydrophobic crown ether, 4,4'-(5')-di-(*tert*-butylcyclohexano)-18-crown-6 (Dtb18C6) (Figure 5). The composition of the aqueous phase seemed to have a dramatic effect on the distributions ratios for $Cs^+$ and $Sr^{2+}$, Addition of HCl, $Na_3$citrate, $NaNO_3$, and $HNO_3$ (at low concentrations due to hydrolysis of $[PF_6]^-$) to the aqueous phase led to a decrease in distribution ratios and also a decrease in the water content of the ILs. The addition of $Al(NO_3)_3$, however, led to an increased water content of the ILs and, surprisingly, increased distribution ratios probably due to a salting-out effect, with the trend ($D_{Sr} > D_{Cs}$) following traditional solvent extractant behavior. Distribution ratios for $Sr^{2+}$ were observed to be as high at 645 at 2M $Al(NO_3)_3$ when using $[C_8mim][PF_6]$ and stripping is possible via the addition of water after loading from $Al(NO_3)_3$ (*15*).

Extraction mechanisms for the partitioning of $Sr^{2+}$ with dicyclohexano-18-crown-6 (DCH18C6) into 1-alkyl-3-methylimidazolium ILs have been thoroughly investigated by Dietz, *et.al.* The extraction mechanism has been shown to change from cation exchange to strontium nitrato-crown ether complex partitioning (solvent extraction) as the hydrophobicity of the IL cation is increased, marked by an increase in alkyl chain length. For example, with $[C_5mim][Tf_2N]$ and $[C_6mim][Tf_2N]$ ILs, a cation exchange mechanism is favored whereas when the hydrophobicity of the IL is increased ($[C_8mim][Tf_2N]$), a mixed cation exchange/solvent extraction mechanism is observed, and increasing further the hydrophobicity of the IL ($[C_{10}mim][Tf_2N]$) results in extraction of metal ions as a neutral complex via a dominant solvent extraction mechanism (*16*).

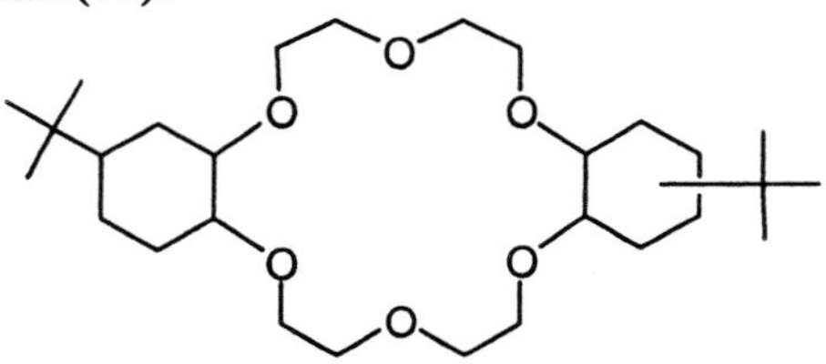

*Figure 5. 4,4'-(5')-di-(*tert-*butylcyclohexano)-18-crown-6 (Dtb18C6)*

## Task Specific Ionic Liquids (TSILs)

The tunable nature of ILs allows for the covalent attachment of extractant functionalities to the cation component in the rational design of ILs aimed at performing specific tasks. TSILs can be viewed as either (i) appending complexing functionality onto the IL components, or (ii) addition of an ionic moiety (IL component) to an extractant to increase affinity with an IL diluent. Incorporation of an extractant functionality into the ILs itself provides opportunities to develop new and novel chemistries with ILs while at the same time bypassing the need to add the extractant itself. In doing so, the favorable properties of the extractant can be maintained, leading to a "loaded" solvent environment with high extractive ability (*17*).

**Ethylene-oxide functionalized ILs.** We have recently attached polyether functionalities to an IL that act as an ethylene oxide (EO-*n*) bridge between two imidazolium cations with the potential benefit of selective extraction behavior observed in crown ethers.

Partitioning studies using radiolabeled $Cs^+$ and $Hg^+$ were performed from nitric acid media into the TSIL, made as the $[(CF_3SO_2)_2N]^-$ salt, which remains relatively hydrophobic. No extractive ability is observed for the $Cs^+$ ion over a wide nitric acid range with the EO-2 TSIL, but the $Hg^+$ ion shows a strong acid dependency, with distribution ratios decreasing with increasing nitric acid concentration. Notably, the distribution of Hg(II) with this TSIL as the extracting phase is higher than that using a 'conventional' IL, or when the IL phase is doped with polyethylene glycol as an extractant, showing that the imidazolium functions in the extractant do play an active role in complexation and indicating that improved and selective TSILs can be designed. Studies are currently underway to prepare TSILs with a longer ethylene oxide chains (Figure 6) in order to provide a larger "pocket" for metal ion binding and to investigate selective $Sr^{2+}$ binding over $Cs^+$ in nuclear waste simulants (18).

**Urea, thiourea, and thioether functionalized ILs.** The Davis group at the University of South Alabama has synthesized a series of TSILs with urea, thiourea, and thioether functionalities covalently appended to the imidazolium cation. Partitioning studies were performed using radiolabeled $Hg^{2+}$ and $Cd^{2+}$ and these $[PF_6]^-$ TSILs as a function of nitric acid concentration. Distribution ratios using the thioether-appended TSIL show comparable ability for extracting both $Cd^{2+}$ and $Hg^{2+}$ into the IL phase at both pH 1 and pH 7 using both the pure TSIL and a 1:1 TSIL:$[C_4mim][PF_6]$ doped system. The urea- and thiourea-appended TSILs show a differing trend with $Hg^{2+}$ being more efficiently extracted than $Cd^{2+}$ over all acid concentrations using 1:1 doped systems. In many cases, the difference in distribution ratios between $Hg^{2+}$ and $Cd^{2+}$ was maintained over all nitric acid concentrations, thus necessitating conditions other than pH dependence for stripping of these metal ions (19).

**CMPO-functionalized TSIL.** The Davis group at the University of South Alabama has also synthesized a phosphonamide IL where the imidazolium cation has been functionalized with a phosphine oxide functional group similar to that seen in CMPO (Figure 7). Preliminary results from the extraction of tri-, tetra-, and hexavalent actinides using the $[PF_6]^-$ TSIL doped in a 1:5 ratio with another IL show high distribution ratios over a wide pH range, no pH dependence, and little selectivity for one actinide over another. Further studies are in progress to understand the coordination environment as well as the extraction mechanism (17).

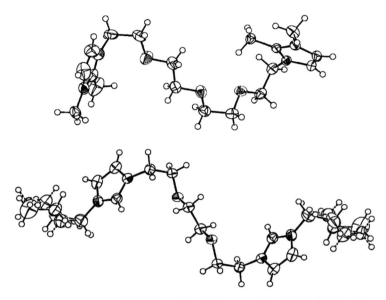

*Figure 6.* Examples of ethylene-glycol functionalized bis-imidazolium IL cations containing two (top) or three (bottom) ether groups, from the crystal structures of the tetraphenylborate salts.

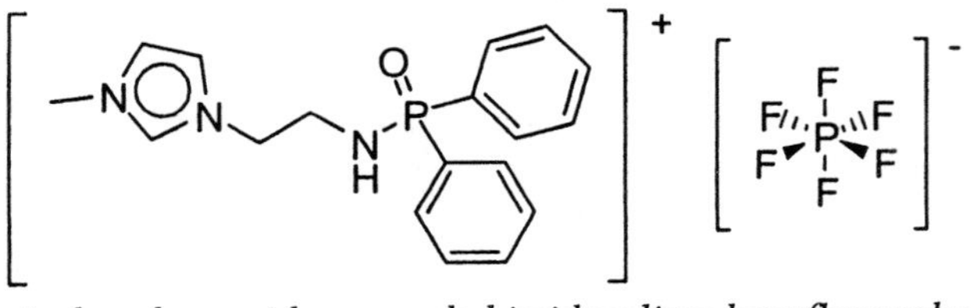

*Figure 7.* A phosphonamide-appended imidazolium hexafluorophosphate IL.

# Aqueous Biphasic Systems (ABS)

**Polymer/salts ABS.** Polyethylene-glycol (PEG) has been well studied for its role in the formation of aqueous biphasic systems via the addition of kosmotropic salts. Here, the ability of any given kosmotropic salt to "salt-out" PEG is based upon the ion's Gibb's Free Energy of Hydration ($\Delta G_{hyd}$). A large negative $\Delta G_{hyd}$ results in less salt being required to form biphasic systems with PEG, yielding better phase divergence with increasing salt concentration. In addition to these ABS, PEG functionalities of various chain length have been appended to resins for applications to Aqueous Biphasic Extraction Chromatography (ABEC™). Both ABS (20) and ABEC™(21) have been applied to the removal of $^{99}$Tc and $^{125}$I from salt solutions and tank waste simulants. In PEG-based ABS, addition of [TcO$_4$]⁻ to a biphasic PEG/salt system results in the partitioning of [TcO$_4$]⁻ to the PEG rich phase without the need for an extractant. Partitioning increases with the kosmotropic nature of the salt and the salt concentration. Partitioning of [TcO$_4$]⁻ on columns filled with ABEC™ resin follows the same theory: [TcO$_4$]⁻ will quantitatively stick to the resin as a [TcO$_4$]⁻-containing kosmotropic salt solution is run down the column. Stripping of the [TcO$_4$]⁻ from the column is easily achieved by passing water down the column.

**Ionic Liquid-based ABS.** As described previously, the use of ILs as direct replacements for organic solvents in organic/aqueous liquid-liquid separations schemes is largely restricted to hydrophobic ILs containing anions that are usually both perfluorinated and expensive. As a result, implementation of these types of ILs on a large scale will be limited due to cost or environmental impact. The potential environmental impact of the [PF$_6$]⁻ anion in ILs was recently established; hydrolysis and degradation at high acidity of this anion leads to the formation of toxic, environmentally non-benign compound, HF. The potential to use hydrophilic ILs in separations schemes is particularly appealing due to the availability of 'greener' anions, cheaper costs, and broader range. Recently, we have shown that aqueous solutions of hydrophilic ILs can be induced to form biphasic systems in contact with concentrated solutions of water-structuring (i.e., kosmotropic) salts, such as in the example of [C$_4$mim]Cl and K$_3$PO$_4$ shown in Figure 8 (22).

The addition of aqueous K$_3$PO$_4$ to aqueous [C$_4$mim]Cl results in the formation of an aqueous biphasic system that has an upper phase rich in IL and a lower phase rich in K$_3$PO$_4$. It is believed that the water-structuring nature of the phosphate anion increases the hydrogen-bonding network in the lower aqueous phase, thus forcing the bulky organic cation to the IL-rich phase. Tracer studies using $^{14}$C-labeled [C$_4$mim]Br, $^{32}$P-labeled H$_3$PO$_4$, and $^{36}$Cl-labeled NaCl confirm phase divergence with increasing K$_3$PO$_4$. Figure 9

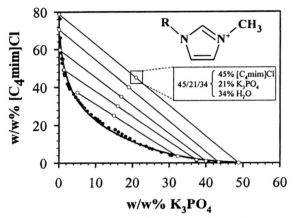

**Figure 8.** *Binodal phase diagram of [C₄mim] Cl/K₃PO₄/aqueous mixtures showing formation of biphase regions with composition.*

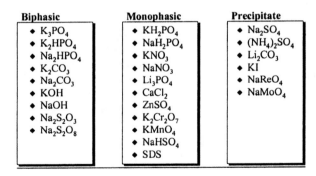

**Figure 9**. *Examples of inorganic salts and an observation of their propensity to form biphases with [C₄mim] Cl IL.*

shows the ability of a range of common inorganic salts to induce biphase formation with [C₄mim]Cl, related directly to the kosmotropic, or water-structuring nature of the respective salts.

This phenomenon is not limited to imidazolium-based ILs, and has also been observed previously with quaternary tetraalkylammonium salts (*23*). Further investigation has shown that this is indeed, a general phenomenon that occurs for other classes of ILs including pyridinium-, phosphonium-, and ammonium-based systems. Hence, the expansion of IL-based separations systems to include hydrophilic ILs is a distinct possibility. In fact, these aqueous biphasic systems possess a sufficient chemical potential between the phases to allow organics, such as short-chain alcohols, and inorganics such as [TcO₄]⁻ to partition to the IL-rich phase without the need for an extractant. Partitioning is observed to increase with both phase divergence as well as the number of carbons in the alcohol alkyl chain. Figure 10 shows partitioning of [$^{99}$TcO₄]⁻ to IL-rich upper phase of IL/salt aqueous biphase of [C₄mim]Cl/K₃PO₄ as a function of tie-line length.

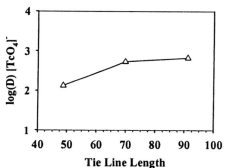

*Figure 10. Preliminary data on the partitioning of [$^{99}$TcO₄]⁻ from aqueous salt solution to IL-rich upper phase of IL/salt aqueous biphase of [C₄mim]Cl/K₃PO₄ as a function of tie-line length (taken from Figure 8)*

Biphasic systems from hydrophobic ILs can potentially be applied to the remediation of nuclear tank wastes due to the highly-salted nature of the supernatant. The liquid supernatant waste in a storage tank is highly alkaline and extremely radioactive due to the presence of $^{137}$Cs as well as [$^{99}$TcO₄]⁻. Studies are underway to explore the potential of forming biphasic systems with hydrophilic ILs when contacted with nuclear waste simulants, the speciation or metathesis that may be occurring, and the use of extractants for the removal of metal ion contaminants.

## Conclusions

The application of ILs to nuclear separations is and will continue to be a growing and diversifying area of research. While ILs are becoming highly studied in areas such as biochemistry, catalysis, and organic synthesis, their potential as solvents for extractions and coordination studies of actinides and fission products is largely untapped. ILs offer many distinct advantages over both traditional nuclear separations solvents such as dodecane and kerosene, as well as the volatile organic compounds of industry, including little or no vapor pressure, non-flammability, and tunable physical properties. ILs have successfully been shown to act as solvent phases for the extraction of metal ions, backbones for TSILs based on well-known extractant functionalities, and the basis for novel aqueous biphasic systems. In addition, ILs provide unique coordination environments as compared to molecular solvents, and sometimes alternate extraction mechanisms. Currently, ILs are expensive commodity chemicals that will continue to become more affordable, and as they become more thoroughly studied as a class of solvents, their impact on the area of nuclear remediation and separations will continue to expand.

## Acknowledgments

This work was supported by funding from the Environmental Management Science Program of the Office of Environmental Management, U.S. Department of Energy, Grants DE-FG07-01ER63266 and DE-FG07-01ER63296, the U. S. Environmental Protection Agency's STAR program (Grant number RD-8314201-0) and the Division of Chemical Sciences, Geosciences, and Biosciences, Office of Basic Energy Research, U. S. Department of Energy (Grant DE-FG02-96ER14673).

## References

1   *Metal-Ion Separation and Preconcentration: Progress and Opportunities*; Bond, A. H.; Dietz, M. L.; Rogers, R. D., Eds.; ACS Symposium Series 716, American Chemical Society: Washington, D.C., 1999.
2   Schulz, W. W.; Horwitz, E. P. *Sep. Sci. Technol.* **1988**, *23* 1191; Mathur, J. N.; Murali, M. S.; Natarajan, P. R.; Badheka, L. P.; Banerji, A. *Talanta* **1992**, *39*, 493.

3   *Ionic Liquids in Synthesis*; Wasserscheid, P.; Welton T., Eds., VCH-Wiley: Weinheim, 2002; *Ionic Liquids: Industrial Applications to Green Chemistry*; Rogers R. D.; Seddon, K. R., Eds.; ACS Symposium Series 818, American Chemical Society: Washington D.C., 2002; *Ionic Liquids as Green Solvents: Progress and Prospects*; Rogers R. D.; Seddon, K. R., Eds.; ACS Symposium Series 856, American Chemical Society: Washington D.C., 2003.

4   Holbrey, J. D.; Reichert, W. M.; Nieuwenhuyzen, M.; Johnston, S.; Seddon, K. R.; Rogers, R. D. *Chem. Commun.* **2003**, 1636.

5   Visser, A. E.; Swatloski, R. P.; Griffin, S. T.; Hartman, D. H.; Rogers, R. D. *Sep. Sci. Technol.* **2001**, *36*, 785.

6   Visser, A. E.; Swatloski, R. P.; Hartman, D. H.; Huddleston, J. G.; Rogers, R. D. In *Calixarenes for Separations*; Lumetta, G. J.; Rogers, R. D.; Gopalan, A. S., Eds.; ACS Symposium Series 757, American Chemical Society: Washington D.C., 1983, p 223-236.

7   Allen, D.; Baston, G.; Bradley, A. E.; Gorman, T.; Haile, A.; Hamblett, I.; Hatter, J. E.; Healey, M. J. F.; Hodgson, B.; Lewin, R.; Lovell, K. V.; Newton, B.; Pitner, W. R.; Rooney, D. W.; Sanders, D.; Seddon, K. R.; Sims, H. E.; Thied, R. C. *Green Chem.* **2002**, *4*, 152.

8   Swatloski, R. P.; Holbrey, J. D.; Rogers, R. D. *Green Chem.* **2003**, *5*, 361.

9   *Principles and Practices of Solvent Extraction*; Rydberg, J.; Musikas, C.; Choppin, G. R. Eds.; Marcel Dekker: New York, 1992; *Ion Exchange and Solvent Extraction*; Marinsky, J. A.; Marcus, Y., Eds.; Marcel Dekker: New York, 1966-1997; Vols. 1-13.

10  Visser, A. E.; Rogers, R. D. *J. Solid State Chem.* **2003**, *171*, 109.

11  Visser, A. E.; Jensen, M. P.; Laszak, I.; Nash, K. L.; Choppin, G. R.; Rogers, R. D. *Inorg. Chem.* **2003**, *42*, 2197.

12  Marcus, Y.; Asher, L. E. *J. Phys. Chem.* **1978**, *82*, 1246.

13  Tsurubou, S.; Mizutani, M.; Kadota, Y.; Yamamoto, T.; Umetani, S.; Sasaki, T.; Le, Q. T. H.; Matsui, M. *Anal. Chem.* **1995**, *67*, 1465.

14  Bradshaw, J. S.; Izatt, R. M.; Bordunov, A. V.; Chu, C. Y.; Hathway, J. K. *Comprehensive Supramolecular Chemistry, Volume 1: Molecular Recognition: Receptors for Cationic Guests*, Pergamon: Oxford, **1996**, 35.

15  Visser, A. E.; Swatloski, R. P.; Reichert, W. M.; Griffin, S. T.; Rogers, R. D. *Ind. Eng. Chem. Res.* **2000**, *39*, 3596.

16  Dai, S.; Ju, Y. H.; Barnes, C. E. *J. Chem. Soc., Dalton Trans.* **1999**, 1201; Dietz, M. L.; Dzielawa, J. A.; Laszak, I.; Young, B.A.; Jensen, M.P. *Green Chem.* **2003**, *5*, DOI: 10.1039/b21507p.

17  Davis Jr., J. H. In *Ionic Liquids: Industrial Applications to Green Chemistry*; Rogers, R. D.; Seddon, K. R., Eds.; ACS Symposium Series 818, American Chemical Society: Washington D.C., 2002, 247-258.

*18* Holbrey, J. D.; Visser, A. E.; Spear, S. K.; Reichert, W. M.; Swatloski, R. P.; Broker, G. A.; Rogers, R. D. *Green Chem.* **2003**, *5*, 129.

*19* Visser, A. E.; Swatloski, R. P.; Reichert, W. M.; Mayton, R.; Sheff, S.; Wierzbicki, A.; Davis Jr., J. H.; Rogers, R. D. *Chem. Commun.* **2001**, 135.

*20* Rogers, R. D.; Bond, A. H.; Bauer, C. B.; Zhang, J.; Rein, S. D.; Chomko, R. R.; Roden, D. M. *Solvent Extr. Ion Exch.* **1995**, *13*, 689.

*21* Rogers, R. D.; Griffin, S. T.; Horwitz, E. P.; Diamond, H. *Solvent Extr. Ion Exch.* **1997**, *15*, 547.

*22* Gutowski, K. E.; Broker, G. A.; Willauer, H. D.; Huddleston, J. G.; Swatloski, R. P.; Holbrey, J. D.; Rogers, R. D. *J. Am. Chem. Soc.* **2003**, *125*, 6632.

*23* Nagaosa, Y.; Sakata, K.; *Talanta*, **1998**, *46*, 647.

Chapter 4

# Task-Specific Ionic Liquids for Separations of Petrochemical Relevance: Reactive Capture of $CO_2$ Using Amine-Incorporating Ions

James H. Davis, Jr.

Department of Chemistry, University of South Alabama, Mobile, AL 36688 and The Center for Green Manufacturing, University of Alabama, Tuscaloosa, AL 35487

## Introduction

Natural gas, largely composed of $CH_4$, is rarely removed from the earth in a state pure enough to trunk directly into pipeline systems for delivery to consumers *(1)*. Depending upon the particular production field, the gas usually contains adulterants which run the gamut from water and heavier hydrocarbons to so-called acid gases. Perhaps more so than the removal of some species, the scrubbing of acid gases – $CO_2$, $H_2S$ and others – from natural gas streams is a process of vital importance. The former is non-combustable and its persistance in a gas stream lowers the fuel value of the entraining gas. In contrast, $H_2S$ burns but is highly toxic and its combustion produces noxious and corrosive sulfur oxides. With supplies of relatively pure "sweet gas" dwindling in the face of increasing world demand, production fields regarded heretofore as being of marginal value are being increasingly looked to as the production sources of the future. These "sour gas" fields will demand unprecedented scales of treatment for the removal of acid gases.

The removal of acid gases from natural gas – "scrubbing" or "sweetening" – is a process which has been practiced since the 1930's (2). A number of technologies have been used to accomplish this purification, some of them based upon the differing physical properties of the gases, but most relying upon chemical agents to effect acid gas capture. In turn, chemical agents are used in either of two ways, as physical solvents or reagents for reactive capture. In either case, the established materials used in these processes themselves introduce complications into the gas removal process. As both gas demand and pressures for increased sour gas utilization increase, so too will the demand for improved scrubbing agents.

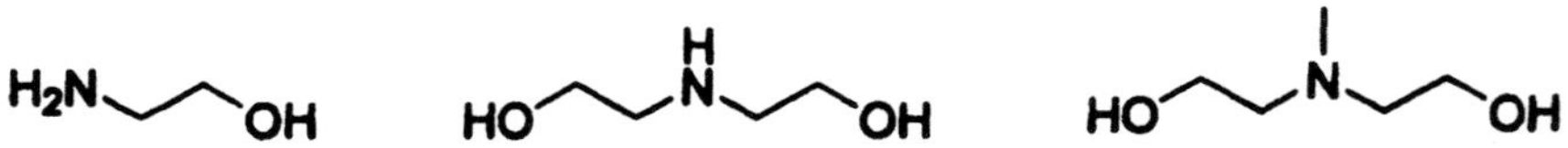

*Figure 1. Alkanolamines in common use for $CO_2$ scrubbing. Left to right: monoethanolamine (MEA); diethanolamine (DEA); methyldiethanolamine (MDEA).*

The most widely used technology for the removal of acid gases is based upon scrubbing by aqueous alkanolamines. Commonly used alkanolamines may be primary, seconday or tertiary amines. The structures of several frequently used materials are depicted in Figure 1. The means by which alkanolamine effect the capture of $CO_2$ is twofold [Scheme 1]. In the case of scrubbing reagents with primary or secondary amines, the $CO_2$ can react with two equivalents of amine, forming ammonium carbamates. In the case of any of the amine groups, reaction of the amine with the water solvent produces equilibrium concentrations of ammonium hydroxides, which react with $CO_2$ to form bicarbonate ion. In the case of $H_2S$, the principal operating mechanism appears to be the formation of hydrosulfide anion.

Regardless of the gas being removed, the purging of the natural gas stream through the scrubbing solution creates problems. In the first place, the exiting gas stream may be rendered acid gas free, but it has been thoroughly charged with water vapor which has to be removed in a subsequent, energy intensive condensation step. At the same, the alkanolamine – which is a neutral compound – may be taken up in some quantity into the gas stream. While not posing problems in the sense that it is both combustable and not especially toxic, its loss represents some economic drag on the scrubbing process. Finally, the extrusion of $CO_2$ from the loaded or spent scrubbing fluid may require the heating of the solution, which of course involves heating the solvating water along with the reagent, another undesireable energy cost.

In terms of their potential utility in gas processing, ionic liquids can in a very real way be thought of as liquid solids. To the degree that solids have been considered or are utilized as sorbents for gas processing in any arena, their chief advantage over liquids stems from their lack of vapor pressure. However, the limited surface area for solids – even materials like porous carbons – makes their capacity generally lower than ideal for many gas processing applications. When the solubility of a particular gas in a certain liquid is higher than its adsorption limit on a specific solid, the liquid would be preferred *except for the potential for cross-contamination created by the liquids' own vapor pressure.* Notably, one of the defining characteristics of ionic liquids is their lack of measureable vapor pressure up to their decomposition point. Given the foregoing considerations, it seemed opportune to us to investigate ionic liquids for the capture of various acid gases.

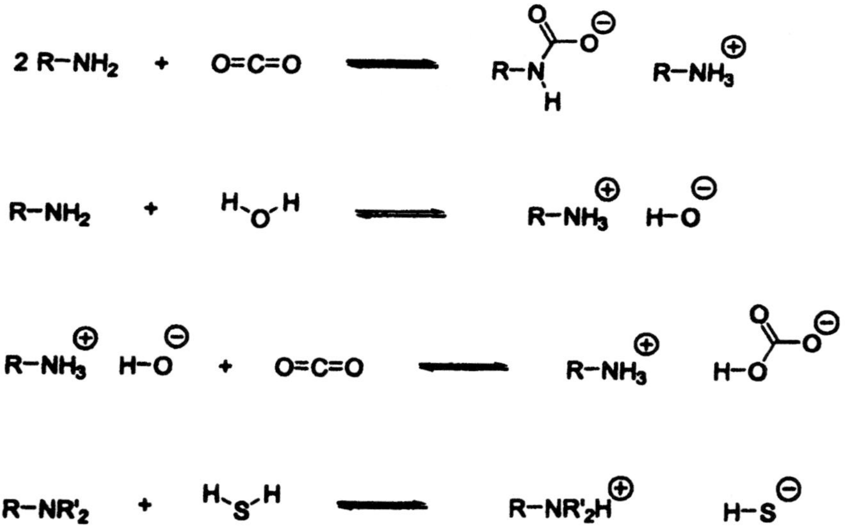

*Scheme 1. Mechanisms of reactive acid gas capture operative in aqueous amine solutions.*

## Task-Specific Ionic Liquids for Reactive Capture

In a paper presented in 2000 at an AIChE symposium on advances in solvent selection, we coined the term "task-specific" ionic liquids to describe a new, non-traditional type of IL under development by us in which the cation, anion, or both of the IL incorporated functional groups capable of imparting specific properties or reactivities to the resulting IL (3).     Among the ILs under development were materials incorporating appended amine groups.   These materials became the initial testbed for our continuing studies of the reactive capture of acid gases with ionic liquids.

Our first TSIL with appended amine groups built upon the work of Herrmann, who had reported the synthesis of an amine-appended imidazolium cation which he used as a precursor to imidazolidene carbene ligands (4). In that instance, the salt was not isolated as the amine free base, and the counterion was bromide, making the salt both high melting and non-basic, impediments to their utilization as ILs for acid gas capture. After developing a protocol for both the generation of the free amine and an anion exchange, we isolated several structurally related amine-appended imidazolidene ILs in modest yields. Using these compounds, we were able to demonstrate their utility as $CO_2$ capture reagents (5). For example, using IL 1 [Figure 2], we were able to capture near the theoretical maximum of $CO_2$ for this salt (0.5 mol / mol). Given the water-free nature of both the IL and the model gas stream, we expected the capture mechanism to be formation of an ammonium carbamate derivative of the IL. The $^{13}$C NMR of the product salt after treatment with $CO_2$ firmly established this to be the case, the expected doubling of all imidazolium ring peaks being observed as well as a signal for the carbamate carbon (5).

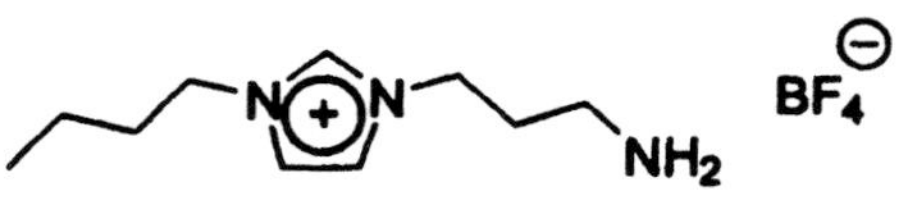

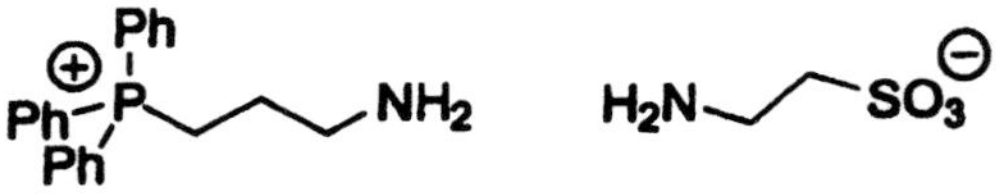

Figure 2. Amine appended TSILs with one (IL 1) and two (IL 2) appended amine groups per ion pair.

As mentioned, the incorporation of one amine group per ion pair coupled with the need for two amine groups to effect $CO_2$ capture by carbamate formation dictates that the maximum capture capacity for salts of type 1 is 0.5 mol $CO_2$ / mol IL. Consequently we began to examine approaches to boost the $CO_2$ capture capacity by boosting the number of amine groups per ion pair. While there are several potential approaches for doing so, we began our effort by focusing on the anion as the carrier for the second amine group. There are several immediate possibilities in this regard, including the conjugate bases of amino acids, a concept suggested by Quinn and Pez in patents concerning the use of certain salt hydrates for $CO_2$ capture *(6)*. However, since carboxylate-based anions tend to be less than ideal in IL formulations, we decided to probe the use of the non-proteogenic amino acid taurine in combination with amine-appended cations. At the same time, we decided to probe the use of a new amine appended cation, this species incorporating a phosphonium head group [Figure 2]. Our rationale for doing so stemmed from both a desire to identify less costly head groups as well as to avoid the potential for imidazolium $C^2$-H deprotonation by the appended amine group. It should be noted that the isolated salt contained 3-6 molecules of water per ion pair which were tenaciously bound, making it formally a molten salt- or ionic liquid hydrate.

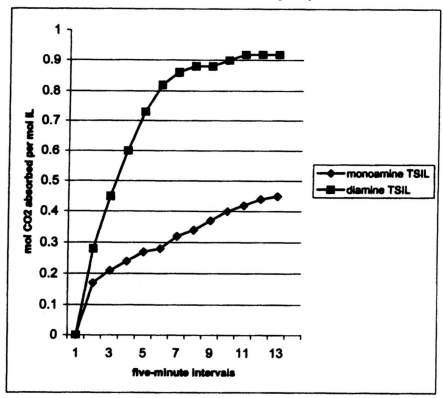

*Figure 3. Measured $CO_2$ uptake capacity of TSILs containing one versus two amine groups per ion pair.*

When exposed to a stream of dry $CO_2$, bis-amine IL 2 proved to be capable of absorbing twice as much $CO_2$ per mole of IL as did IL 1. [Figure 3] In addition, where the viscosity of 1 visibly increased with the uptake of $CO_2$, that of 2 appeared to decrease, a phenomenon we have yet to understand. The [13]C-NMR of the product exhibited two poorly resolved peaks for carbamate carbons (presumably one for taurine-bound carbamate and one for cation bound carbamate), but also exhibited a small third peak in the same vicinity, ascribed to bicarbonate. Observation of the latter was not surprising given the water content of the sequestering salt. Like $CO_2$ loaded 1, decarboxylation is readily accomplished by heating the samples in vacuo overnight. Uptake experiments were repeated on the same sample of 2 for five runs with no observed diminution of $CO_2$ uptake capacity.

## Task-Specific Ionic Liquids for Increased $CO_2$ Physical Solubility

As mentioned, current technology for gas sweetening uses aqueous solutions of alkanolamines. Given that the sequestration mechanism turns on either the direct reaction of the amine group with $CO_2$ or with water to generate hydroxide, the question arises as to what function is served by the alcohol group of these molecules. In short, the alcohol group increases the solubility of the amine in the water and aids in its retention by the scrubbing solution as the natural gas is purged through the system. Unfortunately, the presence of the alcohol in the same molecule as the amine (especially on a β carbon) also renders these molecules susceptible to (under certain pH conditions) the formation of five-membered cyclic carbamates, species from which extrusion of $CO_2$ is not readily accomplished. Over time, this leads to a loss of capture capacity on the part of the scrubbing solution. With amine-appended TSILs, such problems are avoided since the molecules do not incorporate the complicating hydroxyl group. Still, as $CO_2$ is absorbed by the TSILs, each system becomes not just a mixture of ions but rather one of zwitterions. In this case, one can imagine the creation of an extensive network of ionic cross-links which could effect diffusivity of excess $CO_2$ through the system. Such a process appears to be operative, being evidenced in TSILs applied to membranes for $CO_2$ separation becoming rapidly impermeable to CO2 transport (7). And, since the development of membrane based gas separation systems is being widely investigated, we have also begun to develop ionic liquids which we hope will prove to be better physical solvents for $CO_2$, in order to provide a complimentary technology.

It is well known that a number of highly polar aprotic organic solvents have a high capacity for dissolving certain gases such as $CO_2$. While most of these solvents – compounds such as dimethylformamide (DMF), dimethylsulfoxide (DMSO) and N-methylpyrrolidinone (NMP) – have relatively low vapor pressures, those vapor pressures still pose a problem for gas processing. At the same time, elegent work by Brennecke has shown that $CO_2$, compared to several other gases, has a relatively high solubility in "conventional" ionic liquids, and at least one patent has been issued based upon their use in membrane systems *(8)*. Consequently, it is our hope that task-specific ionic liquids in which one of the ions incorporates functionality similar to that in DMF, DMSO, NMP or other solvents might exhibit even higher capacities for the solvation of $CO_2$. To this end, we have been working to prepare such compounds.

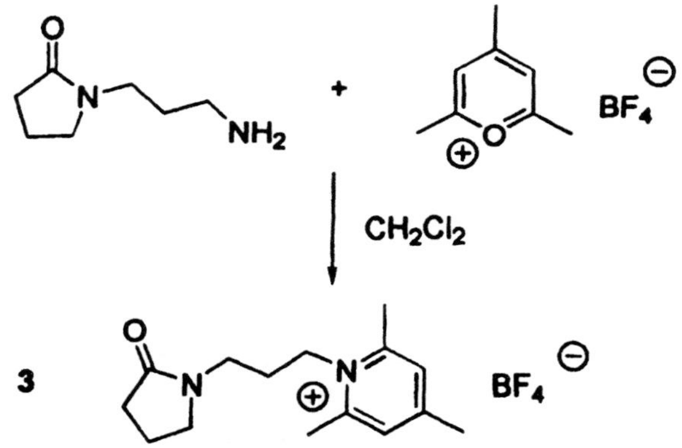

*Figure 4. Synthesis of an NMP-like TSIL from commercially available starting materials.*

One of our first successful forays in this area centers on the synthesis of a pyridinium IL with an appended NMP-like group, 3 [Figure 4]. We believe our synthetic methodology for the assembly of this compound to be unique as an approach to IL synthesis. It is well established that pyrrylium salts will react directly with primary amines, cleanly substituting an R-N fragment for the ring oxygen with the concommitant loss of water. By combining the commercially available compounds 2,4,6-trimethylpyrrylium tetrafluoroborate with N-(3-aminopropyl)-pyrrolydin-2-one, we were able to isolate IL 3 as a viscous liquid in 58% yield. We note that IL 3 is not the first pyrrolidinone-based IL to have been proposed. A 2002 paper by Demberelnyamba reported the synthesis of a new IL by N-alkylation of N-vinyl pyrrolidinone with an alkyl iodide *(9)*. However, we have been unable to repeat this work and note that the N-alkylation of amides is virtually unprecedented. Further, a 1984 paper by Smith specifically descrbes the O-alkylation of N-vinyl pyrrolidinone with oxonium salts (these more powerful alkylating reagents being required to accomplish

56

alkylation on any part of the molecule), and the subsequent decomposition of those species to amino acids upon contact with water *(10)*.

To date, we have not begun to measure the solubility of $CO_2$ in IL 3. However, we have found the solubility of a number of polar molecules (e.g., sugars) to be higher in it than in conventional IL, pointing to the similarity of this molecule to that which it is intended to resemble, NMP. The complete results of our work to prepare new solvent-analog ILs as well as their gas solubility capacities will be reported in due course.

## References

1. *Understanding Natural Gas*. Accessible on the web at: siepr.stanford.edu/about/Natural_Gas.pdf
2. Kohl, A. and Nielsen, R. *Gas Purification*; Gulf: Houston, 1997.
3. Davis, J.H., Jr. and Wierzbicki, A. Proceedings of the Symposium on Advances in Solvent Selection and Substitution for Extraction; AIChE: New York, 2000; Paper 14F.
4. Herrmann, W..A.; Kocher, C.; Goossen, L. J. and Artus, G. R. J. *Chemistry: a European Journal*, 1996, *2*, 1625-1635.
5. (a) USP pending ( filed April 2002). (b) Bates, E. D.; Mayton, R. D.; Ntai, I.; Davis, J. H., Jr.; *J. Am. Chem. Soc.*, 2002; *124*, 926-927.
6. (a) Quinn, R.; Pez, G. P. and Appleby, J. B. USP 5,338,521 (August 16, 1994). (b) Quinn, R. and Pez, G. P. USP 4,973,456 (November 27, 1990).
7. Scovazzo, P. and Davis, J. H., Jr. Unpublished results.
8. (a) Anthony, J. L.; Maginn, E. J. and Brennecke, J. J. Gas Solubilities in 1-butyl-3-methyl imidazolium hexafluorophosphate. In *Ionic Liquids*; Rogers, R. D. and Seddon, K. R., Eds.; American Chemical Society: Washington, D. C., 2002; ACS Symp. Ser. No. 818, pp 260-269. (b) Brennecke, J. F. and Maginn, E. J. USP 6,579,343 (June 17, 2003).
9. Demberelnyamba, D.; Shin, B. K. and Lee, H. *Chemical Commun.*, 2002, 1538-1539.
10. Smith, M. B. and Shroff, H. N. *J. Org. Chem.*, 1984, *49*, 2900-2906.

# Chapter 5

# Ionic Liquids as Alternatives to Organic Solvents in Liquid–Liquid Extraction of Aromatics

G. Wytze Meindersma, Anita (J. G.) Podt,
Mireia Gutiérrez Meseguer, and André B. de Haan

University of Twente, Separation Technology Group, Faculty of Science
and Technology, P.O. Box 217, 7500 AE Enschede, The Netherlands

The ionic liquids 1-ethyl-3-methylimidazolium ethylsulfate
and 1,3-dimethylimidazolium methylsulfate are suitable for
extraction of toluene from toluene/heptane mixtures. The
selectivity for the toluene/heptane separation using [emim]
ethylsulfate increases with decreasing toluene content in the
feed, from 13.4 (95% toluene) to 55.3 (5% toluene) at 40°C
and from 8 to 35 at 75°C. With [mmim]methylsulfate the
selectivity increases from 6 to 67.8 at 40°C and from 11.4 to
55 at 75°C. These selectivities are a factor of 2-3 higher
compared to those obtained with sulfolane as the extractant.
Because ionic liquids have a negligible vapour pressure,
evaporating the extracted hydrocarbons could achieve the
recovery of the ionic liquid. A conceptual process scheme for
the extraction has been set up. Preliminary calculations show
that both the investment costs and the energy costs will be
considerably lower with ionic liquids compared to using
sulfolane as the solvent.

# Naphtha cracker process

Naphtha crackers convert naphtha into ethylene, propylene and other hydrocarbons by thermal cracking. Some cracker feeds contain a considerable amount of aromatic compounds, in the range of 10-25 wt%, that are not converted into olefins in the cracker furnaces, but remain as such in the process stream. The separation of aromatic hydrocarbons (benzene, toluene, ethyl benzene and xylenes) from $C_4$ - $C_{10}$ aliphatic hydrocarbon mixtures is challenging since these hydrocarbons have boiling points in a close range and several combinations form azeotropes. The conventional processes for the separation of these aromatic and aliphatic hydrocarbon mixtures are liquid extraction, suitable for the range of 20-65 wt% aromatic content, extractive distillation for the range of 65-90 wt% aromatics and azeotropic distillation for high aromatic content, >90 wt% [1].

In ethylene cracker plants, industrial separation of aromatic compounds and $C_5^+$ aliphatic hydrocarbons from the cracker products is carried out by extractive or azeotropic distillation of the $C_5^+$ hydrocarbon stream. Typical solvents used are polar components such as sulfolane, N-methyl pyrrolidone (NMP), N-formyl morpholine (NFM) or ethylene glycols. This implicates additional distillation steps to separate the extraction solvent from both the extract and raffinate phases and to purify the solvent, with, consequently, additional investments and energy consumption.

## Objectives

Although separation of aromatic and aliphatic hydrocarbons after the furnace section is industrial practice, selective removal of aromatic compounds from the cracker feed has not yet been considered. However, removing the major part of the aromatic compounds present in the feed to the crackers would offer several important advantages. They occupy a large part of the capacity of the furnaces, put an extra load on the separation section of the $C_5^+$-aliphatic compounds and the presence of aromatic compounds in the feed to the cracker has a negative influence on the thermal efficiency. Furthermore, the presence of aromatic compounds in the feed is undesirable, because they tend to foul the radiation sections (coking of the coils) and the Transfer Line Exchangers. The improved margin will be around $20/ton of feed or $ 48 million per year for a naphtha cracker with a feed capacity of 300 ton/hr. Preliminary calculations showed that extraction with conventional solvents is not an option because additional separation steps are required to purify the raffinate, extract and solvent streams, which would induce high investment and energy costs.

Extraction of aromatics with ionic liquids is expected to require less process steps and less energy consumption than extraction with conventional solvents because ionic liquids have a negligible vapour pressure.

For an economically feasible operation, the amount of aliphatic compounds in the aromatic product stream should be as low as possible. An option is to obtain the pure aromatic hydrocarbons from the aromatic product stream. On the other hand, total removal of the aromatic compounds from the feed is not necessary, as aromatics are formed during the cracking process. Therefore, the purity of both the aromatic and aliphatic product streams is set to 98 wt%.

The extraction of toluene from mixtures of toluene and heptane is used as a model for the aromatic/aliphatic separation.

# Extraction with ionic liquids

The variability of the anion and R-groups in imidazolium or pyridinium cations may be used to adjust the properties of the ionic liquids. Ionic liquids possess a number of properties, which may be of importance in their application as extractive media in liquid/liquid extraction processes. The requirements of a suitable extraction solvent for the separation of aromatic and aliphatic hydrocarbons are:

- High solubility of aromatic hydrocarbons in the IL
- No or low solubility of aliphatic hydrocarbons in the IL
- High separation factor and a high distribution coefficient
- Simple recovery of the IL from both the extract and the raffinate phase

### Selection of suitable ionic liquids

Aromatic hydrocarbons are soluble in some ionic liquids, such as [bmim]BF$_4$, [mmim]PF$_6$, [bmim]PF$_6$, [hmim]PF$_6$, [emim]I$_3$, [bmim]I$_3$, [omim]Cl, [bmim]Cl/AlCl$_3$, [mmim]Tf$_2$N (= [mmim]bis(trifyl)imide or [mmim][(CF$_3$SO$_2$)$_2$N]), [emim]Tf$_2$N, [bmim]Tf$_2$N, [1,2-dimethyl-3-ethyl imidazolium]Tf$_2$N, [emim][C$_2$H$_5$OSO$_3$], [4-methyl-N-butylpyridinium]BF$_4$, while aliphatic hydrocarbons (alkanes and cyclo-alkanes) are not [2 - 11]. This suggests that ionic liquids can be used as extractants for aromatic hydrocarbons from mixtures of aromatic and aliphatic hydrocarbons.

Brennecke and Maginn [12] reported in 2001 that until that date hardly any chemical engineer was involved in the design and development of ionic liquids for practical applications. There are only a few publications concerning extractive separations of aromatic/aliphatic hydrocarbon mixtures using ionic liquids [7, 8, 13 - 16]. Selvan et al. [7] describe the separation of toluene and heptane using [emim]I$_3$ and [bmim]I$_3$ at temperatures of 45°C and 35°C, respectively (Figure 1).

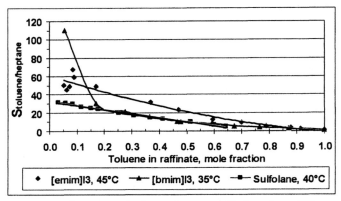

*Figure 1. Selectivity of toluene/heptane, IL data from [7].*

These ionic liquids are in principle suitable for the extraction of aromatic hydrocarbons from a mixed aromatic/aliphatic feed stream, since they show both a high selectivity and a high distribution coefficient at low toluene concentrations, although [bmim]I$_3$ showed a comparable selectivity as sulfolane at toluene concentrations above 17%. As the data at low concentrations of toluene are quite scattered ([emim]I$_3$) or consist of only one point ([bmim]I$_3$), it was necessary to carry out experiments in the low toluene concentration range to determine whether these ionic liquids are indeed suitable for the separation of aromatic and aliphatic hydrocarbons. However, we soon stopped experimenting with these ionic liquids, since they were very corrosive. Our stainless steel stirrers were corroded after only a few extraction experiments. Selvan et al. did not report any corrosion, but they used a Teflon paddle stirrer. Therefore, it is not surprising that we could not find any other references regarding these ionic liquids. Brennecke and Maginn [12] reported that halide containing ILs are certainly corrosive and Swatloski et al. [17] observed that HF formation is possible when using [bmim]PF$_6$. Therefore, we wanted to use halide-free ionic liquids for further testing in our project.

Krummen et al. [9] reported about the measurement of activity coefficients at infinite dilution ($\gamma_i^\infty$) for several solutes in the ionic liquids [mmim]Tf$_2$N, [emim]Tf$_2$N, [bmim]Tf$_2$N and [emim]ethylsulfate. From these data, selectivities ($S_{ij}^\infty = \gamma_i^\infty/\gamma_j^\infty$) and distribution coefficients ($k_i^\infty = 1/\gamma_i^\infty$) can be calculated (Figure 2a and b). From these figures, it is apparent that of these ionic liquids, [emim]ethylsulfate shows the highest selectivity for toluene/n-heptane. The

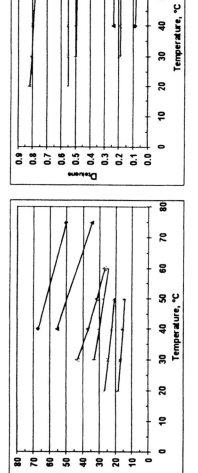

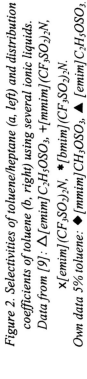

*Figure 2. Selectivities of toluene/heptane (a, left) and distribution coefficients of toluene (b, right) using several ionic liquids. Data from [9]: Δ[emim]C₂H₅OSO₃, +[mmim](CF₃SO₂)₂N, ✕[emim](CF₃SO₂)₂N, ✱[bmim](CF₃SO₂)₂N. Own data 5% toluene: ◆[mmim]CH₃OSO₃, ▲[emim]C₂H₅OSO₃.*

ionic liquid [bmim]⁺[Tf₂N]⁻ shows a lower selectivity and a higher capacity (distribution coefficient) than the ILs with cations [emim]⁺ and [mmim]⁺. The same trend is seen for other aromatic/aliphatic hydrocarbons tested: benzene/cyclohexane and benzene/hexane.

From the data of Krummen et al. [9], it is clear that a shorter R-group is favourable for a high toluene/heptane selectivity, but results in a decrease in capacity. This observation of a higher selectivity with shorter R-groups is in accordance with the findings of Hanke et al. [18]. The existence of π-electrons in orbitals above and below an aromatic ring results in a much stronger electrostatic field around an aromatic molecule compared to a saturated aliphatic molecule. Longer alkyl chains on the imidazolium group make the compound less aromatic in character, resulting in less interaction with the aromatic ring.

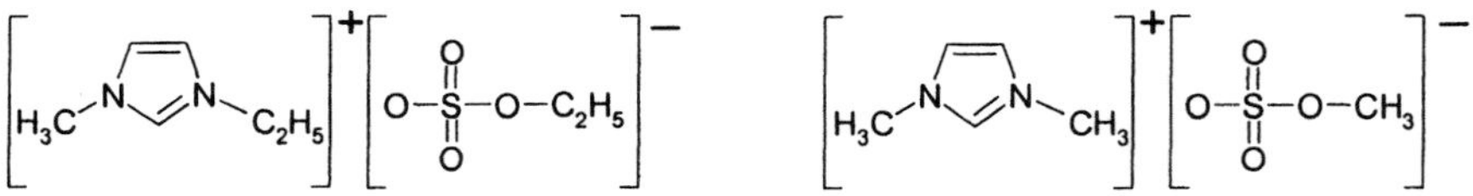

*Figure 3. 1-ethyl-3-methylimidazolium ethylsulfate ([emim]ethylsulfate) and 1,3-dimethylimidazolium methylsulfate ([mmim]methylsulfate).*

Based on the above-mentioned considerations, we decided to use the ionic liquids [emim]ethylsulfate and [mmim]methylsulfate (Figure 3) as extractants for the separation of toluene and heptane. If the same trend is occurring as with the [Tf₂N]⁻-containing ILs, we expect a higher selectivity and a lower capacity using [mmim]methylsulfate than with [emim]ethylsulfate, because both the R-group of the cation and the alkyl chain in the anion are shorter.

Other possibilities would have been to use [mmim]ethylsulfate or [emim] methylsulfate, but these ILs or the starting materials for a simple synthesis were not readily available.

## Experimental Section

### Chemicals

The following chemicals were purchased from Merck: toluene p.a. (min. 99.5 %), n-heptane p.a. (min. 99 %), n-hexane p.a. (min. 99 %), 1-butanol p.a.

(≥ 99.5 %). 1-Methylimidazole for synthesis (> 99 %) was purchased from Merck-Schuchardt. Diethyl sulfate (≥ 99 %) was purchased from Fluka and acetone-$d_6$, 99.5 atom% D, was purchased from Aldrich. The ionic liquid 1,3-dimethylimidazolium methylsulfate (98 %) was purchased from Solvent Innovation, Aachen.

### Preparation of [emim]ethylsulfate

The preparation of the ionic liquid [emim]ethylsulfate was based on the procedure described by Holbrey et al. [19]. Diethyl sulfate (116 mL) was added drop-wise to a mixture of 1-methylimidazole (70 mL) in toluene (400 mL), cooled in an ice-bath under argon. After addition of the diethyl sulfate, the reaction mixture was stirred at room temperature for five hours. Formation of two phases occurred: the toluene solution and a denser ionic liquid phase. The upper, organic phase was decanted and the lower, ionic liquid phase, was three times washed with toluene (total amount of toluene: 200 mL), dried at 75°C under reduced pressure (0 bar) in a rotary evaporator to remove any residual organic solvents and finally in a vacuum exsiccator.

A sample of the ionic liquid was dissolved in acetone-$d_6$ and analysed by $^1$H NMR (Varian 300 MHz).

### Experimental procedure

Before each experiment, the ionic liquids were dried at 75°C under reduced pressure in a rotary evaporator. Liquid-liquid extraction experiments were carried out in jacketed vessels with a volume of approximately 70 mL. The top of a vessel was closed using a PVC cover, through which a stirrer shaft passed. Two stainless steel propellers were used with an electronic stirrer (Heidolph RZR 2051 control). The vessels were jacketed to circulate water from a water bath (Julabo F32-MW) in order to maintain the temperature inside the vessels at either 40 or 75°C.

For each experiment, 20 mL of the ionic liquid and 10 mL of a toluene/n-heptane mixture were placed into the vessel. The temperature and the ratio of toluene in n-heptane were varied. We established that equilibrium was reached within 5 minutes. However, in order to be sure that the two phases were in equilibrium, the extraction experiment was continued for 15 minutes. After stirring, the two phases were allowed to settle for about one hour.

**Analysis**

Samples (approximately 0.5 mL) were taken from both phases. 1-Butanol (0.5 mL) was added to each sample to avoid phase splitting, n-hexane (0.2 mL) to the raffinate phase samples as an internal standard for the GC analysis and n-hexane (0.1 mL) to the extract phase samples.

The concentrations of toluene and n-heptane in the samples were analysed by a Varian CP-3800 gas chromatograph with an Alltech Econo-Cap EC-Wax column (30 m x 0.32 mm x 0.25 μm). However, most of the raffinate phase samples were analysed with a Varian CP-3800 gas chromatograph with a Varian 8200 AutoSampler. All extract phase samples were analysed with a Varian CP-3800 gas chromatograph without an autosampler. Because the ionic liquid has no vapour pressure, it could not be analysed by GC. Therefore, the concentrations of the ionic liquid in both phases were calculated by using a mass balance.

## Results and discussion

Liquid-liquid equilibrium data were collected for mixtures of toluene and heptane at 40 and 75°C using [emim]ethylsulfate and [mmim]methylsulfate.

The distribution coefficient or capacity, $D_i$, is directly calculated from the ratio of the mole fractions in the extract and raffinate phases. The distribution coefficients of toluene and heptane are defined by the ratio of the concentrations of the solute in the extract (IL) phase and in the raffinate (organic) phase, according to formula 1:

$$D_{tol} = C^{IL}_{tol}/C^{org}_{tol} \text{ and } D_{hept} = C^{IL}_{hept}/C^{org}_{hept} \tag{1}$$

The selectivity, $S_{tol/hept}$, of toluene/heptane is defined as the ratio of the distribution coefficients of toluene and heptane:

$$S_{tol/hept} = D_{tol}/D_{hept} = (C^{IL}_{tol}/C^{org}_{tol})/(C^{IL}_{hept}/C^{org}_{hept}) \tag{2}$$

From Figures 4 and 5, it can be seen that the distribution coefficients of toluene, using [emim]ethylsulfate as an extractant, remain slightly above 0.2 over the entire concentration range for both 40 and 75°C. The distribution

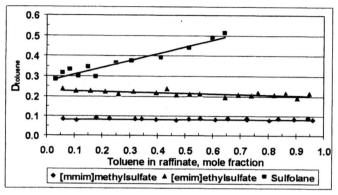

*Figure 4. Distribution coefficient of toluene, T = 40°C.*

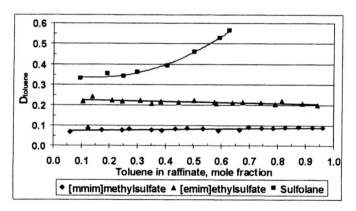

*Figure 5. Distribution coefficient of toluene, T = 75°C.*

coefficient of toluene with [mmim]methylsulfate is about 0.08 over the entire concentration range, also for both temperatures. It is lower than with [emim] ethylsulfate, due to the shorter alkyl chain in the imidazolium cation, as expected from the data of Krummen et al. [9].

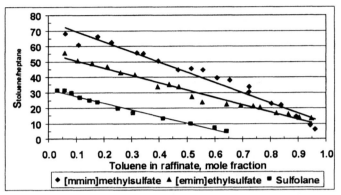

*Figure 6. Selectivity of toluene/heptane, T = 40°C.*

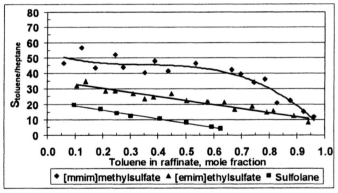

*Figure 7. Selectivity of toluene/heptane, T = 75°C.*

With [emim]ethylsulfate the toluene/heptane selectivity at 40°C increases from 13.4 to 55.3 with decreasing toluene content in heptane, from 95 to 5 v/v% in the feed (Figure 6). The toluene/heptane selectivity at 75°C increases from 8 (95% toluene in the feed) to 35 (10% toluene in the feed) (Figure 7). The selectivity at 75°C is lower than at 40°C, mainly due to higher concentrations of heptane in the extract (IL) phase, resulting in higher distribution coefficients for heptane (0.0043 at 40°C and 0.0057 at 75°C with 5% toluene), while the distribution coefficient of toluene remains about constant: 0.21 at 40°C and 0.22 at 75°C. The higher distribution coefficients of heptane at a higher temperature are in accordance with the findings of Krummen et al. [9]. From his data, distribution coefficients for heptane with [emim]ethylsulfate can be calculated: 0.0051 at 40°C and 0.0069 at 60°C.

With [mmim]methylsulfate, which has both a shorter alkyl chain on the imidazolium ring and a shorter alkyl group in the anion, the selectivity is considerably higher than using [emim]ethylsulfate, which is according to our expectations. The selectivities obtained with this IL range from 6 (95% toluene in the feed) to 67.8 (5% toluene in the feed) at 40°C and from 11.4 to 55 at 75°C (Figures 6 and 7). The data with [mmim]methylsulfate are quite scattered, because the analysis of heptane in the extract phase proved to be very difficult as the concentrations were extremely low: between 0.03 and 0.14 mole% at both temperatures.

For high toluene concentrations, the selectivity with [mmim]methylsulfate at 75°C increases at decreasing toluene content in the feed, but it levels off at concentrations of 60% and lower. This is caused by the distribution coefficient of heptane, which increases only slightly from 0.0014 to 0.0018 from 5% to 60% toluene in the feed and increases from 0.0018 to 0.0075 for 60% to 95% toluene in the feed. The distribution coefficients of toluene remained at about the same level of 0.08 over the entire concentration range.

Both the distribution coefficients of toluene and the toluene/heptane selectivities with [emim]ethylsulfate in our extraction experiments are higher than the values calculated by Krummen et al. [9] from activity coefficients at infinite dilution. The distribution coefficients of toluene at 40 °C are: 0.214 and 0.187, respectively (Figure 2b). We measured a toluene/heptane selectivity at 40°C of 55.3 at 5 v/v% toluene in the feed, compared to 36.4 from the data of Krummen et al. (Figure 2a). The differences in these values are probably due to the different methods, because Krummen et al. use VLE measurements to obtain activity coefficients at infinite dilution and calculate the capacity and selectivity from the activity coefficients. We have used LLE experiments and calculated the distribution coefficients and selectivities directly from the concentrations of the components in the extract and raffinate phases with formulas (1) and (2).

The distribution coefficients of toluene in sulfolane decrease from 0.50 (65% toluene in the feed) to 0.28 (5% toluene in the feed) at 40°C and from 0.56 (65% toluene) to 0.33 (10% toluene) at 75°C (Figures 4 and 5). These distribution coefficients are much higher than those obtained with the two ILs tested. However, the toluene/heptane selectivities of the ILs are a factor of 2-3 higher than those with sulfolane. The toluene/heptane selectivity with sulfolane increases from 4.6 (65% toluene in the feed) to 31 (5% toluene in the feed) at 40°C and from 3.7 (65% toluene) to 19 (10% toluene) at 75°C, as can be seen in Figures 6 and 7. A higher selectivity means less extraction stages, but a lower distribution coefficient requires a higher solvent to feed ratio.

Because ionic liquids have very low vapour pressures, it was possible to recover them after the experiments by drying the extract phase at 75°C under reduced pressure (0 bar) in a rotary evaporator. From NMR-analysis of the product, it was concluded that toluene and n-heptane were completely removed from the ionic liquid. Analysis by NMR of both the original ionic liquid and the regenerated ionic liquid showed no differences between the two samples. This means that no degradation of the ionic liquid after repeated recycling could be detected. This confirms the assumption that the regeneration and recycling of the ionic liquids are indeed simple.

The average concentration of the ionic liquid [emim]ethylsulfate in the raffinate phase, which also contains some toluene, is low: 0.9 mole% at 40°C and 2.5 mole% at 75°C. The average concentration of [mmim]methylsulfate in the raffinate is 0.12 mole% at 40°C and 1 mole% at 75°C. Removal of the ionic liquid from the raffinate phase is possible by using water, since these ionic liquids are water-soluble.

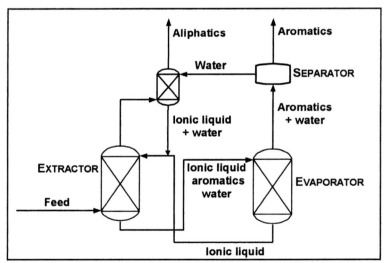

*Figure 8. Conceptual process scheme for extraction of aromatic hydrocarbons from aliphatic hydrocarbons with ionic liquids.*

A preliminary process design has been set up (Figure 8) and a flow-sheeting model was developed for this extraction to simulate the process. The raffinate phase from the extractor, containing aliphatic hydrocarbons and a very small amount of ionic liquid, is washed with water in order to remove the ionic liquid, which is recycled to the extractor. The extract phase from the extractor, containing the ionic liquid, the extracted aromatic hydrocarbons and the water from the water wash, is heated by low-pressure steam in an evaporator, where the volatile aromatic hydrocarbons and the water are being evaporated. The ionic liquid is recycled to the extractor. The top stream of the evaporator will then undergo a phase separation, where the water is separated from the organic compounds and can be recycled to the water wash.

From Figure 8, it is apparent that there are only a few process steps for the extraction of aromatics when ionic liquids are used, because regeneration of the ionic liquid is easily done by evaporation. Preliminary calculations indicate that the investment costs for extraction with ionic liquids are about 50% lower compared to an extraction with sulfolane, because fewer process steps for the purification of the product streams and for the regeneration of sulfolane are required. Also the energy costs will be lower, because low-pressure steam can be used for the evaporation, while the sulfolane process requires high-pressure steam due to the high temperature, 190°C, in the solvent stripper.

Optimisation of this process must still be carried out. For instance, the optimum between the solvent to feed ratio vs. the number of stages has to be found: a higher S/F ratio means fewer stages, but higher costs of the solvent. Also, the washing of the raffinate with water needs to be optimised further.

## Conclusions

The ionic liquids 1-ethyl-3-methylimidazolium ethylsulfate and 1,3-dimethylimidazolium methylsulfate are suitable for extraction of toluene from toluene/heptane mixtures. The selectivity for the toluene/heptane separation using [emim]ethylsulfate increases with decreasing toluene content in the feed: from 13.4 (95% toluene) to 55.3 (5% toluene) at 40°C and from 8 to 35 at 75°C. With [mmim]methylsulfate the selectivity increases from 6 to 67.8 at 40°C and from 11.4 to 55 at 75°C. These selectivities are a factor of 2-3 higher compared to those obtained with sulfolane as the extractant. However, the distribution coefficients of toluene in the ionic liquids are lower than in sulfolane.

Since ionic liquids have a negligible vapour pressure, evaporating the extracted hydrocarbons from the ionic liquid phase could achieve an easy recovery of the ionic liquid. NMR analysis showed no degradation of the ionic liquids after repeated regeneration and recycling of the ionic liquids.

A conceptual process scheme for the extraction has been set up. Preliminary calculations show that both the investment costs and the energy costs will be considerably lower compared to extraction using sulfolane as the solvent.

## Acknowledgements

The authors thank Peter Wasserscheid, RWTH Aachen, for helpful discussions about the application of suitable ionic liquids, Henny Bevers, Separation Technology Group, University of Twente, for analytical support, and DSM Research and Novem for their financial support.

## References

1. Weissermel, K.; Arpe, H.-J., *Industrial Organic Chemistry*, 4th Completely Revised Edition, Wiley-VCH, Weinheim, D., 2003, pp 313-336.
2. Dupont, J.; Consorti, C.S.; Spencer, J., Room Temperature Molten Salts: Neoteric "Green" Solvents for Chemical Reactions and Processes, *J. Braz. Chem. Soc.* **2000**, 11 (4) 337-344.
3. Domańska, U.; Marciniak, A., Solubility of 1-Alkyl-3-methylimidazolium Hexafluorophosphate in Hydrocarbons, *J. Chem. Eng. Data* **2003**, 48 (3) 451-456.
4. Holbrey, J.D.; Reichert, W.M.; Nieuwenhuyzen, M.; Sheppard, O.; Hardacre, C.; Rogers, R.D., Liquid clathrate formation in ionic liquid-aromatic mixtures, *Chem. Commun.* **2003**, 4, 476-477.
5. Blanchard, L.A.; Brennecke, J.F., Recovery of organic products from ionic liquids using supercritical carbon dioxide, *Ind. Eng. Chem. Res.* **2001**, 40 (1) 287-292 (+ 40, 2550).
6. Letcher, T.M.; Soko, B.; Ramjugernath, D.; Deenadayalu, N.; Nevines, A.; Naicker, P.K., Activity Coefficients at Infinite Dilution of Organic Solutes in 1-Hexyl-3-methylimidazolium Hexafluorophosphate from Gas-Liquid Chromatography, *J. Chem. Eng. Data* **2003**, 48, 708-711.
7. Selvan, M.S.; McKinley, M.D.; Dubois, R.H.; Atwood, J.L. Liquid-Liquid Equilibria for Toluene + Heptane + 1-Ethyl-3-methylimidazolium Triiodide and Toluene + Heptane + 1-Butyl-3-methylimidazolium Triiodide, *J. Chem. Eng. Data* **2000**, 45, 841-845.
8. Letcher, T.M.; Deenadayalu, N., Ternary liquid-liquid equilibria for mixtures of 1-methyl-3-octyl-imidazolium chloride+benzene+an alkane at T=298.2K and 1atm, *J. Chem. Thermodyn.* **2003**, 35, 67-76.

9.  Krummen, M.; Wasserscheid, P.; Gmehling J., Measurement of Activity Coefficients at Infinite Dilution in Ionic Liquids using the Dilutor Technique, *J.Chem. Eng. Data* **2002**, 47, 1411-1417.

10. Heintz, A.; Kulkokov, D.V.; Verevkin, S.P., Thermodynamic Properties of Mixtures Containing Ionic Liquids. 1. Activity Coefficients at Infinite Dilution of Alkanes, Alkenes, and Alkylbenzenes in 4-Methyl-*n*-butylpyridinium Tetrafluoroborate Using Gas-Liquid Chromatography, *J. Chem. Eng. Data* **2001**, 46, 1526-1529.

11. Heintz, A.; Kulikov, D.V.; Verevkin, S.P., Thermodynamic Properties of Mixtures Containing Ionic Liquids. 2. Activity Coefficients at Infinite Dilution of Hydrocarbons and Polar Solutes in 1-Methyl-3-ethyl-imidazolium Bis(trifluoromethyl-sulfonyl) Amide and in 1,2-Dimethyl-3-ethyl-imidazolium Bis(trifluoromethyl-sulfonyl) Amide Using Gas-Liquid Chromatography, *J. Chem. Eng. Data* **2002**, 47, 894-899.

12. Brennecke J.F.; Maginn, E.J, Ionic Liquids: Innovative Fluids for Chemical Processing, *AIChE Journal* **2001**, 47 (11) 2384-2389.

13. Huddleston, J.G.; Willauer, H.D.; Swatloski, R.P.; Visser, A.E.; Rogers, R.D., Room temperature ionic liquids as novel media for 'clean' liquid-liquid extraction, *Chem. Commun.* **1998**, 16, 1765-1766.

14. Shyu, L.-J.; Zhang, Z.; Zhang, Q. (Akzo Nobel), Process for the extraction of an aromatic compound from an aliphatic phase using a non-neutral ionic liquid, PCT Int. Appl. WO 2001/40150 A1, **07-06-2001**.

15. Arlt, W; Seiler, M.; Jork, C.; Schneider, T. (BASF), Ionic liquids as selective additives for the separation of close-boiling or azeotropic mixtures, PCT Int. Appl. WO 2002/074718 A2, **26-09-2002**.

16. Gmehling, J.; Krummen, M., Einsatz ionischer Flüssigkeiten als selektieve Lösungsmittel für die Trennung aromatischer Kohlenwasserstoffe von nichtaromatischen Kohlenwasserstoffen durch extractieve Rektifikation und Extraktion, DE 101 54 052 A1, **10-07-2003**.

17. Swatloski, R.P.; Holbrey, J.D.; Rogers, R.D., Ionic liquids are not always green: hydrolysis of 1-butyl-3-methylimidazolium hexafluoroborate, *Green Chem.* **2003**, 5 (4), 361-363.

18. Hanke, C.G.; Johansson, A.; Harper, J.B.; Lynden-Bell, R.M., Why are aromatic compounds more soluble than aliphatic hydrocarbons in dimethylimidazolium ionic liquids? A simulation study, *Chem. Phys. Lett.* **2003**, 374, 85-90.

19. Holbrey, J.D.; Reichert, W.M.; Swatloski, R.P.; Broker, G.A.; Pitner, W.R.; Seddon, K.R.; Rogers, R.D., Efficient, halide free synthesis of new, low cost ionic liquids: 1,3-dialkylimidazolium salts containing methyl- and ethylsulfate anions, *Green Chem.* **2002**, 4, 407-413.

# Chapter 6

# Use of Ionic Liquids in Oil Shale Processing

## Mihkel Koel

Tallinn University of Technology, Ehitajate, 5, Tallinn 19086, Estonia

Containing sufficient amount of organic matter, oil shale utilization to produce petroleum-like oils and gaseous products could be feasible under certain conditions. Liquid products could be as an alternative synthetic fuel source or as an even more valuable petro-chemical feedstock for the chemical industry. There is a place for utilization of ionic liquids either as solvents for extraction or reaction media with very good properties for catalyst solubilization in the processing of oils and gaseous products. This paper considers the potential for the extraction of shale oil with ionic liquids and anticipated problems of recycling the ionic-fluid extractant.

The world's vast reserves of oil shale are many times greater than the proven remaining known reserves of crude oil and natural gas combined. As such, shale oils will play an important role in future energy supply. Conversion of solid fuels to chemicals feedstocks and cleaner-burning liquid or gaseous fuels is becoming a desirable goal to pursue. Considerable attention has been given to processes for the gasification and/or liquefaction of solid fuels to produce petroleum-like oils and gaseous products as well as further upgrading of produced oils for fuels and chemicals feedstock. Given the size of the shale deposits, there exists an urgent need for the development of efficient processes– processes which are capable of providing useful products in an economical way – in order to make oil shale utilization feasible (*1*).

The engineering science of oil shale processing fits well with studies of other alternatives to petroleum raw material sources such as lignocellulosic biomass. This biomass also has a very complex structure, and new chemical technologies, in particular novel solvent systems, are needed for its processing. Estonian oil shale has a content rich in heteroatoms (oxygen, nitrogen, sometimes sulphur and heavy metals). This makes the task of shale process development even more challenging than that of many coals.

Estonia has a uniquely strong, historical oil shale industry: 95 % of its electricity is generated in oil shale fired power stations and 15 % of liquid fuels are shale oil. There are two types of oil shale in the Baltic Basin: 1) the lower Ordovician Dictyonema shale found in the Estonian deposit. The inorganic part of this shale is primarily clay (76 %) with a small proportion of carbonate (3 %). The organic content is below 20 %, and it is not in current industrial use; 2) The Middle Ordovician Kukersite oil shale found in the Estonian deposit. The inorganic part of this shale is mineralogically complex, composed primarily of carbonates (41 %). The organic content is approximately 36 %. This is the historic shale resource of the Estonian oil shale industry. Most of oil shales (among them Estonian) have a quite inert, kerogen matrix and give very little extractable mass with different solvents under normal laboratory conditions. The search for new -especially environmentally benign- processes and technologies is a very important present concern.

## Liquefaction of oil shale

The organic part of oil shale – kerogen - is a cross-linked macromolecular system. The term "liquefaction" means the structural degradation of carbonaceous material typically, but not necessarily, accompanied by addition of hydrogen to the molecular structure of the material. Hydrocarbon semi-solids and liquids often are further converted, structurally degraded, altered and/or hydrogenated. The essence of an oil shale liquefaction process is the structural degradation of, and/or the addition of hydrogen to, a kerogen material, with heteroatom removal being an important consideration.

Classical methods for the liquefaction of solid fuel include pyrolysis, solvent extraction, direct hydrogenation and indirect hydrogenation. The kerogen molecule is thermally broken down at 500-600 °C. The most widely spread method for oil shale processing is pyrolysis or retorting (*2*). There are other processes than pyrolysis at high temperature to convert kerogen into soluble products.

- Solvent extraction and modified solvent extraction which utilizes a hydrogen donor solvent system with catalyst at temperatures 300-400 °C and pressures 70-200 atm.
- Indirect liquefaction involves reacting solid fuel with steam and oxygen at high temperature to produce synthesis gas consisting primarily of hydrogen, carbon monoxide and methane, and then catalytically reacting the hydrogen and carbon monoxide to synthesize hydrocarbon liquids by the Fischer-Tropsch process.
- Direct liquefaction process involves the hydrogenation of shale particles at high pressure of hydrogen and using a solid catalyst.
- Modern processes under intensive study are supercritical water gasification (SCWG) process and the hydrothermal upgrading (HTU) process. The first can be classified as an indirect liquefaction process, and the second is the direct liquefaction process.

It is worthwhile to focus on these latter, less-classical approaches to liquefaction in terms of their details. In the SCWG process, the product is fuel gas containing more than 50 % hydrogen on a mole basis. Crude oil is the target product in HTU, and hydrogen in SCWG (3). In SCWG, the reaction generally takes place at temperatures above 600°C and at the pressures at or higher than that typical at or just above the critical point of water. At temperatures higher than 600 °C, water becomes a strong oxidant, and oxygen in water can be transferred to the carbon atoms of the carbonaceous material. As a result of the high density, carbon is preferentially oxidized into $CO_2$ but small amounts of CO are also formed. The hydrogen atoms of water and of the kerogen are set free and form $H_2$. The gaseous product consists of more than 50 % hydrogen on a mole basis, $CO_2$ in second quantity at about 33 %, and the rest including $CH_4$ and CO.

The HTU process is a liquefaction process carried out under sub-critical water conditions, the product contains crude, organic compounds, gases, and water (4). The HTU process is a promising liquefaction process- it can be used for the conversion of a broad range of carbonaceous feedstocks. Because the process is especially designed for wet materials, the drying of feedstocks is not necessary. Treated in water in a temperature range of 300–350°C and a pressure range of 120–180 bar, kerogen is depolymerized to a hydrophobic liquid product so-called "crude". Gases are also produced, consisting of $CO_2$, $H_2$, methane and CO. Other product includes water and organic compounds. There are studies on liquefaction of the organic matter of Kukersite oil shale (5). At 350 °C the oil yield was up to 85 % of the total organic matter. The extracts were characterised by a high content of aliphatic compounds, especially by olefins with a double bond in the middle of the chain. The obtained crude oil is heavy, having a high density and molecular weight and containing a great deal of asphaltenes (6).

After completion of the reaction, a substantial portion of the produced fluids, including gases and easily removable liquids, may be recovered from any solid materials in the reaction mixture by the use of conventional solid/gas and solid/liquid separation techniques. The recovered hydrocarbon liquids may then be further treated by filtration, centrifugation, distillation, solvent extraction etc. Unlike petroleum, crude oil from these processes has a very complex structure, and, thus, new chemical technologies, especially novel solvent systems, are needed for its processing.

## Ionic liquids for liquefaction of oil shale

There are only few papers available about using ionic liquids for treating coals or oil shales. All these are using chloroaluminate(III) based ionic liquids. The first example is almost 30 years old. It is shown that Green River oil shale can be degraded using a sodium chloride saturated tetrachloroaluminate melt (7). This solvent is acting as Lewis base. At a temperature 320 °C, 58 % conversion of the organic carbon present was achieved.

Newman, *et al* (8) used mixture of pyridinium chloride and AlCl₃ , which is liquid at near room temperature (30-40 °C), as a catalysts for Friedel-Craft alkylation of coal with 2-propanol. Comparing the solubility's of the alkylated and non-treated coals in various solvents showed that the alkylated coal was in all cases considerably more soluble. Friedel-Craft acylation alters the coal's primary structure and oxyalkylation alters its secondary structure making it potentially more susceptible to further modification, but with minimum damage to the sub-units of the coal macromolecule. In other studies different coals were acylated by acetyl chloride in presence of ionic liquid – pyridinium chloride-aluminum chloride, and coal solubility was increased up to 17 % (9).

Chloroaluminate (III) ionic liquids have investigated as media for the dissolution of kerogen for oil exploration by Seddon and coworkers (10) at the University of Sussex between 1990 and 1995, and this was concluded in (11). Kerogen can be dissolved by heating of the sample in acidic ionic liquid for one-minute periods in a microwave oven. By different spectroscopic means it was found that products (liquid and solid) were substantially altered compare to initial oil shales.

Another trial was done by Sutton (12) studying the solubilisation of Insoluble Organic Matter (IOM) from sediments and bacteria. Less than 1 % of sedimentary IOM and 5 % of Kimmeridge Clay IOM was soluble in DCM following ionic liquid treatment, whilst alkyl chains were lost from the insoluble portion which also increased in aromaticity. The ionic liquid was found to non-quantitatively promote ether cleavage, protonation and rearrangement reactions. This ionic liquid dissolution procedure has not provided the substantial progress in elucidating the molecular character of IOM promised by earlier reports.

The difference that treatment with ionic liquids can offer is well demonstrated on the example of cracking of polyethylene to gaseous alkanes and low-volatile cyclic alkanes in chloroaluminate (III) ionic liquids (*13*). Yields of liquids as high as 95 % with HDPE have been obtained. The cracking of polyethylene, using chloroaluminate (III) ionic liquids, gives a significantly different distribution of hydrocarbon products. The major products of the reaction are C3–C5 gaseous alkanes (such as isobutane) and branched cyclic alkanes, all of which are useful feedstocks, and this selectivity towards low molecular weight feedstocks— the best possible products from this cracking reaction – gives an additional advantage.

Koel *et al* (*14*) conducted oil shale treatment with 1-butyl-3-methylimidazolium based ionic liquids: 1-butyl-3-methylimidazolium cation and $PF_6^-$ anion ([bmim]·[$PF_6$]) and with different mol % of $AlCl_3$ ([bmim][Cl*$AlCl_3$]). These ionic liquids are easy to prepare and this makes them as a good choice for treatment of oil shales. The hexaflourophosphate ionic liquid appears to be a poor solvent for either oil shale at room temperature or at elevated temperatures, but below 200 °C where the thermal degradation of oil shale kerogen starts. The yield of soluble products was only 0.07 % for the Dictyonema shale and 0.02 % for the Kukersite shale, which is on the same level as solvent extraction. This means that the hexaflourophosphate ionic liquid itself does not have catalytic activity to solubilise kerogen. For the acidic chloroaluminate ionic liquid, the extraction yield generally increased with extraction temperature, suggesting that this ionic liquid does not function as a thermal agent only but has catalytic activity and is solubilising kerogen. For Kukersite increasing extraction temperature increases yield of soluble product in case of acidic IL substantially (up to 4 % from total mass of shale taken at 175 °C). For basic ionic liquid there was only small increase in the recovery of kerogen at temperature 175 °C.

Different behaviour was seen in case of Dictyonema, where the increase in yield of soluble products was estimated on case of basic IL (up to 1.7 % from total mass of shale taken); acidic IL gives slight increase but yield remains lower than 0.5 %. A rationale for this might be in the different mineral matrix of this oil shale, and it points to uniqueness of different oil shales.

An ionic liquid which itself is a Lewis acid or a solvent for a Lewis acid could be applied for acid-catalysed, oil-shale conversion for producing liquid and gaseous products at lower temperatures than that of pyrolysis. If these acid catalyst systems can be readily regenerated their use will be not limited. Ionic liquids could be used to overcome the difficulties in catalytic hydrogenation serving as a means to bring hydrogen gas in contact with the kerogen, especially if a solid catalyst is used. All the experiments described above share a step involving quenching of the reaction by addition of water. This way the ionic liquid is destroyed and there are no data about potential continuous processes. The future direction of these studies should be towards regeneration/recycling of process ionic liquids which could make the kind of treatment economically and environmentally interesting for industry.

The bench scale experiments are already a new source of data to refine the models for the macroscopic structure of kerogen (*15*). It is important to employ complementary, solution based analytical methods in order to learn more about kerogen. Successful dissolution of kerogen preserving structurally important fragments may therefore have a substantial impact upon fundamental geochemistry and the petrochemical industry.

## Ionic liquids for upgrading of shale oil

Whatever process is used for liquefaction of kerogen, the crude oil from that is very complex and needed further processing/separation to get fuels and feedstock for chemicals, where high-value chemicals are the best way to increase the overall value of shale oil. Ionic liquids need to be investigated with a view to upgrading technologies and look for one-step conversion processes that are economic in small-scale applications also. The applications of ionic liquids in a range of reactions continues to expand and the list of studies performed covers a broad range of areas: oligomerization and polymerization, hydrogenation, dimerization and telomerization of dienes, carbonylation, oxidation and radical reactions, Heck, Suzuki, Stille, Sonogashira, Negishi, and Ullmann coupling reactions, allylation reactions, and others. Their successful use as solvents has been demonstrated for a wide range of organic reactions including acid catalyzed reactions and transition metal catalyzed transformations (*16,17*). The main potential interest in using ionic liquids results in the possibility to modify the properties of salts in large extent. From a chemical point of view, this leads to enhancement of reaction rates and improvement of chemo- and regioselectivities relative to other organic solvents (*18*).

Although the exploration of 1,3-dialkylimidazolium-based ionic liquids in catalysis is in its exploratory stage, these salts have already stepping in to industrial catalytic processes. Indeed, IFP (Rueil-Malmaison, France) has launched a commercial process for the dimerization of butenes to isooctenes (Difasol process) (*19*), and this new process provides significant benefits over the existing homogeneous Dimersol X process.

Catalytically active salts which were effective for breaking apart hydrocarbon molecules in case of kerogen liquefaction can be used for crude oil upgrading also. Expected operating temperatures will be much lower than used in current conversion methods. University of Regina work has shown that heavy oil conversion can occur at a surprisingly low temperature, whereas current technology uses temperatures in the 400–550 °C range (20).

Existing experience with room temperature ionic liquids based on alkylmethylimidazolium hexafluorophosphate suggests that such systems may be easily adapted to conventional liquid–liquid extraction practice. The good correlation between the distribution of the solutes in the ionic liquid system and their distribution in an n-octanol–water system is a useful base for design of extraction systems (21). The search for task-specific ionic liquids for use in extraction systems is going on to find the appropriate combination of physical properties and performance geared towards industrial application. Results on extraction of metals (22) can be developed for crude oil applications as well as there are examples of removal of sulfur from diesel fuel. The results indicate that future technical sulfur content specifications should be achievable with multistage extraction using ionic liquids (23). Already, some patents are claimed in this removal of sulphur compounds (24). It is actual and very fast developing area where next steps would be on extracting oxygen and nitrogen containing compounds from crude oils.

Another advantage of ionic liquids could be find in their large effect in microwave assisted chemical transformations as it was already used in case of oil shale dissolutions (11). Small amount of ionic liquids can insure an efficient absorption of microwave energy and a good distribution of heat. Reactions can proceed in a much faster way than in conventional organic solvents (25).

## Recovery of ionic liquids

Separating the conversion product from the ionic liquid without losing the catalytic activity of the catalyst solubilized in ionic liquid or catalytic activity of ionic liquid, also the recovery of ionic liquid in extraction process itself improves the economics of the process.

In principle all methods of separation like distillation, gas extraction, crystallization and liquid-liquid extraction should be viewed in light of the special properties of ionic liquids. In this area activity has just begun and only sparse results are available. Interestingly, the low vapor pressures of ionic liquid systems offer some advantages and suggest novel methods of solvent recovery or solute separation.

If the products to be removed from an ionic liquid have relatively high vapor pressures then simple distillation or a wiped film evaporator should be highly effective in removing these components. Since the ionic liquid is essentially non-volatile, this would be an easy separation and should require relatively few equivalent stages or effective area. Thanks to the low vapor pressure of the ionic liquids, distillation can be envisioned without azeotrope formation (26). Thus, for volatile compounds that do not decompose at temperatures near their boiling points (ionic liquids are stable at least to 150 °C, mostly 250 °C), a flash or an easy distillation column would be the separation method of choice. However, it is often limited because of the general thermal instability of organometallic catalysts. One obviously additional benefit of distillation is that it does not require the addition of any new compounds.

Another alternative for the removal of volatile components from an IL is gas stripping or extraction. Simply bubbling air or some other inert gas through the IL mixture should strip out the volatile components. The disadvantage of this technique is that removing volatile products from the gas by condensation may require significant refrigeration costs. On the other hand, like distillation, there will not be any ionic liquid contamination of the product in the gas phase.

A departure from gas stripping is using a fluid in its' supercritical state. Supercritical carbon dioxide (scCO$_2$) can bring breakthrough in technology of extraction with ionic liquids. The advantages of using CO$_2$ as an extraction medium include low cost, nontoxic nature, recoverability, and ease of separation from reaction products. Brennecke et al (27) demonstrated that CO$_2$ and [bmim]·[PF$_6$] could be used together as extraction system. Supercritical CO$_2$ dissolves in the liquid to facilitate extraction, but the ionic liquid does not dissolve in carbon dioxide, so pure product can be recovered. A variety of polar and nonpolar aliphatic and aromatic substrates was recovered effectively from the IL by batch wise extraction with scCO$_2$ (28).

Existing results suggest a promising future for biphasic IL/scCO$_2$ systems. A primary advantage of such systems is the solubility or stability of organometallic or enzymatic catalysts in ILs and their negligible solubility in scCO$_2$. On the other hand, many organic reactants and products are reasonably soluble in scCO$_2$. The development of continuous-flow catalytic systems in which both the IL/catalyst and the CO$_2$ can be recycled would be an important advance en route to large-scale commercial applications. There have been reported examples indicating that the continuous flow supercritical fluid–ionic liquid biphasic system provides a method for continuous flow homogeneous catalysis, with built in separation of the products from both the catalyst and the reaction solvent, and that this is possible even for relatively low volatility products (29,30).

One option for separation of organic solutes whose melting points are significantly lower is crystallization of the ionic liquid solvent or for solutes with high melting point the precipitation from the ionic liquid. The liquid remaining would still contain some IL because of possible eutectic. This way crystallization remains always problematic for effective separation or recovery ionic liquids.

If the target product to be removed from the ionic liquid is not volatile or if it is thermally labile then the natural choice might be liquid-liquid extraction.

Moreover, if there are multiple organic compounds in the IL mixture that have differing polarities or functionalities, then liquid-liquid extraction may afford the opportunity for some selectivity in the extraction. But there are several restrictions relating to solvent-choice for extraction from an ionic liquid:

- The ionic liquid and the solvent must be immiscible.
- The solubility of the solvent in the IL-rich phase and the solubility of IL in the solvent-rich phase. The solubility of the solvent in the IL-rich phase may have positive or negative consequences. On the other hand, any solubility of the ionic liquid in the solvent-rich phase is almost always a disadvantage.
- The partition coefficient of the product between the IL-rich phase and the solvent-rich phase must favor partition into the solvent.

Thus, liquid-liquid extraction offers some opportunities for selective extraction of the organic solute from ionic liquid but introduces additional issues associated with contamination of the extract phase with the ionic liquid and subsequent downstream separation and waste treatment challenges. In addition, use of conventional volatile organic liquid solvents for the separation step is contrary to the overall objective to reduce the use of VOC solvents and their subsequent environmental pollution problems. Requiring the liquid extractant to be environmentally benign, as well, would severely limit the choices of potential extraction solvents.

## Conclusion

Ionic liquids provide a unique medium for developing new chemistry and a potential solution for processing oil shale, which is providing a variety of challenges to turn them into practical fluids for industrial use. To achieve the full potential of this new class of solvents in petrochemistry will require serious participation of chemical engineeres in the design and development practical applications with ionic liquids.

## Acknowledgements

Prof. Charles H. Lochmüller of Duke University {USA} is recognised for valuable discussions and friendly advice. Estonian Science Foundation is acknowledged for support (grant No. 5610).

# References

1. Brendow,K.; Global Oil Shale Issues and Perspectives. *Oil Shale* **2003**, *20*, 81-92.

2. Russell, P.L. *Oil Shales of the world. Their origin, occurrence and exploitation;* Pergamon Press; 1990.

3. Antal Jr., M.J.; Allen, S.; Schulman, D.;. Xu, X.; Divilio, R.; Biomass gasification in supercritical water. *Ind. Eng. Chem. Res.* **2000**, 39, 4040-4048.

4. Rocha J.D.; Brown S.D.; LoveG.D.; Snape C.E.; Hydropyrolysis: a versatile technique for solid fuel liquefaction, sulphur speciation and biomarker release. *J.Anal.Appl.Pyrol.* **1997**, *40*,91-109.

5. Nappa L., Klesment I., Vink N., Kailas K.; Low Temperature Decomposition of Organic Matter of Oil Shale by Solvent Extraction (in Russian). *Eesti TA Toim., Keemia* **1982**, *31*,17-24.

6. Luik H., Klesment I.; Liquefaction of Kukersite Concentrate at 330-370°C in Supercritical Solvents. (in Russian). *Eesti TA Toim., Keemia* **1985**, 34, 253-263.

7. Bugle, R.C.; Wilson, K.; Olsen, G.; Wade Jr., L.G.; Osteryoung, R.A.; Oil-shale kerogen: low temperature degradation in molten salts. *Nature* **1978**, 274,578-580.

8. Newman, D.S.; Winans,R.E.; McBeth, R.L.; Reactions of Coal and Model Coal compounds in Room Temperature Molten Salt Mixtures. *J.Electrochem.Soc.* **1984**, *131*,1079-1083.

9. Newman, D.S.; Kinstle, T.H.; Thambo, G.; The acylation of coal and model coal compounds in room temperature molten salts; in Mamontov, G.; Hussey, C., Eds.; *Proc. Intern. Symp.on Molten Salts*, 1987.

10. Pattell Y. The dissolution of kerogeen; M.Phil Thesis 1994, U.of Sussex, U.K.; Dutta L. Dissolution of Liassic Kerogen and coal; M.Phil Thesis 1995, U.of Sussex, U.K.

11. Patell, Y.; Seddon, K.R.; Dutta, L.; Fleet, A.; The dissolution of kerogen in ionic liquids, in R.D.Rogers, K.R.Seddon, S.Volkov, Eds.; *Green Industrial Applications of Ionic Liquids*, Kluwer Academic Press 2003, NATO Science ser.II: Math.Phys.Chem.92; p 499-510.

12. Sutton, P. A.; The Quantitative Isolation of 'Insoluble Organic Matter' (IOM) from Sediments and Bacteria, and its Attempted Dissolution using the Ionic Liquid 1-Ethyl-3-Methylimidazolium Chloride-Aluminium (III) Chloride; PhD thesis 2000, U. of Plymouth, U.K.

13. Adams, C. J.; Earle, M.J.; Seddon, K.R.; Catalytic cracking reactions of polyethylene to light alkanes in ionic liquids. *Green Chem.***2000**, *2*,21-23.

14. Koel, M.; Hollis, W.K.; Rubin, J.B.; Lombardo, T.J.; Smith, B.F.; Ionic liquids for oil shale extraction, in Rogers, R.D.; Seddon, K.R.; Volkov, S., Eds.; *Green Industrial Applications of Ionic Liquids,* Kluwer Academic Press 2003, NATO Science ser.II: Math.Phys.Chem.92; p 193-208.

15. Lille Ü.; Current knowledge on the origin and structure of Estonian kukersite kerogen. *Oil Shale* **2003**, *20*,253-263.

16. Wasserscheid, P.; Keim, W.; Ionic Liquids: New "solutions for transition metal catalysis. *Angew. Chem. Int. Ed. Engl.***2000**, *39*, 3772-3789.

17. Dupont, J.; de Souza, R.F. ; Suarez, P.A.Z.; Ionic Liquid (Molten Salt) Phase Organometallic Catalysis, *Chem. Rev.* **2002**, *102,* 3667-3692.

18. Olivier-Bourbigou H.; Magna L. ; Ionic liquids: perspectives for organic and catalytic reactions. *J.Mol.Cat.A: Chem.* **2002**, *182*, 419–437

19. Chauvin,Y.; de Souza, R. F.; Olivier, H. ; *U.S. Patent* **1996**, No. 5723712.

20. Saskatchewan Research Council Publication 2000; Number P-110-492-G-00

21. Huddleston,J.G.; Willauer,H.D.; Swatloski,R.P.; Visser,A.E.; Rogers,R.D.; Room temperature ionic liquids as novel media for 'clean' liquid–liquid extraction. *Chem. Commun.* **1998**, 1765-1766.

22. Visser,A.E.; Swatloski,R.P.; Reichert, W.M.; Mayton,R.; Sean Sheff,S.; Wierzbicki,A.; Davis, Jr.,J.H.; Rogers,R.D.; Task-specific ionic liquids for the extraction of metal ions from aqueous solutions. *Chem. Commun.* **2001**, 135–136.

23. Bösmann, A; Datsevich, L; Jess, A; Lauter, A; Schmitz, C; Wasserscheid, P.; Deep desulfurization of diesel fuel by extraction with ionic liquids, *Chem. Commun.* **2001**, 2494-2495.

24. O'Rear,D.J.; Boudreau, L.C.; Driver, M.S.; Munson, C.L. ; *WO Patent* **2002**, No. 2002034863; Schoonover, R.E.; *U.S.Patent* **2003**, No. 2003085156; Schucker, R.C.; Baird, W.C., Jr.; *U.S. Patent* **2001**, No. 6274026.

25. Leadbeater, N.E.; Torenius, H.M.; A Study of the Ionic Liquid Mediated Microwave Heating of Organic Solvents. *J.Org.Chem.* **2002**, *67*, 3145-3148.

26. Keim, W.; Vogt, D.; Waffenschmidt, H.; Wasserscheid, P.; New Method to Recycle Homogeneous Catalysts from Monophasic Reaction Mixtures by Using an Ionic Liquid Exemplified for the Rh-Catalyzed Hydroformylation of Methyl-3-pentenoate. *J. Catal.* **1999**, *186*, 481-484.

27. Blanchard, L. A.; Hancu, D.; Beckman, E. J.; Brennecke, J. F.; Ionic liquid/$CO_2$ biphasic systems: New media for green processing. *Nature* **1999**, *399*,28-29.

28. Blanchard, L.A.; Brennecke, J.; Recovery of organic products from ionic liquids using supercritical carbon dioxide. *Ind.Egn.Chem.Res.* **2001**, *40*,287-292.

29. Sellin, M.F.; Webb, P.B.; Cole-Hamilton, D.J.; Continuous flow homogeneous catalysis: hydroformylation of alkenes in supercritical fluid–ionic liquid biphasic mixtures. *Chem. Commun.* **2001**, 781–782.

30. Brown, R.A.; Pollet, P.; McKoon, E.; Eckert, C.A.; Liotta, C.A.; Jessop, P.G.; Assymmetric hydrogenation and catalyst recycling using ionic liquid and supercritical carbon dioxide. *J. Am. Chem. Soc.* **2001**, *123*, 1254-1255.

## Chapter 7

# Deep Desulfurization of Fuels by Extraction with Ionic Liquids

**Andreas Jess[*] and Jochen Eßer**

Department of Chemical Engineering, University of Bayreuth,
Universitätsstraße 30, D–95447 Bayreuth, Germany
[*]Corresponding author: Jess@uni-bayreuth.de

In recent years, much attention has been given to deep desulfurization of diesel oil and gasoline, since exhaust gases
containing SOx cause air pollution and acid rain. Moreover, a
lower sulfur content than today - still several 100 ppm S in
many countries - would allow the use of new engines and catalytic systems. Therefore, the S-level in fuels will be drastically
decreased in Europe and other countries in 2005 down to at
least 50 ppm. The current process of catalytic hydrodesulfurization (HDS) is limited for such ultra-low sulfur fuels, respect-
tively the expenses are high to meet these future requirements.
Alternative processes are therefore desirable. This paper presents a potential alternative, the (reversible) extraction of organic S-compounds by ionic liquids (ILs), e. g. by butylmethyl-
imidazolium(BMIM)-chloroaluminate, but also by halogen-free
ILs like BMIM-octylsulfate. In the presented experiments, extraction of model oils (dibenzothiophene derivatives mixed
with dodecane) as well as of a real diesel oil were investigated.
The results show the excellent and selective extraction
properties of ILs, especially with regard to those S-compounds,
which are very hard to remove by HDS. In addition, organic N-
compounds like indole are also extracted with an even much
higher efficiency. The application of mild process conditions
(ambient pressure and temperature) and the fact that hydrogen
is not needed, are additional advantages of this new concept in
comparison to HDS. The estimation of the basic parameters of
a technical extraction process with ILs will be given.

# Introduction

Crude oils contain significant amounts of sulfur in form of various organic compounds (Fig. 1). Depending on their origin, crude oils strongly vary in their S-content. Crudes from South America may contain up to 5 wt.-% sulfur, and those from North Africa as little as 0.2 %. The average content is typically 1 %.

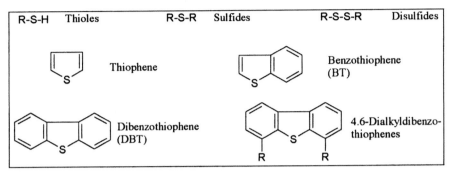

**Fig. 1**: Typical sulfur compounds in liquid fuel

During about the last three decades much attention has been given to the desulfurization of fuels like diesel and gasoline, since exhaust gases containing SOx cause air pollution and acid rain. The S-limit was therefore gradually decreased, e. g. in Germany from 5000 ppm (1975) down to today's value of 350 ppm; fuels with even lower S-contents are already available due to tax benefits. Due to these restrictions, $SO_2$-emissions from gasoline and diesel today only contribute about 2 % to the total emissions (Germany). Nevertheless, the S-limit will be set Europe-wide in 2005 to 50 ppm. In addition, in some countries (Sweden, Germany) the majority of fuels will then be even "S-free" (by definition < 10 ppm S). These additional restrictions mainly aim at the reduction of CO-, $NO_x$-, and particulate emissions:

- Vehicles with gasoline engines usually carry three-way catalytic converters. For the further reduction of the above named emissions as well as of fuel consumption, the automotive industry will introduce direct injection and lean fuel engines, which need a very S-sensitive so-called storage DeNOx-catalyst. To keep a high degree of efficiency of the catalysts, they have to be regenerated periodically. In case of S-levels > 10 ppm, the regeneration cycles are too frequently, which could nullify the better efficiency of such engines [1].

- For diesel motors the (cancerogenic) particulate emissions need to be reduced further. This can only be achieved with novel catalysts and particle filters, but these systems are only suitable for (practically) S-free diesel oil.

By 2010, the S-limit will therefore probably have reached in many countries both for gasoline and diesel oil a value of 10 ppm (or even less, e. g. with the introduction of new technologies like fuel cells). This means that the degree of S-conversion/separation that is needed to meet this limit is 99.9 % (assuming a S-content in the raw feed of 1 % and a target S-content of 10 ppm) compared to today's value of ("only") about 97 % (to reach 350 ppm) in European refineries.

## Current Technology for desulfurization of fuels

State of the art of desulfurization is hydrodesulfurization (HDS). Thereby, the S-compounds are converted on catalysts based on CoMo or NiMo to $H_2S$ and the corresponding hydrocarbon according to (example thiole):

$$R - S - H + H_2 \longrightarrow R - H + H_2S$$

Typical reaction conditions are 350 °C and 30 to 100 bar. High pressure reactors and vessels etc. are needed, resulting in high investment costs. $H_2S$, which is separated from the desulfurized oil, is then subsequently converted by catalytic oxidation with air into elementary sulfur. Hydrogen, which is fed into the HDS-reactor together with preheated oil, is only consumed to a small extent in the trickle bed reactor, and so the hydrogen is recycled (after separation) back into the reactor (typically with recycle rates > 50 !) [2].

The S-compounds, which remain in the diesel oil, e. g. after desulfurization to 350 ppm S (S-limit in Europe up to 2005), are above all derivatives of thiophene (Fig. 1). Mercaptanes and (di)sulfides are much more reactive and therefore practically nonexistent in partially desulfurized oils. Particularly refractory are dibenzothiophene (DBT), methyldibenzothiophene (MDBT), and above all 4,6-dimethyldibenzothiophene (4,6-DMDBT). The reaction rate to convert these S-components by HDS decreases in the named order by factors of about 2 and 10, respectively [3, 4]. (*Remark: The desulfurization rate can therefore be approximated by a reaction second order with respect to the total amount of S. This expresses, that the rate of desulfurisation superproportionally decreases with rising degree of desulfurization by the increasing portion of the remaining, less reactive compounds like 4,6-DMDBT*). So in return, the expenses also drastically increase both with respect to investment (reactor size, pressure) and production costs (energy, hydrogen consumption and recycle).

In order to improve the current HDS-technology, the focus of research is the development of more active catalysts based on noble metals. In general, they are not S-resistant for S-contents > 100 ppm [5], are therefore only applicable for deep desulfurization, and of cource the above mentioned expenses remain. Alternative processes are therefore desirable, if possible without the need of hydrogen, high pressure and/or high temperature.

## Alternative processes for deep desulfurization

Possible alternative processes for deep desulfurization are adsorption and absorption (extraction). (*Biological and other more "exotic" methods discussed in the literature are beyond the focus of this paper.*)

A new adsorption technology for gasoline was developed by Phillips (*S Zorb SRT-Gasoline*) [6]. The first commercial unit came on stream in 2001. The new technology uses a specific sorbent (not specified in the open literature) that attracts S-compounds and removes the sulfur. The S-loaded sorbent is continuously withdrawn from the fluid bed reactor and transferred to a regenerator, where the sulfur is removed as $SO_2$ by oxidation with air. A similar process for diesel is currently under development, but obviously this is more problematic.

The 2nd option is extraction of S-compounds by an appropriate liquid extracting agent. In principle, an ideal agent should have the following properties:

- high partition coefficient for S-compounds, above all for dibenzothiophenes,
- regeneration should be easy, i. e. S-extraction should be reversible and so the agent should have no or at least a very low vapour pressure,
- the agent should be absolutely indissoluble in oil, and in addition hydrocarbons should not or only to a small extent be soluble in the agent, and
- the agent should feature a high thermal stability, be non-toxic and environmentally benign.

Ideal candidates are ionic liquids (ILs) [7-11]. As shown below, ILs have excellent (and reversible) extraction properties for S- (but also for N-compounds), no vapour pressure, and are - if chosen carefully - insoluble in oil. The basic concept of such an extraction process is given in Fig. 2.

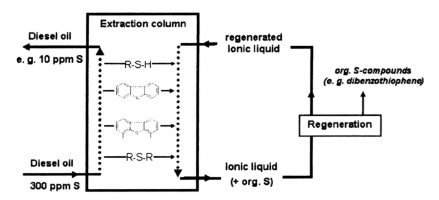

**Fig. 2:** Concept of deep desulfurization of diesel by extraction with ILs (possible methods of regeneration: re-extraction with low-boiling hydrocarbons, destillation - if needed under reduced pressure - or with supercritical $CO_2$)

Subsequently, experimental results of the extraction of S-compounds from diesel oil by ILs is described. In addition, the basic parameters of a technical extraction process are estimated.

## Deep desulfurization of fuels using ionic liquids

Ionic liquids (ILs) are low melting (< 100 °C) salts, which represent a new class of solvents. Up to now, ILs have mainly been studied with respect to biphasic homogeneous catalysis. The range of known and available ILs has been expanded so that many different candidates are accessible today. Fig. 3 shows some ILs, which have been up to now investigated in our group (Uni. Bayreuth) in cooperation with the group of P. Wasserscheid (University Aachen) [7-11].

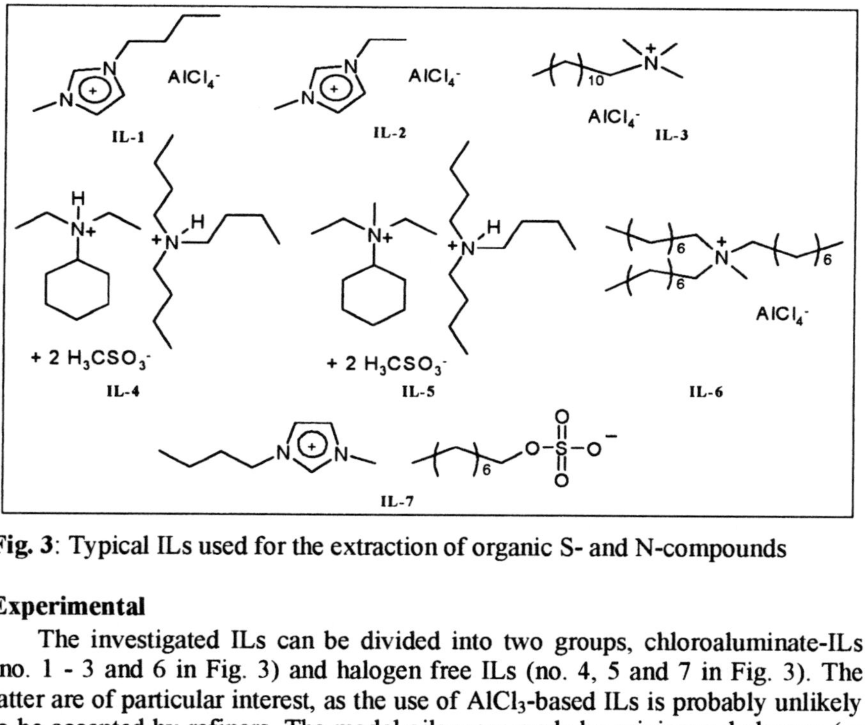

**Fig. 3**: Typical ILs used for the extraction of organic S- and N-compounds

### Experimental

The investigated ILs can be divided into two groups, chloroaluminate-ILs (no. 1 - 3 and 6 in Fig. 3) and halogen free ILs (no. 4, 5 and 7 in Fig. 3). The latter are of particular interest, as the use of $AlCl_3$-based ILs is probably unlikely to be accepted by refiners. The model oils were made by mixing n-dodecane (or an oil from FLUKA with 18 % aromatic hydrocarbons) with the following S-compounds: dodecanethiol, DBT, and a mixture of DBT, MDBT and 4,6-DMDBT.

All ILs form biphasic systems with the model oils at room temperature. For the extraction experiments, the IL was added to the model oil in a mass ratio of typically 1/1 up to 1/5. First of all, the oil-IL-mixture was stirred at room temperature in a schlenk tube or a comparable vessel. It was found that the extraction process proceeds very quickly, i. e. the final sulfur distribution was reached in less than 5 min (thermodyn. equilibrium). After stirring, two phases quickly separate, and the oil could be analyzed with respect to the S-content etc. To confirm that the extraction is reversible and can therefore be described by a partition coefficient (Nernst's law), the S-compound was re-extracted from the S-loaded IL with fresh dodecane. (*Remarks: Technically, re-extraction would certainly not be done with dodecane, but with more efficient re-extraction agents like cyclohexane [10] and others, which are currently under investigation [11]. In addition, the schlenk tubes used up to now for the micro-scale experiments can not be considered to be a typical or least of all an optimal extraction vessel. Therefore extraction with respect to industrial scale extraction will also be tested in a lab-scale mixer-settler unit with 10 stages.*)

All S-contents of the oil presented in this paper were determined by elementary analysis (Antek Elemental Analyzer 9000, pyro-chemiluminescence N-detector, pyro-fluorescence S-detector). Repeated measurements indicate an average error of this method of about 1 ppm (both for S and N). The S-content of the IL was determined based on a respective mass balance.

Leaching of the ILs into the oil (also never observed) could easily be controlled and monitored for all experiments by the elementary N-analysis, as in contrast to the model oils (dodecane and also aromatic hydrocarbons like benzene and methylnaphthaline) all the investigated ILs (see Fig. 2) contain nitrogen.

The extraction of N-compounds (model system indole-dodecane) was also investigated. N-compounds, which are present in (raw) diesel oil in a range comparable to S-compounds, strongly inhibit the catalytic hydrodesulfurization process. It could therefore be interesting to combine a HDS unit with an upstream unit for separation of N-compounds (details will be given below).

To determine the parameters of an extraction column or a mixer-settler system, the respective triangular diagrams were calculated based on experiments with the system *1-methylnaphthaline/dodecane/BMIM-octylsulfate* and the system *FLUKA oil/DBT/BMIM-octylsulfate*.

Eventually, the extraction of real predesulfurized diesel oil was investigated.

**Results and discussion of extraction experiments with Ionic Liquids**

A typical example of an extraction experiment is shown in Fig. 4. The partition coefficient with BMIM-chloroaluminate is 3.4 $(g_S/g_{IL})/(g_S/g_{oil})$, indicating that DBT can be easily extracted from the oil. The respective re-extraction experiment approved this partition coefficient. A similar extraction with dodecanethiol (with BMIM-chloroaluminate) gives a partition coefficient of 8.4 $(g_S/g_{IL})/(g_S/g_{oil})$, indicating that the extraction properties of the IL is not limited to aromatic S-compounds. (*Remark: For a technical deep desulfurization, this result is nonrelevant, because such reactive S-compounds like thiols are already converted by HDS, i. e. pre-desulfurized diesel is practically free of thiols [4].*)

In order to check the feasibility of a technical deep desulfurization process by extraction with IL, the desulfurized oil from the first extraction step was again treated with fresh IL. This process was repeated up to four times to reach five extraction steps (or theoretical plates in a cross current operation).

The almost linear relationship of log(S) *vs.* number of extraction steps underlines that the extraction can be described in a wide range of S-concentration by a constant partition coefficient (Nernst's law), as shown in Fig. 5 for the IL BMIM-octylsulfate, which was again confirmed by re-extraction experiments.

The extraction efficiency of the chloroaluminate-IL (Fig. 4) is higher than the one of the halogen-free IL BMIM-octylsulfate, although in the latter case still a partition coefficient of about 2 is reached (Fig. 5). As expected, the degree of desulfurization is increased by increasing the IL to oil-ratio; thereby the partition coefficient does not change. In case of a 1 to 1 ratio (by mass), only 5 extraction steps are needed to reach a S-content of about 4 ppm (starting from 500 ppm S).

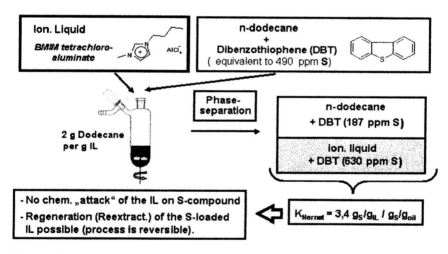

**Fig. 4**: Typical extraction experiment with model diesel oil

With respect to deep desulfurization of diesel oil, it is highly important, that the S-compounds, which are specifically hard to convert (desulfurize) by HDS, can be extracted by ILs. Therefore, an experiment with a model diesel oil containing a mixture of dodecane with DBT (9 ppm S), MDBT (133 ppm S) and 4,6-DMDBT (301 ppm S) was conducted. The result clearly indicates that in contrast to the differences in selectivity to convert DBT-derivatives by HDS, the selectivity of the IL (here BMIM-octylsulfate) to extract MDBTs and above all 4,6-DMDBT is practically identical to the base case of DBT-extraction (Tab. 2).

As already mentioned, the extraction of N-compounds can also play an important role in a combination of extraction with IL (for deep desulfurization)

with an up-stream HDS unit (for desulfurization down to e. g. 100 ppm S). An extraction experiment with indole and BMIM-octylsulfate as extracting agent resulted in a partition coefficient of 340 $(g_N/g_{IL})/(g_N/g_{oil})$, which is - compared to the coefficient of DBT of about 2 $(g_S/g_{IL})/(g_S/g_{oil})$ - an amazingly high value.

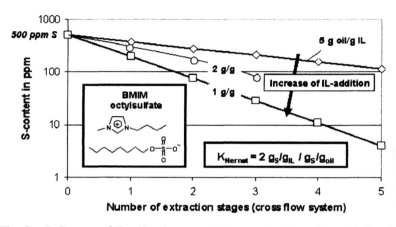

**Fig. 5**:  Influence of IL-oil ratio on multistage extraction of model diesel oil

**Table 2**:  **Comparison of selectivities of extraction and of hydrotreating** (*data from [3], ** data from [4])

| S-compounds present in pre-desulfurized diesel oil | Selectivity of hydrotreating (CoMo-catalyst) | Selectivity of IL-extraction (1-methyl-3-n-butyl-imidazoliumoctylsulfate) |
|---|---|---|
| **DBT** | Reaction rate: 100 % (standardized) | $K_{Nernst}$ = 100 % (standardized) |
| **4-MDBT** | 10 %* to 50 %** | 90 % |
| **4,6-DMDBT** | 1 %* to 20** % | 89 % |

Finally, the desulfurization by extraction of a real pre-desulfurized diesel oil (from MIRO-refinery in Karlsruhe, Germany; S-content of about 375 ppm S) was investigated with different ILs (Fig. 6). It must be emphasized that the ratio of the diesel to IL (5 g oil/g IL) was high, and therefore the final S-contents after e. g. 3 extraction steps (cross current mode) were not much lower than about 200 ppm.

In case of the real diesel oil, the partition coefficients are only about half of the values with model diesel oils. The reason for this is up to now not clear and will be the object of further investigations. Nevertheless, the experiments show, that also real oil can be desulfurized by extraction with ILs.

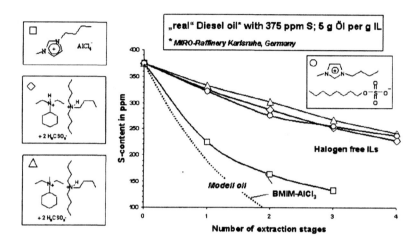

**Fig. 6**: Desulfurization of predesulfurized "real" diesel oil by extraction with IL

## Engineering aspects of desulfurization by extraction with ILs

So as an interim result, the following can be stated: ILs do effectively extract S- and N-compounds. The S-loaded IL can be regenerated by re-extraction or probably by other means like distillation or by supercritical carbon dioxide. The optimization of regeneration of the loaded IL was not yet studied in detail; investigations of this important step with respect to a technical extraction process are presently done.

For the estimation of the design of a technical extraction column (or a mixer-settler system), three main problems/questions are still left:

- To what extent are hydrocarbons also soluble in ILs ("cross solubility")?
- Is the (desulfurized) diesel oil "contaminated" with IL?
- What is the number of theoretical plates of a technical extraction unit?

The respective experiments and answers will be given below.

**Cross-solubility of ionic liquids in hydrocarbons and vice versa**

The triangular diagram of the system *n-dodecane-methylnaphthaline-BMIM-octylsulfate* is shown in Fig. 7 [11]. A mixture of IL and pure methylnaphthaline is monophasic. For a content of less than about 50 % of this aromatic hydrocarbon (in the oil), a biphasic system is established (at least for an IL-content of less than 90 % in the whole mixture). Within the biphasic area, an IL-rich phase and an oil phase is established. The IL-rich-phase contains both the paraffinic and the aromatic hydrocarbon to a relative small extent, whereby the extraction selectivity for the latter is higher.

This cross solubility of hydrocarbons in the IL is an unwanted effect, and should be decreased by further investigations in order to find more suitable ILs. So during regeneration of the oil- and S-loaded IL, a S-rich oil phase would be produced, and has to be further processed to minimize the losses of diesel oil, e. g. by recycling this stream back to the hydrotreater (see below).

It is important to state, that the oil phase (with a slightly lower naphthaline-content than the fresh oil), is free of any IL, which was proven by elementary N-analysis during all experiments with dodekane, FLUKA oil and for comparison also with benzene: The N-content (respectively the IL-content) in the oil phase is at least less than 1 ppm, i. e. below the detection limit of the elementary N-analyzer. This important fact is supported by numerous other experiments.

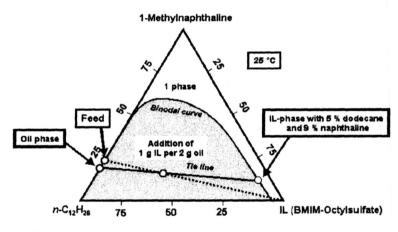

**Fig. 7:** Cross solubility of hydrocarbons in BMIM-octylsulfate

**Number of theoretical plates of an extraction process with ILs**

Triangular diagrams of the system *oil-S-compound-IL* are given in Fig. 8. The oil used (FLUKA, white spirit) had a content of 18 % aromatic hydrocarbons, which is typical for diesel oil (rest paraffines and 5 ppm S), and a boiling range from 180 to 220 °C (1 bar). For clearness, only a "cut-out" of the whole triangular diagram of the system *DBT-BMIM-octylsulfate* is shown in the lower half of Fig. 8 (S-content up to 750 ppm; for a better understanding of the cut-out see the upper part of Fig. 8). In this system, the cross solubility of oil in the IL was only 5 % compared to the model system outlined in Fig. 7, and seems to be independent of the oil to IL-ratio used in extraction. So in case of a high ratio of oil to IL, the portion of the oil transferred into the IL would be small.

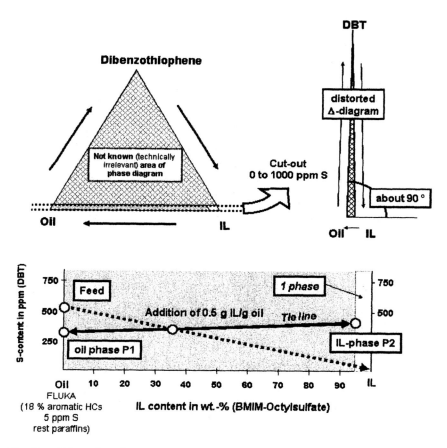

**Fig. 8**:  Triangular diagram (cut-out) of the system *oil-DBT-BMIM-octylsulfate*

To determine the number of theoretical plates of a column or the numbers of mixers/settlers, a relative complicate construction using several tie lines (the first is shown in Fig. 8) and a pole (outside the triangular diagram) is needed. The result of such a calculation for the system described in Fig. 8 is a number of theoretical plates of 6 (for counter current operation of an extraction column and a mixer-settler system) and about 4 for a system with cross current mode.

Much easier (and better to understand) is the determination of the number of theoretical plates by assuming, that the cross solubility of hydrocarbons in the IL can be neglected (at least for this calculation). So the biphasic system is reduced to an IL-phase with dissolved org. S-compounds (but without hydrocarbons) and an oil phase (with a certain residual S-content), i. e. only organic S is exchanged between the IL- and the oil-phase. The resulting standard procedure to calculate the theoretical plates based on this simplification is shown in Fig. 9.

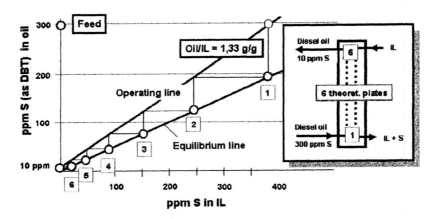

**Fig. 9**: Number of theoretical plates of an extraction column (counter current, 25 °C, IL: BMIM-octylsulfate, calculation based on data with model diesel oil, simplified solution neglecting the cross solubility of hydrocarbons in the IL)

Based on the equilibrium line (Nernst's law) and the operating line, which depends on the mass balance of sulphur (DBT), the S-content in the feed, the target S-content in the final product (here 10 ppm), and the IL to oil ratio (here 1,33 g/g), the number of theoretical plates can be determined by a step function as shown in Fig. 9. The resulting number is 6, which is equivalent to the complicate accurate method using the triangular diagram as roughly described before. According to informations from an industrial partner, a number of less than 10 is acceptable for typical extraction columns; for a value of less than about 3, a mixer-settler would probably be used.

**Integration of extraction unit for deep desulfurization in a refinery**

Several options do exist to integrate a process of deep desulfurization by extraction with ILs into a refinery (Fig. 10):

- The extraction process is installed downstream of the hydrotreater in order to separate the DBT-derivatives from the pre-desulfurized product stream of the hydrotreater. Depending on the IL to oil-ratio needed in the extraction unit, a certain portion of the oil is dissolved in the IL. In Fig. 13, a ratio of 1 was assumed. The loaded IL is then fed to a regeneration unit, and subsequently recycled into the extraction. The oil with the S-compounds from IL-regeneration can then be processed in other units of the refinery, e. g. in thermal cracking, coking, as co-feed in partial oxidation of heavy oils or in a power plant. Alternatively, this S-rich oil from IL-regeneration can also be recycled and added to the feed of the HDS reactor. By this means, the refractory S-compounds like 4,6-DMDBT are converted in the end to $H_2S$ after several cycles (HDS and extraction) and a long total residence time in the HDS, respectively. By this means, the loss of diesel oil would be minimized.

- Alternatively or in addition to a S-extraction, a small up-stream N-extraction unit can be installed to separate mainly the N-compounds. This would relief the HDS unit, because N-compounds strongly inhibit the HDS reactions: According to [4], the activity of a classical CoMo-catalyst decreases by about 30 %, if only 100 ppm N is present. The partition coefficient for the extraction of N-compounds by ILs is so high, that the IL to oil ratio needed for extraction would be very small. So in return, practically all N- and only a very small amount of S-compounds and above all of hydrocarbons would be dissolved. So the portion of the raw diesel oil dissolved in the IL would be so small, that the N-rich oil from the regeneration of the loaded IL can easily be used as a co-feed in several units of the refinery (see above).

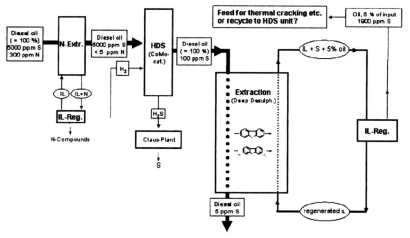

**Fig. 11**: Desulfurization of pre-desulfurized real diesel oil by extraction with IL

## Summary and outlook

The experiments on extraction of S- and N-compounds from oil by ILs indicate, that such a process could be an attractive alternative to common catalytic hydrotreating, after all for deep desulfurization down to values of 50 ppm S or even lower. The results show the excellent and selective extraction properties of ILs, especially with regard to S-compounds, which are very difficult to remove by HDS (e. g. 4,6-DMDBT). The application of very mild process conditions (ambient pressure and temperature) and the fact that no hydrogen is needed, are additional advantages compared to traditional HDS.

Nevertheless, further investigations are still needed in order to develop such a new extraction process based on ILs:

- For a better and optimized extraction process (e. g. compared to BMIM-octyl-sulfate), ILs should be found with a lower cross solubility for hydrocarbons and an even higher partition coefficient for S-compounds.

- The thermodynamic properties of selected extraction systems need to be studied in more detail in order to determine properly all the data for a basic design of a technical process (triangular diagrams etc).

- The regeneration of the loaded IL is a critical process step, and will therefore be intensively studied in the future ($scCO_2$, re-extraction with short-chain hydrocarbons, distillation – if needed under vacuum, etc.).

Finally, it must be emphasized that extraction of polar organic compounds is probably not only limited for deep desulfurization. Other applications are currently investigated: Separation of Cl- and O-compounds, and the extraction of flavours and odors, i. e. the separation of high-value fine chemicals.

## References

(1) König, A.; Herding, G.; Hupfeld, B.; Richter, Th.; Weidmann, K., *Topics Catalysis* **2002**, *16/17*, 23–31.

(2) Song, Ch.; Ma, X., *Appl. Catal. B.* **2003**, *41*, 207–238.

(3) Zhang, Q.; Ishihara, A.; Kabe, T., *Sekiyu Gakkaishi* **1996**, *39*, 410–417.

(4) Schmitz, C. Ph.D. Thesis, University Bayreuth **2003**.

(5) Babich, I.V.; Moulijn, J.A., *Fuel* **2003**, *82*, 607–631.

(6) Gislason, J., *Hydrocarb. Eng.* **2002**, *7*, 39–42.

(7) Jess, A.; Wasserscheid, P. *Z. Umweltchem. Ökotox* **2002**, *14*, 145–154.

(8) Bösmann, A.; Datsevitch, L.; Jess, A.; Lauter, A.; Schmitz, C.; Wasserscheid, Patent WO 03/037835 A2 (2.11.2001).

(9) Bösmann, A.; Datsevich, L.; Jess, A.; Lauter, A.; Schmitz, C.; Wasserscheid, P., *J. Chem. Soc. Chem. Commun.* **2001**, 2494–2495.

(10) Eßer, J., Diploma Thesis, RWTH Aachen **2002**.

(11) Eßer, J., PhD Thesis, University Bayreuth (in preparation).

Chapter 8

# Opportunities for Membrane Separation Processes Using Ionic Liquids

Thomas Schäfer, Luis C. Branco, Raquel Fortunato, Pavel Izák, Carla M. Rodrigues, Carlos A. M. Afonso, and João G. Crespo

REQUIMTE–Department of Chemistry, Faculdade de Ciências e Tecnologia, Universidade Nova de Lisboa, 2829–516 Caparica, Portugal

An overview will be given on the application of membrane separations involving ionic liquids. In particular, the use of pervaporation and supported liquid membranes for the recovery of solutes will be presented and discussed in detail, revealing the opportunities that exist for the still recent area of coupling membrane separation processes and ionic liquids.

Membrane separation techniques can excel conventional downstream processes because separation of target solutes is achieved in many applications without any phase transition, and selectivity is exceeding that of conventional processes owing to particular solute-membrane interactions. In most cases, membrane processes do not involve any additional extraction aid such as organic solvents, can be operated at ambient temperature, and have in general a

© 2005 American Chemical Society

much lower energy consumption than comparable conventional separation techniques. They therefore are considered a "clean", environmentally benign technology.

A large variety of membrane processes are nowadays known, their differences being determined by the function the membrane fulfils, rather than the membrane material that is employed. A traditional classification distinguishes between membrane processes that employ porous membranes (microfiltration, ultrafiltration) with the separation principle governed by a size-exclusion effect, and those applying so-called "non-porous" or "dense" membranes (pervaporation, vapour permeation, gas separation) where the separation occurs on the basis of a solution-diffusion mechanism (1). Membranes may also serve as a base for creating a large interfacial area for mass transport, as is the case in membrane contactors, or host selective phases immobilised within the membrane pores through which the solute transport takes place (supported liquid membranes).

This classification of processes solely on the basis of the membrane intrinsic porosity can be misleading with regard to the function of the membrane, which should always be evaluated in view of the size and physico-chemical properties of the solutes to be recovered. It is easily comprehensible that the order of magnitude of "porous" or "dense" is different for gas molecules than for macromolecules such as proteins. Not surprisingly, a number of significant membrane processes fall in between the regime of "porous" and "non-porous", depending on how far and in what way the membrane in fact interacts with the target solutes (nanofiltration).

As is the case with ionic liquids, membranes can to some degree be tailor-made for individual applications with regard to membrane material (organic or inorganic), structure and porosity (regular or asymmetric), and composition (homogenous or composites, surface modification). Combining both ionic liquids and membranes therefore literally suggests itself not only with regard to benignity, but also in view of process versatility.

The literature on the use of membrane processes involving ionic liquids is still scarce. The use of room-temperature ionic liquids (RTILs) for separation of gases by using supported liquids membranes (SLM) was mentioned as early as 1988 for the separation of gases, using tetramethylammonium fluoride tetrahydrate as the immobilised phase supported in a poly(trimethylsilylpropyne) (PTMSP) membrane (2,3). More recently, SLMs were studied with success using more recent generations of ionic liquids for liquid-gas and gas-gas separations (4) as well as the recovery of organic compounds from solvents (5). Pervaporation was reported to successfully recover solutes from ionic liquids using different types of membranes (6), and for augmenting the conversion during biocatalysed esterifications by selectively removing water (7).

In order to highlight the potential of membranes using ionic liquids, the two until today in literature most cited membrane processes will be discussed in detail in the following: pervaporation and supported liquid membranes. The aim will be to not only call attention to the advantages of membrane processes combined with ionic liquids, but to also emphasise pitfalls – which more often than not bear the potential of significant progress, once they are understood and overcome.

## Pervaporation

Pervaporation is a membrane separation process that employs dense, non-porous membranes for the selective separation of dilute solutes from a liquid bulk solvent into a solute enriched vapour phase (8). The separation concept of pervaporation is based on the molecular interaction between the feed components and the dense membrane material, unlike some pressure-driven membrane separation processes involving porous membranes, such as microfiltration, whose general separation concept is primarily based on size-exclusion.

The principle of pervaporation is illustrated in Figure 1.

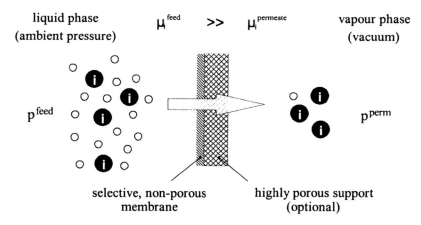

*Figure 1. Principle of pervaporation.*

A non-porous membrane separates a liquid feed, commonly close to atmospheric pressure, from a downstream compartment in which a vacuum is applied. In order to warrant high fluxes while maintaining a high mechanical stability of the membrane, composite membranes are often employed which consist of an, as thin as possible, selective membrane layer cast on a highly porous non-selective support (Figure 1).

When the feed contacts the membrane, the solutes (denoted as $i$ in Figure 1) adsorb on and subsequently absorb in the membrane surface due to solute-polymer interactions. Preferential solute-polymer interactions imply that the solvating power of the polymer is higher for the solutes than for the bulk solvent. The sorption of these solutes at the membrane surface creates a solute concentration gradient across the membrane, resulting in a diffusive net flux of solute across the membrane polymer. In pervaporation, any solute having diffused toward the membrane downstream surface is ideally instantaneously desorbed and subsequently removed from the downstream side of the membrane. This can be achieved either by applying a vacuum (vacuum pervaporation), or by passing an inert gas over the membrane downstream surface (sweeping-gas pervaporation) such that the diffusive net flux across the membrane is maximal. The vapour so desorbed is commonly recovered by condensation at an adequate temperature.

The driving force in pervaporation is thus the gradient of the concentration, or more precisely of the chemical potential, of a solute between the bulk phases of each side of the membrane, while the transport is governed by a solution-diffusion mechanism such that the solute flux is described in its most general and convenient form by eq. 1 as

$$J_i = \frac{D_{i,m} S_{i,m}}{\ell_m}\left(c_{i,f} - c_{i,P}\right) = \frac{L_{i,m}}{\ell_m}\left(c_{i,f} - c_{i,P}\right) \tag{1}$$

with $J_i$: flux of $i$ with the dimension being quantity of solute (in moles or mass) per unit membrane area and unit time; $D_{i,m}$: solute diffusion coefficient in the membrane; $S_{i,m}$: solute sorption (partitioning) coefficient in the membrane; $\ell_m$: membrane thickness; $L_{i,m}$: membrane permeability for solute $i$. Hence, whilst the driving force is identical in pervaporation and evaporation, it is the membrane permeability L that infers an additional selectivity of the pervaporation process in view of processes governed by the vapour-liquid equilibrium alone. It should be noted that, for mere convenience, concentrations are commonly used in order to calculate pervaporation fluxes. Data on solute mass transport so obtained, however, are only valid for the reference system in which they were determined and may not be readily extrapolated. For this purpose one would revert to using the chemical potential.

The selectivity of the pervaporation process toward two solutes $i$ and $j$ is expressed as their ratio of permeabilities, or their separation factor,

$$\alpha_{i,j} = \frac{L_i}{L_j} = \frac{c_{i,P}/c_{i,f}}{c_{j,P}/c_{j,f}} \qquad (2)$$

In practice, flux and selectivity can often be found counteracting: increasing the flux of a membrane often results in a loss of selectivity, and *vice versa*. The optimum configuration of a pervaporation is therefore found by optimising not only membrane characteristics, but also the condensation strategy and, if possible, the optimum feed temperature and feed fluid dynamics.

## Recovery of Solutes by Pervaporation

In order to show the potential for recovering solutes by pervaporation, laboratory-scale pervaporation experiments were carried out using a range of binary solutions of a common ionic liquid, 1-butyl-3-methylimidazolium hexafluorophosphate [$C_4$MIM][$PF_6$], and model solutes all of which differing strongly in their physical-chemical properties, namely: water, chlorobutane, ethyl hexanoate, and naphthalene. Different dense hydrophilic and hydrophobic polymeric membranes, namely polyoctylmethylsiloxane (POMS) polyether block amide (PEBA) and polyvinylalcohol (PVA) were chosen for the solute recovery based on their selectivity for the individual model solute. All experiments were carried out at 323.15 K and a permeate pressure of 10 Pa, using a standard laboratory pervaporation set-up (9) with an effective membrane area of 0.01 $m^2$. The feed volume used was 110 $cm^3$ throughout all experiments.

The degree of recovery and the partial solute fluxes are depicted in Figure 2. All solutes were recovered to a degree higher than 99.2 % (limit of the analytical sensitivity) of their initial feed concentration (6). No ionic liquid was detected in any of the permeates, with the detection limit being 74 $\mu g \cdot kg^{-1}$ using inductively coupled plasma (ICP) spectroscopy and phosphorus as the reference atom for the hexafluorophosphate anion of [$C_4$MIM][$PF_6$], as could be expected owing to the non-measurable vapour pressure of the ionic liquid. Partial fluxes using the membranes described were stable throughout the operation, suggesting that the ionic liquid either did not penetrate the membrane polymer, or that it penetrated the membrane but did not affect solute transport. This is opposed to observations made using Nafion® membranes and 1-butyl, 3-methylimidazolium trifluoromethane sulfonate, where the ionic liquid swell the membrane polymer and thus affected solute transport significantly (10).

Water exhibited the highest flux, using a hydrophilic PVA-composite membrane. Because of their hydrophilicity, PVA-membranes are highly permeable for water, while hindering the permeation of hydrophobic molecules. This is particularly interesting for reversible condensation reactions or biocatalytic esterifications carried out in ionic liquids, during which the selective and continuous removal of the water formed shifts the equilibrium to higher yields of the target product, as will be discussed in more detail below.

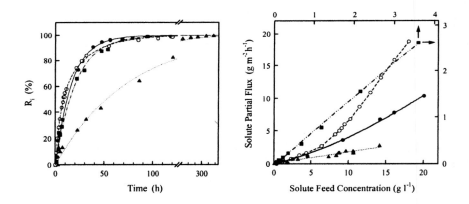

*Figure 2. left: Recovery of water by a PVA (4μm)-composite membrane (O); chlorobutane (●) and ethyl hexanoate (■) by POMS (25 μm)-composite membrane; naphthalene by a homogeneous PEBA (30 μm)-membrane (▲). right: Partial fluxes of the solutes as a function of their respective feed concentration; for graphical reasons, the x- and y-axes for ethyl hexanoate are depicted on the top and on the right of the graph (indicated by arrows). (Reproduced from reference 6. Copyright 2001Royal Chemical Society.)*

Also the low-volatile model solute, naphthalene, was recovered quantitatively at 323 K, 168 K below its boiling point of 491 K using a homogenous PEBA-membrane. A homogenous membrane was chosen for this separation because in composite membranes the pressure loss in the porous support can be sufficient to cause undesired condensation of the low-volatile solute on the membrane downstream face. Although the rate of recovery was lower than that of the other solutes, owing to a smaller driving force, this example illustrates the potential that pervaporation has even for the recovery of high-boiling compounds.

It should be pointed out that the choice of the right membrane material and operating conditions are crucial for the pervaporation to possess an advantage over evaporation techniques, as otherwise no advantage may be observed (11). Under the operating conditions reported here, POMS, for example, showed an about 2.7 times higher selectivity for ethyl hexanoate than for chlorobutane, which could be expected as POMS has been reported to be particularly suitable for the recovery of esters (9).

Despite these promising results, however, a thorough evaluation and comparison of process efficiency and performance should only be made on the basis of thermodynamic variables such as activities or the chemical potential, rather than on the more common and more readily accessible concentrations. Here the need for thermodynamic data on solute activities in different ionic liquids, as they are long available for conventional solute-solvent systems, becomes imperative for the separation process design (12). If the separation task involves a mere purification of the ionic liquid, or in the practically rare case of binary solutions with the solvent being an ionic liquid, pervaporation does not offer any advantage over evaporative methods. Owing to the non-measurable vapour pressure of the ionic liquid, in this case the separation process selectivity is not relevant and the pervaporation membrane constitutes only an additional resistance to the overall mass transport.

## Enhancing Reversible Reaction Conversion with Pervaporation

During reversible reactions in which water is one of the products, such as condensations, esterifications, and eliminations, the reaction equilibrium may be favourably shifted by removing water using and adequate separation technique. Pervaporative water removal (hydrophilic pervaporation) is particularly interesting inasmuch as the target solute, water, not only partitions favourably into the hydrophilic membrane, but also diffuses fastest owing to its small molecular size. Hence, while the pervaporative recovery of organic compounds is in general primarily sorption-controlled, hydrophilic pervaporation is mostly transport (diffusion) controlled (1).

Using a similar experimental set-up as mentioned before, an esterification reaction according to Figure 3 was carried out in [C$_4$MIM][BF$_4$] with and without pervaporative removal of the water produced. [C₄MIM][BF₄] was chosen for its relatively low viscosity, as well as its ability to easily dissolve the crystalline borneol. Water was removed during esterification using a PVA-based membrane (PERVAP® 2205, Sulzer, Germany) suitable for applications in presence of organic acids.

*Figure 3. Esterification of (-)-borneol with acetic acid in [C₄MIM][BF₄] in presence of para-toluenesulfonic acid at 333 K, yielding (-)-bornyl acetate and water.*

The resulting increase of the conversion of borneol was two-fold from 22 to 44 %, clearly confirming the suitability of pervaporation for shifting the reaction equilibrium also in ionic liquids (13).

It should be pointed out that especially for the removal of compounds from biotransformations carried out in ionic liquids, pervaporation is the only separation technique that can be applied without either degrading the biocatalyst or interfering with the bioconversion: it can be efficiently operated at a moderate temperature, and does not require any extraction aid detrimental to the performance of the biocatalyst that, in turn, can be reused without loss of activity (7). The target product, such as esters formed by esterification, may then be recovered using an appropriate hydrophobic membrane, with an example being the pervaporation of ethyl hexanoate utilising a POMS-composite membrane (Figure 2).

## Supported Liquid Membranes (SLM)

If a liquid is immobilised in the pores of a membrane, a so-called "supported liquid membrane" (SLM) is obtained (Figure 4). The immobilisation occurs owing to capillary forces. Favourable van-der-Waals interactions between the liquid and the membrane ideally prevent the liquid from partitioning into the adjacent feed and receiving phase during operation. The porous membrane serves, hence, in this case as a framework for the liquid extractant, across which selective transport occurs.

The principal difference to using the extractant in a liquid-liquid separation or as an emulsion liquid membrane is the fact that in an SLM the extractant remains confined within the membrane pores, such that the need for any phase separation after extraction can be avoided. A potential advantage of SLMs is furthermore that by choosing the appropriate immobilised solvent-membrane combination, loss of extractant to either the feed or the receiving phase should be minimum. The transport of a target solute across an SLM is determined, similarly to as was described for pervaporation, by a partitioning of the target solute into the immobilised liquid, across which it is subsequently transported according to a concentration gradient (more precise: a gradient of the chemical potential) of the target solute between the feed phase and the receiving phase.

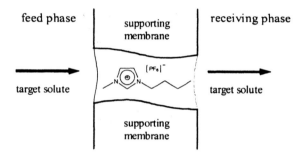

*Figure 4. Principle of a supported liquid membrane (SLM).*

SLMs have been widely studied on the laboratory-scale for diverse applications such as recovery of metal-ions, gases, contaminants from effluents, as well as fermentation products. However, their industrial breakthrough has been impaired by the lack of stability of the SLM owing to evaporation of the immobilised liquid, or its dispersion or dissolution into the adjacent phases. Because ionic liquids lack a measurable vapour pressure, and may be immiscible with various solvents including water, they appear an attractive alternative for stable SLM. In the following, two case studies employing SLMs based on ionic liquids will therefore be presented: the recovery of amino acid derivates from and to aqueous solutions, and the selective transport of isomeric amines from and to an organic solvent.

**Transport of Amino Acid Derivates**

Amino acid derivates belong to the nowadays so-called "nutraceuticals", components of beneficial value for products from both the pharmaceutical and food industry, requiring therefore a selective downstream processing that also complies with the delicacy and final utilisation of these compounds.

Prior to conducting transport studies using SLMs, the partitioning of three amino acid esters, namely: proline benzyl ester (e-Pro), phenylalanine methyl ester (e-Phe), and phenyglycine methyl ester (e-Phg), from aqueous solution to $[C_8MIM][PF_6]$ was investigated. This particular ionic liquid was chosen due to its low water solubility (14). It was found that the affinity of $[C_8MIM][PF_6]$ for each of the three esters varied, with the respective equilibrium partitioning coefficients P decreasing from $P_{e-Pro} = 9.84$ to $P_{e-Phe} = 0.88$ and $P_{e-Phg} = 0.33$, suggesting a promising application for SLM (16).

$[C_8MIM][PF_6]$ was subsequently immobilised in a porous PVDF flat membrane (FP Vericel™, Gelman Sciences) of 140 μm thickness to create an SLM according to a procedure described elsewhere (14). The SLM was contacted with the aqueous feed and receiving phases, and the transport of the esters across the SLM measured. Surprisingly, the SLM so prepared did not show any selectivity at all to any of the amino acid esters. Thorough transport studies involving tritiated water revealed that the formation of water microenvironments within the ionic liquid was responsible for this phenomenon: with the SLM being comparably thin, the ionic liquid rapidly equilibrated with the adjacent aqueous phases with regard to its water content, forming water microenvironments through which the main transport of the amino acid esters occurred, and thus by-passing the selective interaction with the ionic liquid. This observation was not made during transport studies using the bulk liquid membranes as they are orders of magnitude thicker, thus delaying a similar formation of water microenvironments.

Although not a successful application of SLM, this case study proves that in order to evaluate the potential of a SLM for a given separation task, the overall transport under actual operating conditions needs to be considered, as partitioning experiments alone might be misleading.

**Selective Transport of Isomeric Amines**

An SLM was prepared by immobilising $[C_4MIM][PF_6]$ in a porous PVDF flat membrane (FP Vericel™, Gelman Sciences), separating the feed and the receiving phase both of which consisted of diethyl ether as the bulk solvent.

The transport studies were then conducted with a feed mixture of isomeric amines, selective separation of which is a challenge for any separation process. The feed consisted of an equimolar mixture of a secondary and a ternary amine, namely diisopropylamine (DIIPA), and triethylamine (TEA). The SLM was operated in the continuous mode for up to 14 days (5).

In contrast to what had been observed in the previous example, a selective enrichment of DIIPA was observed in relation to TEA with the ratio between the two amines being shifted from 1:1 in the feed toward 4:1 in the receiving phase. Such a selectivity between two isomeric compounds can be considered excellent for any separation process (17).

With the diffusion coefficients of the isomeric amines being expectedly similar within the ionic liquid owing to their very similar molar volume, it could be proven by $^1H$ NMR spectroscopic analysis that the reason for this selective enrichment is in fact a preferential interaction of $[C_4MIM][PF_6]$ with DIIPA: addition of either DIIPA or TEA to neat $[C_4MIM][PF_6]$ resulted in a drift of the chemical shift of the H(2) and H(4) protons of the $[C_4MIM]$ cation that was more pronounced for DIIPA than for TEA (Figure 5). A possible explanation of this higher affinity can be both the higher basicity of DIIPA with regard to TEA, as well as the fact that the latter is sterically more hindered when approaching the $[C_4MIM]$ cation (5,15).

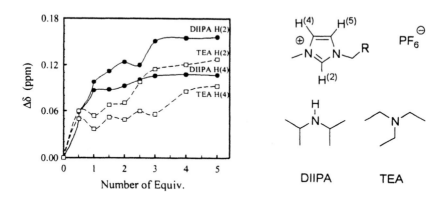

*Figure 5. Drift of the chemical shift of H(2) and H(4) of the imidazolium-ring of $[C_4MIM][PF_6]$ as promoted by addition of either DIIPA or TEA. (Reproduced from reference 15. Copyright 2002 Wiley-VCH.)*

The selectivity so observed could be maintained throughout the 14 days of operation of the SLM (Figure 6), which is a remarkable operational stability (18, 19). It can hence be seen that SLMs can be a very promising alternative to conventional extraction techniques. However, the appropriate combination of membrane and ionic liquid needs to be found for a given separation task and a given bulk solvent in order to warrant a high stability of the supported liquid membrane and hence long operation times.

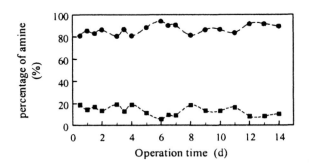

*Figure 6. Stability of the SLM during 14 days of operation with the relative percentage of both amines remaining at a constant ratio of about 4:1. (Reproduced from reference 5. Copyright 2002 Wiley-VCH.)*

## Conclusions and Outlook

Pervaporation and Supported Liquid Membranes are benign membrane processes that in combination with ionic liquids are capable of opening up new opportunities for downstream processing.

Pervaporation can be employed in an energy-efficient way for removing selectively volatile compounds from ionic liquids for recovery, or favourably shifting equilibria during reversible reactions. Its mild operating conditions qualify pervaporation in particular for temperature-sensitive applications such as bioconversions.

Supported Liquid Membranes based on ionic liquids possess a high potential for the selective transport of compounds between two organic phases, such as the resolution of isomeric amines. Their use between two aqueous phases apparently is not adequate owing to the formation of water microenvironments in the ionic liquid phase. A potential application of great interest, however, is the use of Supported Liquid Membranes in gas-liquid, and gas-gas separations, where the lack of a measurable vapour pressure of the ionic liquid is the principal advantage over conventional solvents.

# References

1.  Mulder M. *Basic Principles of Membrane Technology*; Kluwer Academic Publishers: Dordrecht, 1996.
2.  Quinn, R.; Pez, G.P.; Appleby, J.B. U.S. Patent 4,780,114, 1988.
3.  Quinn, R.; Appleby, J.B.; Pez, G.P. New facilitated transport membranes for the separation of carbon dioxide from hydrogen and methane. *J. Membrane Sci.* **1995** *104(1-2)* 139-146.
4.  Scovazzo, P.; Visser, A.E.; Davis, J.H.; Rogers, R.D.; Noble, R.D.; Koval, C. In: *Ionic Liquids: Industrial Applications for Green Chemistry*; Rogers, R.D., Seddon, K.R., Eds.; ACS Symposium Series 818; American Chemical Society: Washington, DC, 2002; pp 69-87.
5.  Branco, L.C.; Crespo, J.G.; Afonso, C.A.M. Highly selective transport of organic compounds by using supported liquid membranes based on ionic liquids. *Angew. Chem. Int. Ed.* **2002,** *41(15)*, 2771-2773.
6.  Schäfer, T.; Rodrigues, C.M.; Afonso, C.A.M.; Crespo, J.G. Selective recovery of solutes from ionic liquids by pervaporation - a novel approach for purification and green processing. *Chem. Commun.* **2001,** 1622-1623.
7.  Gubicza, L.; Nemestóthy, N.; Fráter, T.; Bélafi-Bakó, K. Enzymatic esterification in ionic liquids integrated with pervaporation for water removal. *Green Chem.* **2003,** *5(2)*, 236-239.
8.  Böddeker, K.W. Terminology in Pervaporation. *J. Membrane Sci.* **1990,** *51(3)*, 259-272.
9.  Schäfer, T.; Bengtson, G.; Pingel, H.; Böddeker, K.W.; Crespo, J.P.S.G. Recovery of aroma compounds from a wine-must fermentation by organophilic pervaporation. *Biotechnol. Bioeng.* **1999,** *62(4)*, 412-421.
10. Doyle, M.; Choi, S.K.; Proulx, G. High-temperature proton conducting membranes based on perfluorinated ionomer membrane-ionic liquid composites. *J. Electrochem. Soc.* **2000,** *147 (1)*, 34-37.
11. Fadeev, A.G.; Meagher, M.M. Opportunities for ionic liquids in recovery of biofuels. *Chem. Commun.* **2001,** 295-296.
12. Krummen M.; Wasserscheid P.; Gmehling J. Measurement of activity coefficients at infinite dilution in ionic liquids using the dilutor technique. *J. Chem. Eng. Data* **2002,** *47(6)*, 1411-1417.

13. Izák, P.; Afonso, C.A.M.; Crespo, J.G.; Department of Chemistry, Universidade Nova de Lisboa, Portugal, *unpublished results*.

14. Fortunato, R.; Afonso, C.A.M.; Reis, M.A.M.; Crespo, J.G. Supported liquid membranes using ionic liquids: study of stability and transport mechanisms. *J. Membrane Sci.* **2004**, *in press*.

15. Branco, L.C.; Crespo, J.G.; Afonso, C.A.M. Studies on the selective transport of organic compounds by using ionic liquids as novel supported liquid membranes. *Chem. Eur. J.* **2002**, 17(8), 3865-3871.

16. Fortunato, R.; González-Muñoz, M.J.; Kubasiewicz, M.; Luque, S.; Alvarez, J.R.; Afonso, C.A.M.; Coelhoso, I.M.; Crespo, J.G. Liquid membranes using ionic liquids: influence of water on solute transport. *submitted for publication*.

17. Wankat, P. *Equilibrium-Staged Separations*; Prentice Hall: New Jersey, 1998.

18. Chiarizia, R. Stability of supported liquid membranes containing long-chain aliphatic amines as carriers. *J. Membrane Sci.* **1991**, *55*, 65-77.

19. Danesi, P.R. Reichley-Yinger, L., Rickert, P. Lifetime of supported liquid membranes: the influence of interfacial properties, chemical composition and water-transport on the long term stability of the membranes. *J. Membrane Sci.* **1987**, *315*, 117-145.

Chapter 9

# Ionic Liquids: Highly Effective Medium for Enantiopure Amino Acids via Enzymatic Resolution

Sanjay V. Malhotra and Hua Zhao

Department of Chemistry and Environmental Science, New Jersey Institute of Technology, Newark, NJ 07102

Various room-temperature ionic liquids (RTILs) were prepared and investigated in the enzymatic resolution reactions for obtaining enantiopure amino acids. RTILs can boost the enantioselectivity of enzymes when an appropriate concentration of ionic liquid is used. The preliminary results from this study indicate that ionic liquids are suitable alternative media over organic solvents for achieving high enantiopurity of amino acids.

## Introduction

As one of the most important Active Pharmaceutical Ingredients (APIs), α-amino acids are vital intermediates for many pharmaceutical and biological applications. For example, *L*-(+)-homophenylalanine ((S)-2-amino-4-phenylbutanoic acid) is a vital component of angiotensin-converting enzyme (ACE) and renin inhibitors (1, 2). Many ACE inhibitors such as Benazepril, Enalapril and Lisinopril, have been intensely studied as a medicinal target for the treatment of hypertension and heart failure (3). Therefore, obtaining enantiopure α-amino acids has always been of interest and a challenge to chemists.

Enzymatic resolution of amino acid derivatives is one of the simplest and most efficient methods of synthesizing enantiomerically enriched amino acids. Traditionally, mixtures of organic solvents and water mixture have been used as reaction media. For example, the kinetic resolution of different amino acids by enzyme alcalase has been studied in acetone-water (4), acetonitrile-water (4), ethanol-water (4), 1-propanol-water (4), tetrahydrofuran-water (4), dioxane-water, t-butanol-water (5), 2-methyl-2-propanol-water (6), and DMF (7), etc.

In recent years, ionic liquids have gained a lot of attention as green solvents in organic synthesis and other chemical processes. Ionic liquids with melting point at room temperature and below (as low as −96 °C) can now be produced, which is an important reason that ionic liquids are becoming a more attractive substitute for volatile and toxic organic solvents (8) and considered environmentally friendly. Also, ionic liquids have many favorable properties, e.g., they are as good solvents for a wide range of inorganic, organic and polymeric materials, have adjustable polarity, and catalytic effects, etc. Therefore, they have been investigated as reaction media in many organic and organometallic syntheses (9). More recently, ionic liquids have been used in the studies of enzymatic systems, such as lipase catalyzed kinetic resolution of 1-phenylethanol in ionic liquids (10), enzymatic catalysis in the formation of Z-aspartame in ionic liquids (11), as catalysts in alcoholysis, ammoniolysis and perhydrolysis reactions by lipase in ionic liquids 1-butyl-3-methylimidazolium tetrafluoroborate or hexafluorophosphate (12), etc. Markedly enhanced enantioselectivity was achieved in ionic liquids for the lipase-catalyzed transesterifications (13).

Recently we achieved some promsing results on the enzymatic resolution of amino acid ester using room-temperature ionic liquids (14, 15). With the foreseeable advantages of using ionic media to substitute organic solvents, a systematic kinetic study of the enzymatic solution of amino acids was reported in this chapter. These results further prove that the RTILs are promising media for the enzymatic resolution of amino acids.

## Material Preparation

Ionic liquids 1-ethyl-3-methylimidazolium tetrafluoroborate ($[EMIM]^+[BF_4]^-$) (I), $N$-ethyl pyridinium tetrafluoroborate ($[EtPy]^+[BF_4]^-$) (II) and *N*-ethyl pyridinium trifluoroacetate ([EtPy]⁺[CF₃OO]⁻) **(III)** were prepared using methods modified from the literature (16). The detail preparation procedures was report in our recent article (17).

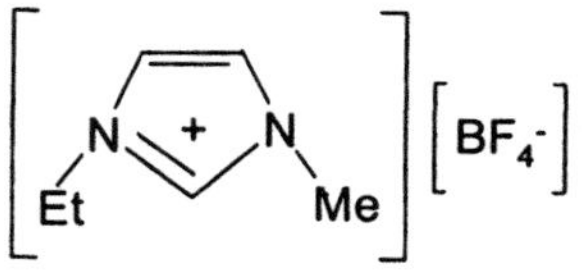

*Figure 1. Structure of three room-temperature ionic liquids.*

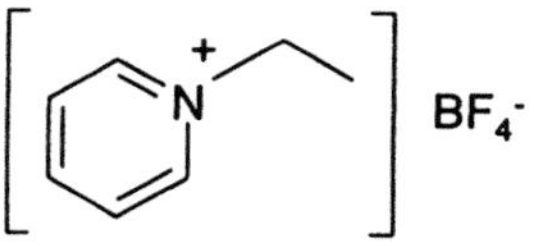

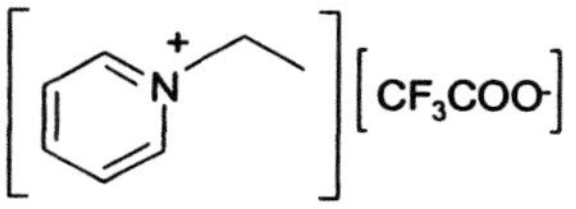

The enzyme Novo alcalase used in this study was produced from *Bacillus licheniforms* by Novozymes A/S and distributed by Sigma-Aldrich as a brown liquid with a specific activity of 2.4 AU/g$^\dagger$ for hydrolysis of dimethyl casein at 50 °C and pH 8.3. In this study, Novo alcalase was used as supplied.

The *N*-acetyl homophenylalanine ethyl ester was prepared following procedures reported earlier (Figure 2) (18). Other amino acid esters were purchased from Sigma-Aldrich. The enantiomeric excess (*ee*) was calculated from the specific optical rotation which was measured by AUTOPOL IV

---

$^\dagger$ The major enzyme component in Novo alcalase is subtilisin Carlsberg. According to Novozymes, one Anson Unit [AU] is the amount of enzyme which, under standard conditions, digests haemoglobin at initial rate liberating per min an amount of tricholoroacetic acid (TCA) soluble product, which gives the same color of phenol reagents as 1 mequiv of tyrosine. Thus, 1 AU = 1000 U, 1U = 1 mmol of *L*-tyrosine methyl ester hydrolyzed per min.

polarimeter (from Rudolph Research Analytical Company). Also, the *ee* measurements were confirmed by HPLC with Chiralpak WH column. The yield calculation was based on the total racemic mixture, therefore, the yield of any one of the enantiomers has a 50% maximum value.

## General Method of Kinetic Resolution

Racemic amino acid ester (0.5) g was dissolved in 60 ml water (85%, v/v) and ionic liquid. To this 0.5 g $NaHCO_3$ was added at a pH ~ 8, followed by addition of 2 ml Novo alcalase. The reaction mixture was gently stirred at 25 °C for 24 h under $N_2$ atmosphere. The *N*-acetyl *D*-ester was extracted by ethyl acetate three times. 6 M HCl was added to the aqueous solution until the pH was lowered to 2-3. The remaining ethyl acetate in the aqueous solution was evaporated and the solution was further concentrated until the precipitate appeared. Removal of water gave *N*-acetyl *L*-acid. The *N*-acetyl *D*-acid was obtained by hydrolyzing *N*-acetyl *D*-ester in 6 M NaOH solution for 2 h. The *N*-acetyl group in the *L*- and *D*-acid was taken off by refluxing in 3 M HCl for 3 h. Figure 2 shows the scheme of enzymatic resolution of homophenylalanine ethyl ester.

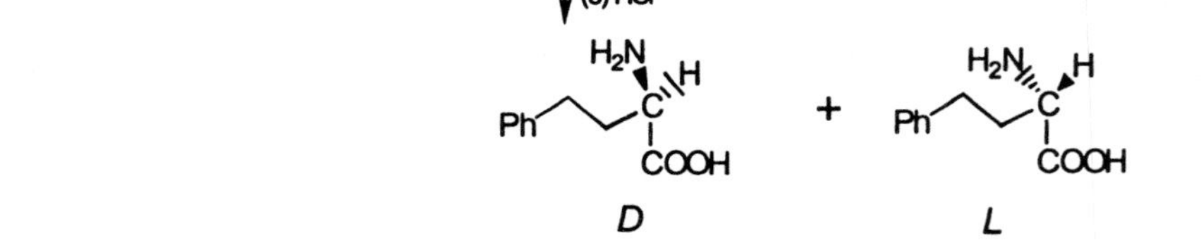

*Figure 2. Scheme of enzymatic resolution of homophenylalanine ethyl ester.*

# Results and Discussion

## HPLC Studies on the Kinetic Resolution of Homophenylalanine Ethyl Ester

Chiral homophenylalanine product samples were analyzed by HPLC with a chiral column (Chiralpak WH). High resolution was achieved under the following HPLC operation conditions:

Column:          Chiralpak WH[†]
Mobile phase:    30% $CH_3OH$, 70% $H_2O$, 0.75 mM $CuSO_4$
Flow Rate:       1.5 ml/min
UV:              230 nm
Temperature:     50 °C

The enzymatic resolution of homophenylalanine ethyl ester was carried out in acetonitrile and water mixture (66% v/v water) by using the enzyme BL-alcalase. The HPLC profile of the L-isomer sample is shown in Figure 3. The reaction conditions were followed by the figure. The L-isomer has a much larger peak compared with the D-isomer. The ee of the L-isomer is 92.4% according to the integration of the HPLC peaks.

When ionic liquid $[EtPy]^+[BF_4]^-$ was used with 83% (v/v) water content, a high ee (97.4%) was observed in the kinetic resolution of homophenylalanine ethyl ester (Figure 4). This indicates that in low concentrations of ionic liquid, the enzyme BL-alcalase shows high enantioselectivity in the kinetic resolution reaction.

## Solvent Effect on the Kinetic Resolution of Homophenylalanine Ethyl Ester

The effect of solvents are significant on the activity and enantioselectivity of enzymes. The kinetic resolution of different amino acids by enzyme alcalase has been studied in various organic solvents (4).

---

[†] The Chiralpak WH column was produced by Chiral Technologies, INC (France) with an internal diameter 4.6 mm, column length 250 mm, partical size 10 μ and adsorbent of ligand exchange.

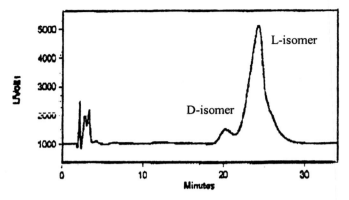

*Figure 3. HPLC profile of kinetic resolution in acetonitrile and water mixture (0.5 g ester, 0.5 g NaHCO₃, 20.4 ml acetonitrile, 39.6 ml water, 1.0 ml BL-alcalase, 25 °C, 24 hours).*

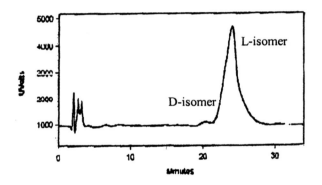

*Figure 4. HPLC profile of kinetic resolution in [EtPy]⁺[BF₄]⁻ and water mixture (0.5 g ester, 0.5 g NaHCO₃, 10.2 ml [EtPy]⁺[BF₄]⁻, 49.8 ml water, 1.0 ml BL-alcalase, 25 °C, 24 hours).*

In this study, three ionic liquids, $[EMIM]^+[BF_4]^-$, $[EtPy]^+[BF_4]^-$ and $[EtPy]^+[CF_3COO]^-$, were compared to the organic solvent acetonitrile. The water content in the mixed solvent seems to be a strong factor influencing the enantioselectivity of the enzyme *BL*-alcalase. Figure 5 illustrates that the water content influences the kinetic resolution of *N*-acetyl homophenylalanine ethyl ester using enzyme *BL*-alcalase. It was observed that a change in concentration of acetonitrile did not significantly affect the *ee* of the L-isomer though the yield changed. However, lower *ee* is expected at high concentration of organic solvent because it may inhibit the activity of enzymes. Although in this case pure water is a good solvent for enzymatic resolution, it is not always a suitable medium for the resolution of amino acids (4). The results in Figure 5, are similar to those shown by Kijima et al (4); who showed that the *ee* declines sharply if the water content is above 70% (v/v) for the kinetic resolution of *DL*-tyrosine ethyl ester in acetonitrile-water mixture.

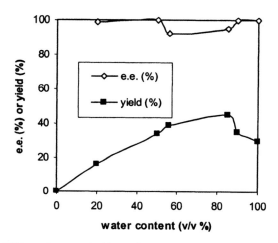

*Figure 5. Effect of acetonitrile and water ratio on the kinetic resolution of DL-homophenylalanine ethyl ester (0.5 g ester, 60 ml mixing solvent, 0.5 g NaHCO₃, 1.0 ml BL-alcalase, 25 °C, 24 hours).*

However, ionic liquids have a different pattern of affecting the kinetic resolution of amino acids. Figure 6, 7 and 8 show the kinetic resolution of *N*-acetyl homophenylalanine ethyl ester in three different ionic liquids $[EMIM]^+[BF_4]^-$, $[EtPy]^+[BF_4]^-$ and $[EtPy]^+[CF_3COO]^-$ respectively. The *ee* of the *L*-isomer declines with the increase of ionic liquid concentration. A high concentration of the ionic liquid results in high ionic strength of the reaction media, which may denature the enzyme and thus decrease the enantioselectivity of the enzyme. Also, an increased concentration of ionic liquid may increase the non-enzymatic hydrolysis of the ester. The effect of ionic liquids on reaction pattern (products *L* & *D* isomer formation) appears to be similar to those in case of acetonitrile as solvent. This indicates that ionic liquids can influence the enantioselectivity and activity of the enzyme *BL*-alcalase at a range of concentration. However, the ionic liquid [EtPy]⁺[CF₃COO]⁻ shows more promising applications among those three ionic solvents since both *ee* and yield of the desired product maintain high values in the presence of 85% (v/v) water (Figure 8).

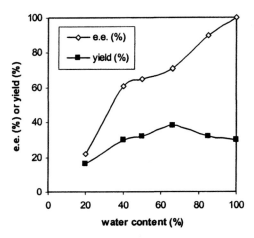

*Figure 6. Effect of ionic liquid [EMIM]⁺[BF₄]⁻ and water ratio on the kinetic resolution of DL-homophenylalanine ethyl ester (0.5 g ester, 60 ml mixing solvent, 0.5g NaHCO₃, 1.0 ml BL-alcalase, 25 °C, 24 hours).*

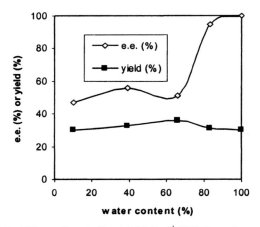

*Figure 7. Effect of ionic liquid [EtPy]⁺[BF₄]⁻ and water ratio on the kinetic resolution of DL-homophenylalanine ethyl ester (0.5 g ester, 60 ml mixing solvent, 0.5g NaHCO₃, 1.0 ml BL-alcalase, 25 °C, 24 hours).*

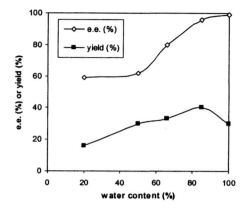

*Figure 8. Effect of ionic liquid [EtPy]⁺[CF₃COO]⁻ and water ratio on the kinetic resolution of DL-homophenylalanine ethyl ester (0.5 g ester, 60 ml mixing solvent, 0.5g NaHCO₃, 1.0 ml BL-alcalase, 25 ℃, 24 hours).*

## Time Effect on the Kinetic Resolution

Since *L*-homophenylalanine ethyl ester is preferred for hydrolysis by the enzyme *BL*-alcalase compared with the *D*-ester, longer reaction time leads to the higher product yield. Kinetic resolutions in ionic liquid $[EtPy]^+[BF_4]^-$-water (Figure 9) indicates that high optical purity was achieved in these cases regardless of reaction time, and increasing product yield was obtained with longer reaction time.

## Temperature Effect on the Kinetic Resolution

Temperature plays an important role in controlling the product optical purity and yield. The effect of temperature on the kinetic resolution of homophenylalanine ethyl ester in ionic liquid $[EtPy]^+[BF_4]^-$ and water mixture was illustrated in Figure 10. Maximum optical purity and yield occur at an optimum temperature 25 °C. It is interesting that at 0 °C, both *ee* and yield were very low while using an ionic liquid as a reaction medium. The possible reason could be that the ionic liquid might have a strong interaction with the enzyme active site at low temperature, and therefore, inhibit the substrates to attach with the enzyme. In an ionic liquid, the enzyme is more sensitive to temperature changes. At a higher temperature than 50 °C, the enzyme *BL*-

120

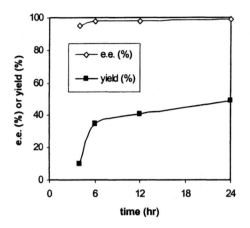

*Figure 9. Effect of reaction time on the kinetic resolution of DL-homophenylalanine ethyl ester in [EtPy]$^+$[BF$_4$]$^-$ + water (0.5 g ester, 85% H$_2$O, 0.5g NaHCO$_3$, 1.0 ml BL-alcalase, 25 °C).*

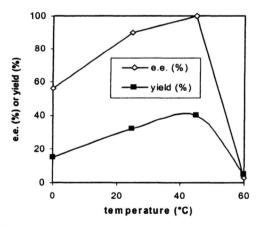

*Figure 10. Effect of reaction temperature on the kinetic resolution of DL-homophenylalanine ethyl ester in [EtPy]$^+$[BF$_4$]$^-$ + water (0.5 g ester, 85% H$_2$O, 0.5g NaHCO$_3$, 1.0 ml BL-alcalase, 6 hrs).*

alcalase loses its enzymatic activity significantly, indicating denaturation of the enzyme by heat.

## Comparison of IL and Acetonitrile on the Resolution of Amino Acids

The data described above for the resolution of *N*-acetyl *DL*-homophenylalanine ethyl ester indicates that at low concentration of ionic liquid, the enzymatic resolution reaction may be boosted by using ionic liquid due to the spatial structure and possible interaction with substrate and enzyme. Due to pharmaceutical interest, various amino acids have been studied using this reaction strategy and the results are shown in Table 1.

As Table 1 shows, higher yield and *ee* in the enzymatic resolution of esters are obtained in ionic liquid, compared to the use of acetonitrile as co-solvent. Ionic liquids improve the optical purity and yield. This indicates that using ionic liquid as a co-solvent in the resolution reaction may increase the enantioselectivity and activity of the enzyme. Especially, two *L*-amino acids (L-serine and *L*-4-chlorophenylalanine) were not achievable in acetonitrile and water solvent using alcalase. Intrestingly however, the resolution reactions for these two compounds are successful using ionic liquid [EtPy⁺][CF₃COO⁻].

# Conclusions

Kinetic resolutions of the homophenylalanine ethyl ester in three ionic liquid ([EMIM]⁺[BF₄]⁻, [EtPy]⁺[BF₄]⁻, and [EtPy]⁺[CF₃COO]⁻) solutions are compared with the results in acetonitrile. Systematic studies have been conducted on the kinetic resolution of *N*-acetyl homophenylalanine ethyl ester by varying reaction parameters, such as co-solvent concentration, reaction time and reaction temperature. The enzyme usually shows high enantioselectivity at a low concentration (15%) of ionic liquid in water. Among the ionic liquids studied, [EtPy]⁺[CF₃COO]⁻ gives the best results in achieving high *ee* and yield. The results indicate that ionic liquids could be an ideal substitute for organic solvents in the kinetic resolution of amino acid esters.

*Table 1.* Enantioselective resolution of *N*-acetyl amino acids in the ionic liquid [EtPy⁺][CF₃COO⁻] (0.5 g ester, 0.5g NaHCO₃, 15% ionic liquid or acetonitrile in water, 3.0 ml alcalase, 25 °C, 24 hours)

| Racemic Ester (each with N-acetyl) | Acetonitrile | | $[EtPy]^+[CF_3COO]^-$ | |
| --- | --- | --- | --- | --- |
| | ee (%) | yield (%) | ee (%) | yield (%) |
| Alanine ethyl ester | 63 | 31 | 86 | 33 |
| Serine methyl ester | — | — | 90 | 35 |
| Threonine methyl ester | 92 | 15 | 97 | 36 |
| Methionine methyl ester | 83 | 30 | 89 | 29 |
| Homophenylalanine ethyl ester | 95 | 35 | 93 | 38 |
| 4-Chlorophenylalanine ethyl ester | — | — | 96 | 39 |
| Norleucine methyl ester | 18 | 32 | 88 | 30 |

— means "non-reactive".

# References

1. Johnson, A.L.; Price, W.A.; Wong, P.C.; Vavala, R.F.; Strump, J.M. *J. Med. Chem.*, **1985**, *28*, 1596-1602.
2. Hayashi, K.; Nunami, K.; Kato, J.; Yoneda, M.; Kubo, M.; Ochiai, T.; Ishida, R. *J. Med. Chem.*, **1989**, *32*, 289-297.
3. Ondetti, M.A.; Cushman, D.W. *J. Med. Chem.*, **1981**, *24*, 355-361.
4. Kijima, T.; Ohshima, K.; Kise, H. Facile optical resolution of amino acid esters via hydrolysis by an industrial enzyme in organic solvents. *J. Chem. Tech. Biotechnol.* **1994**, *59*, 61-65.
5. Chen, S-T.; Chen, S-Y.; Hsiao, S-C.; Wang, K-T. *Biotechnology Letters.* **1991**, *13 (11)*, 773-778.
6. Chen, S-T.; Huang, W-H.; Wang, K-T. *J. Org. Chem.* **1994**, *59*, 7580-7581.
7. Chen, S-T.; Wang, K-T.; Wong, C-H. *J. Chem. Soc. Chem. Commun.* **1986**, 1514-1516.
8. Seddon, K. R. *J. Chem. Tech. Biotechnol.* **1997**, *68*, 351-356.
9. Zhao, H.; Malhotra, S.V. *Aldrichimica Acta.* **2002**, 35 (3), 75-83.
10. Schofer, S. H.; Kaftzik, N.; Wasserscheid, P.; Kragl, U. *Chem. Commun.* **2001**, *5*, 425-426.
11. Erbeldinger, M.; Mesiano, A. J.; Russell, A. J. *Biotechnol. Prog.* **2000**, *16 (6)*, 1129-1131.
12. Lau, R. M.; van Rantwijk, F.; Seddon, K. R.; Sheldon, R. A. *Org. Lett.* **2000**, *2 (26)*, 4189-4191.
13. Kim, K-W.; Song, B.; Choi M-Y.; Kim, M. J. *Org. Lett.* **2001**, *3(10)*, 1507-1509.
14. Zhao, H.; Luo, R.G. and Malhotra, S.V. *Biotech. Prog.* **2003**, *19(3)*, 1016-1018.
15. Zhao, H. and Malhotra, S.V. *Biotech. Lett.* **2002**, *24(15)*, 1257-1259.
16. Holbrey, J. D.; Seddon, K. R. *J. Chem. Soc., Dalton Trans.* **1999**, 2133-2139.
17. Zhao, H.; Malhotra, S.V. and Luo, R.G. *Phy. and Chem. of Liquids.* **2003**, *41(5)*, 487-492.
18. Zhao, H.; Luo, R. G.; Wei, D.; Malhotra, S. V. *Enantiomer.* **2002**, *7(1)*, 1-3.

# Applications

# Chapter 10

# Biphasic Acid Scavenging Utilizing Ionic Liquids: The First Commercial Process with Ionic Liquids

## Matthias Maase and Klemens Massonne

BASF AG, Chemicals Research and Engineering, GCI/P–M 311, D–67056 Ludwigshafen, Germany

Since the end of the 90s ionic liquids have faced a tremendous interest by the scientific community (1). They turned out to be novel materials providing unique new properties that simply had not been available before. Surprisingly no successful industrial application had been reported for quite a long time. In 2003 BASF revealed that they have been running the first commercial process (BASIL™) that makes use of ionic liquids in a dedicated way since beginning of 2002 (2). BASIL™ stands for Biphasic Acid Scavenging utilizing Ionic Liquids. Since the end of 2003 BASIL™ is ready for licensing.

# Introduction

The handling of solids is a true challenge in large scale industrial processes although it might not be obvious to academic groups. Perhaps it is even more challenging than the handling of volatile or highly reactive species since the problems related to solids especially to suspensions often do not occur in a reliable way. But what is wrong with suspensions?

- They often make reaction mixtures much more viscous, leading to insufficient mixing of the reactants.
- Heat transfer in large vessels containing suspensions is often less than satisfactory, so hot spots may occur.
- This in turn favours side reactions and lowers the yield of the desired product.
- It is difficult to transport suspensions through piping or to store them in tanks. In both cases high flow rates or agitation is necessary to prevent the suspension from settling and plugging the system.
- Separation can be an important issue as well. If a solid is formed during a reaction as an unwanted by-product it has to be removed afterwards.
- Reactors that are designed to deal with suspensions are usually more complex and expensive than those suitable for pure liquids.
- Finally if a solid is generated during a reaction scale up is by far not easy. One can hardly predict how the solid looks like when it is prepared in a 10 m$^3$ reactor rather than in a 100 ml flask. This can easily cause serious problems, for example if the solid that could be filtered off in a minute in the lab now takes two days on the large scale just because particles are several orders of magnitude smaller than expected.

# Acid Scavenging

One example in which industrial chemists have to deal with suspensions is acid scavenging. Browsing organic textbooks one will find numerous reactions which liberate acids that have to be scavenged in order to prevent decomposition of the product or other side reactions. Some examples are esterifications, silylations or phosphorylations of alcohols. Hydrochloric acid (HCl) is the acid most commonly formed. Usually, tertiary amines such as triethylamine are added to scavenge the acid. With the acid, these substances form solid salts which turn the reaction mixture into a suspension.

## The Synthesis of Alkoxyphenylphosphines

Alkoxyphenylphosphines are important raw materials in the production of BASF's Lucirines®, substances that are used as photoinitiators to cure coatings and printing inks by exposure to UV light. HCl is formed during the synthesis of diethoxyphenylphosphine (Scheme 1).

$$\text{(PhPCl}_2\text{)} + 2\ \text{ROH} + 2\ \text{R}_3\text{N} \longrightarrow \text{(PhP(OR)}_2\text{)} + 2\ \text{R}_3\text{N} * \text{HCl}$$

*Scheme 1: Synthesis of Dialkoxyphenylphosphines*

Scavenging with a tertiary amine results in a thick, non-stirrable slurry (Figure 1). The problems mentioned earlier significantly lower the yield and capacity of the process.

*Figure 1: Slurry that is formed when a tertiary amine is used as an acid scavenger*
© *BASF Aktiengesellschaft 2002.*

## Ionic Liquids (Dis)solve the problem

If an acid has to be scavenged with a base, the formation of a salt cannot be avoided, but why not form a liquid salt - a so-called ionic liquid? The properties of ionic liquids are:

- they are salts, consisting 100% of ions
- they are liquid at temperatures below 100 °C.

Since they are rather polar materials they often do not mix with solvents or nonpolar product molecules. Using the BASF product 1-methylimidazole as an acid scavenger, an ionic liquid is formed: 1-methyl-imidazolium chloride (Hmim Cl), which has a melting point of about 75 °C (Scheme 2).

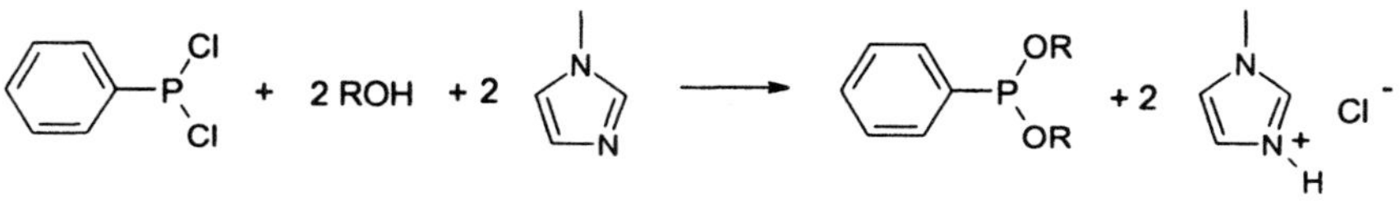

*Scheme 2: 1-methylimidazole as acid scavenger*

After the reaction two clear liquid phases occur (Figure 2) that can easily be separated. The upper phase is the pure product – no solvent is needed anymore - the lower the pure ionic liquid (3). Hmim Cl as an ionic liquid has a great advantage over the classical dialkylated systems: they can be switched on and off just by protonation and deprotonation. This is crucial when recycling and purification of the ionic liquids is considered. To distinguish the switchable ionic liquids - the "Hmims" – from the conventional ones Michael Freemantle used the term "smart ionic liquids" in his C&EN paper on the BASIL™ process (2).

## Lessons Learned: Why have Ionic Liquids been successful with BASIL™ ?

There are several success factors being responsible that BASIL™ was established in routine production so quickly.

*Figure 2: The BASIL™ process. After the reaction two clear liquid phases are obtained - the upper being the pure product the lower being the ionic liquid Hmim Cl*
© *BASF Aktiengesellschaft 2002*

- First of all there was an existing problem – the formation of unwanted solids - to which the unique properties of ionic liquids offered a tailor-made solution.
- The lead structures of ionic liquids - the imidazoles - have been known from academic research already so industry had not to reinvent the wheel.
- The material (1-methylimidazole) was available in large quantities within the company since it is an existing BASF product.
- The process has been improved dramatically.
- There were clear economical benefits from BASIL™ being higher chemical and higher space-time-yields.

Finally one should not forget to mention that luckily the chemistry worked as well. Further investigations revealed that BASIL™ is not restricted to phosphorylation chemistry but is a general solution to all kinds of acid scavenging.

## BASIL™ is more than just Acid Scavenging

Methylimidazole is doing a perfect job by scavenging the acid. Looking closer at it one will find that methylimidazole also helps in setting the acid free. In other words: it acts as an nucleophilic catalyst (4). We found that the phosphorylation reaction is complete in less than a second. Having eliminated the formation of any solids and having increased the reaction rate new reactor concepts were possible. We were now able to do the same reaction that has been done in a large vessel in a little jet reactor that has the size of your thumb. Doing so the productivity of the process has been rised by a factor of $8 \cdot 10^4$ to $690000$ kg m$^{-3}$ h$^{-1}$.

This progress can be attributed to the existence of a new class of novel, promising materials: ionic liquids.

References:

1. Freemantle, M. *Chem. Eng. News* **1998**, *76*, 32; *Green Industrial Applications of Ionic Liquids*; Rogers, R.D., Seddon K.R., Volkov, S., Eds., Kluwer: Dordrecht, 2002; P. Wasserscheid, W. Keim, *Angew.*

*Chem. Int. Ed.* **39**, *2000*, 3772-3789; *Ionic Liquids in Synthesis*; Wasserscheid, P.; Welton T., Eds., VCH–Wiley: Weinheim, 2002; *Ionic Liquids: Industrial Applications to Green Chemistry*; Rogers R. D.; Seddon, K. R., Eds.; ACS Symposium Series 818, American Chemical Society: Washington D.C., 2002; *Ionic Liquids as Green Solvents: Progress and Prospects*; Rogers R. D.; Seddon, K. R., Eds.; ACS Symposium Series 856, American Chemical Society: Washington D.C., 2003.

2. Freemantle, M. *Chem. Eng. News* **2003**, *81*, 9; WO 03/062171 (BASF AG).

3. Seddon, K.R. *Nature Mater.* **2003**, *2*, 363; Rogers, R.D.; Seddon, K.R. *Science,* **2003**, *302*, 792.

4. Chojnowski, J.; Cypryk, M.; Fortuniak W. *Heteroatom. Chem.* **1991**, *2*, 63.

# Chapter 11

# Ionic Liquids at Degussa: Catalyst Heterogenization in an Industrial Process

## B. Weyershausen[*], K. Hell, and U. Hesse

R&D Department, Oligomers/Silicones, Degussa AG Goldschmidtstrasse 100, 45127, Essen, Germany
[*]Corresponding author: telephone: 49 + (201) 1731655; fax: 49 + (201) 1731839; email: bernd.weyershausen@degussa.com

The use of ionic liquids in hydrosilylation reactions allows for an easy recovery and the subsequent reuse of standard hydrosilylation catalysts. Remarkably, the recovered catalyst/ionic liquid solution does not need to be purified or treated before its reuse. Employing this method a variety of organomodified polydimethylsiloxanes was synthesized.

## Introduction

### Hydrosilylation

One of the most fundamental and elegant methods of laboratory and industrial synthesis of organomodified polydimethylsiloxanes is the transition metal catalyzed hydrosilylation of CC-double bond containing compounds with SiH-functional polydimethylsiloxanes (*1*). Regardless of the broad applicability of the hydrosilylation reaction its technical use still suffers from substantial difficulties. Usually, the reaction is homogeneously catalyzed which means that

© 2005 American Chemical Society

after completion of the reaction the catalyst either remains within the product or has to be costly removed. From an economic and ecological point of view in homogenous catalysis the separation, respectively, the immobilization and the reuse of the expensive precious metal catalyst in a subsequent reaction represent serious problems (2). Therefore, there has been no lack of attempts to reduce the amount of catalyst, which in most cases leads to relatively long reaction times. A technical process based on homogenous catalysis is economically efficient only if in combination with acceptable reaction times the catalyst losses can be kept as small as possible. Hence, there is a demand for processes which allow for the recycling of the catalyst without catalyst losses and the lowest possible stress for the products. In the past intensive work has been done on immobilization, heterogenization and anchoring of homogenous catalysts for an easy catalyst separation from the products and recovery of the catalyst. Besides e.g. the extraction of the catalyst or its adsorption at ion exchangers multiphasic reactions represent another possibility for the separation of product and catalyst phase. In the last years biphasic reactions employing ionic liquids have gained increasing importance (3). The ionic liquid generally forms the phase in which the catalyst is dissolved and immobilized.

## Ionic Liquids

Ionic liquids are salts melting at low temperatures, and they represent a novel class of solvents with non-molecular ionic character. In contrast to a classical molten salt, which is a high-melting, highly viscous and very corrosive medium, an ionic liquid is already liquid at temperatures below 100 °C and is relatively low-viscous (4, 5). In most cases ionic liquids consist of combinations of cations such as ammonium, phosphonium, imidazolium or pyridinium with anions such as halides, phosphates, borates, sulfonates or sulfates. The combination of cation and anion has a great influence on the physical properties of the resulting ionic liquid. By careful choice of cation and anion it is possible to finetune the properties of the ionic liquid and provide a tailor-made solution for each task (Figure 1), and that is why ionic liquids are often referred to as designer solvents or materials. With many organic product mixtures ionic liquids form two phases from which arises the possibility of carrying-out biphasic reactions for the separation of the homogenous catalyst (6, 7, 8). Ionic liquids are not any longer a class of esoteric compounds, but they were already used in a multitude of different applications (Figure 2) and since 1999 they have been commercially available (9).

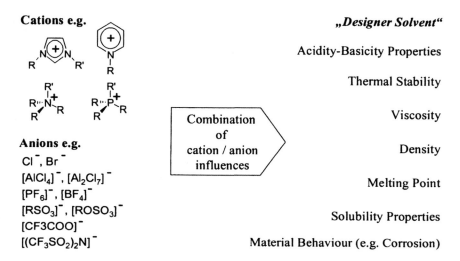

Figure 1. Properties of ionic liquids.

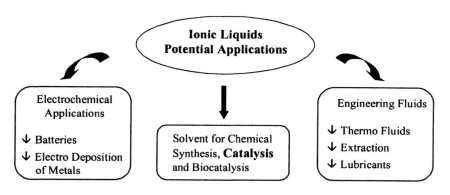

Figure 2. Potential applications of ionic liquids.

For the first time in 1972, Parshall used an ionic liquid for the immobilization of a transition metal catalyst in a biphasic reaction setup (10). He described the hydrogenation of CC-double bonds with PtCl$_2$ dissolved in tetraethylammonium chloride associated with tin dichloride ([Et$_4$N][SnCl$_3$], m.p. 78 °C) at temperatures between 60 and 100 °C. "A substantial advantage of the molten salt medium −over conventional organic solvents− ... is that the product may be separated by decantation or simple distillation". The use of ionic liquids as novel media for transition metal catalysis started to receive increasing attention, when in 1992 Wilkes reported on the synthesis of novel non-chloroaluminate, room temperature liquid salts with significantly enhanced stability towards hydrolysis, such as tetrafluoroborate salts (*11*). Some of the first succesful examples for catalytic reactions in non-chloroaluminate ionic liquids include the rhodium-catalyzed hydrogenation (*12*) and hydroformylation (*13*) of olefins. However, very often catalyst leaching into the organic and/or the product phase is observed, as the transition metal catalyst is not completely retained in the polar ionic liquid phase. Current approaches to circumvent this problem include the modification of the catalyst, respectively, the modification of the ligands (*13*) and the development of task specific ionic liquids (*14*).

## Organomodified Polydimethylsiloxanes

The hydrosilylation of CC-double bond containing compounds with SiH-functional polydimethylsiloxanes is a widely applied reaction in industrial synthesis for the production of organosilicon compounds (Figure 3) on technical scale (*1*).

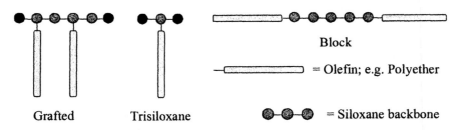

*Figure 3. Organomodified polydimethylsiloxanes (OMS) offer a high synthetical and structural flexibility.*

Apart from our general investigations of the hydrosilylation reaction, we became interested in ionic liquids and their potential to be used in hydrosilylation reactions as a means for catalyst heterogenization. In particular, we aimed at the synthesis of polyethersiloxanes. Polyethersiloxanes constitute an important class of surface active compounds which find use in a broad range of industrial applications (*15*) such as:

- PU foam stabilizers
- PU foam cell openers
- Defoamer oils
- Emulsifiers

- Mar and slip additives
- Wetting agents
- Dispersants
- Cosmetics

## Results and Discussion

Herein, we report on a novel process for the synthesis of organomodified polydimethylsiloxanes employing ionic liquids for the heterogenization and/or immobilization of the precious metal catalyst. The advantage of this novel hydrosilylation process is that standard hydrosilylation catalysts can be used without the need of prior modification to prevent catalyst leaching. To the best of our knowledge this is the first example of a hydrosilylation of olefinic compounds using ionig liquids (Scheme 1). However, a method for the transition metal catalyzed hydroboration and hydrosilylation of alkynes in ionic liquids has been recently described (*16*).

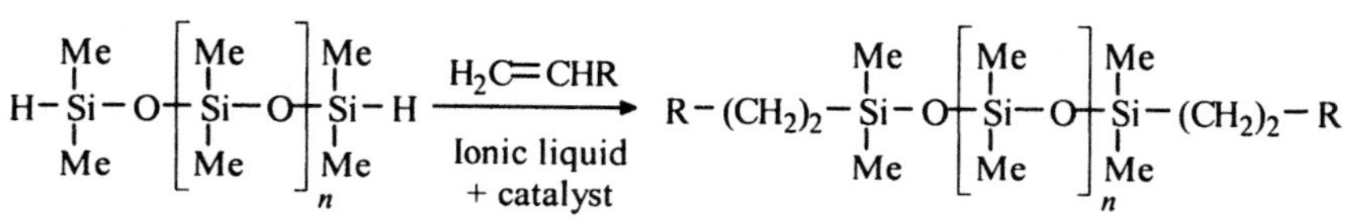

*Scheme 1. Hydrosilylation of olefins (e.g. polyethers).*

Employing the novel hydrosilylation process using ionic liquids a broad range of different organomodified polydimethylsiloxanes was synthesized. The procedure is rather simple and can be described as one pot synthesis (Figure 4).

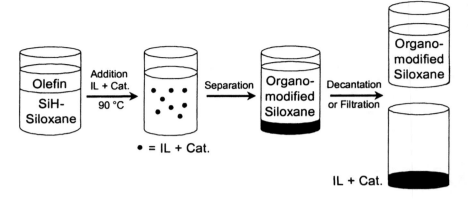

*Figure 4. Schematic representation of the hydrosilylation process using ionic liquids.*

Typically, the reaction is performed in a liquid-liquid biphasic system where the substrates and products (upper phase) are not miscible with the catalyst/ionic liquid solution (lower phase). The SiH-functional polydimethylsiloxane and the olefin are placed into the reaction vessel and heated up to 90 °C. Then the precious metal catalyst (20 ppm) and the ionic liquid (1 %) are added. After complete SiH-conversion the reaction mixture is cooled to room temperature and the products are removed from the reaction mixture by either simple decantation or filtration (in case of non-room temperature ionic liquids). The recovered catalyst/ionic liquid solution can be reused several times without any significant change in catalytic activity. A treatment or workup of the catalyst/ionic liquid solution after each reaction cycle is not necessary. The metal content of the products was analyzed by ICP-OES (*Inductively coupled plasma optical emission spectroscopy*) and the chemical identity of the organomodified polydimethylsiloxane was verified by NMR-spectroscopy ($^1$H-, $^{13}$C- and $^{29}$Si-NMR).

For our investigations we chose SiH-functional polydimethylsiloxanes with different chain lengths and functionality patterns and polyethers with different ethylene (EO) and propylene (PO) oxide contents, as well as one small non-polyether olefin (AGE) in order to evaluate the influence of the hydrophilicity/hydrophobicity of the substrates and corresponding products on the catalytic performance of the various catalyst/ionic liquid solutions and even more importantly, on the separation behaviour of the ionic liquid at the end of the reaction. A clean separation of the ionic liquid from the products is the necessary condition to examine the partitioning of the catalyst between the two phases. Naturally, without a clean separation and a much better solubility of the catalyst in the ionic liquid phase than in the product phase a complete retention of the catalyst in the ionic liquid and the desired recovery of the catalyst cannot be achieved. Figure 5 depicts some of the various polydimethylsiloxanes, olefins, ionic liquids and catalysts used in our investigations; only the ionic liquids and catalysts are shown which gave satisfactory results.

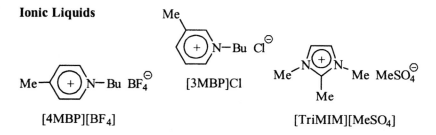

**SiH-functional Polydimethylsiloxanes**

**Olefins**

400EO : x = 0 ; y = 7.5
540 : x = 3 ; y = 5.8
400PO : x = 7.4 ; y = 0

Allyl glycidyl ether (AGE)

**Catalysts**

Pt-92  Di-μ -chloro-dichloro-(cyclohexene)diplatinum (II)
$H_2PtCl_6$  Hexachloroplatinic acid

**Ionic Liquids**

[3MBP]Cl

[4MBP][BF$_4$]

[TriMIM][MeSO$_4$]

*Figure 5. Raw materials.*

The hydrophobicity of the organosilicon products increases with increasing chain length and decreasing SiH-functionalization of the SiH-functional polydimethylsiloxane. The polyethers are the more hydrophobic the higher the content of propylene oxide is. Now, one would expect that the ionic liquid due to its ionic nature separates easier from the organosilicon products the more hydrophobic the latter are. During our studies this anticipation was proven to be right. Table 1 summarizes the results of our investigations.

**Table 1. Results of hydrosilylation reactions of olefins with SiH-functional polydimethylsiloxanes using ionic liquids.**

| Entry No. | Catalyst [20 ppm] | IL (1 %) | SiH-Siloxane | Olefin | SiH-Conv. % / h | Detect. Metal Cont. [ppm] |
|---|---|---|---|---|---|---|
| 1 | Pt-92 | [TriMIM][MeSO$_4$] | $n = 18$ | 400EO | > 99/5 | < 1 |
| 2 | H$_2$PtCl$_6$ | [3MBP][Cl] | $n = 18$ | 540 | > 99/3 | < 1 |
| 3 | H$_2$PtCl$_6$ | [4MBP][BF$_4$] | $n = 18$ | 400PO | > 99/1 | < 1 |
| 4 | H$_2$PtCl$_6$ | [TriMIM][MeSO$_4$] | $n = 18$ | 400PO | 93/5 | < 1 |
| 5 | H$_2$PtCl$_6$ | [TriMIM][MeSO$_4$] | $n = 78$ | AGE | > 99/3 | < 1 |
| 6 | Pt-92 | [TriMIM][MeSO$_4$] | $n = 28$ | AGE | > 99/1 | < 1 |
| 7 | H$_2$PtCl$_6$ | [TriMIM][MeSO$_4$] | $m = 5$ | 400PO | > 99/1 | < 1 |
| 8 | H$_2$PtCl$_6$ | [TriMIM][MeSO$_4$] | $m = 5$ | AGE | > 99/3 | 2 |

It turned out that for successful recovery of the catalyst and its reusability it is crucial to find an appropriate combination of a catalyst and an ionic liquid which has to be harmonized with the hydrophilicity/hydrophobicity of the product. First of all, not every catalyst is soluble in each ionic liquid and secondly, not every ionic liquid separates as facile as desired from the product phase. However, we were able to identify at least one suitable catalyst/ionic liquid combination for the synthesis of each polyethersiloxane (see Table 1). In some cases more than one catalyst/ionic liquid combination gave good results (see entries No. 3 and 4, Table 1). It is notworthy that all of the polyethersiloxanes synthesized in this way exhibit very different polarities.

Furthermore, in two model reactions the concentration of the catalyst/ionic liquid solution was increased to examine its effect on the reaction time. As the catalyst is recovered at the end of the reaction, this would not necessarily cause additional costs as long as at the same time the reaction time would be decreased to an appropriate extent and the catalyst life time would be long enough. To our surprise in case of the reaction of a α,ω-SiH-functional polydimethylsiloxane ($n$ = 18) with the hydrophobic polyether 400PO using a hexachloroplatinic acid/*N*-butyl-3-methylpyridinium tetrafluoroborate solution only little changes of the reaction time were observed when increasing the concentration of the catalyst/ionic liquid solution (entries 1-4, Table 2).

**Table 2. Effect of catalyst concentration in hydrosilylation reactions using ionic liquids.**

| Entry No. | Catalyst [ppm] | IL | SiH- Siloxane | Olefin | SiH- Conv. %/h | Metal Cont. [ppm] |
|-----------|----------------|----|----------------|--------|----------------|-------------------|
| 1 | H$_2$PtCl$_6$ [20] | [4MBP][BF$_4$] | $n$ = 18 | 400PO | > 99/5 | < 1 |
| 2 | H$_2$PtCl$_6$ [50] | [4MBP][BF$_4$] | $n$ = 18 | 400PO | > 99/3 | < 1 |
| 3 | H$_2$PtCl$_6$ [100] | [4MBP][BF$_4$] | $n$ = 18 | 400PO | > 99/2 | < 1 |
| 4 | H$_2$PtCl$_6$ [1000] | [4MBP][BF$_4$] | $n$ = 18 | 400PO | > 99/1 | < 1 |
| 5 | Pt-92 [20] | [TriMIM][MeSO$_4$] | $n$ = 18 | 400EO | > 99/5 | < 1 |
| 6 | Pt-92 [50] | [TriMIM][MeSO$_4$] | $n$ = 18 | 400EO | > 99/2 | 8 |
| 7 | Pt-92 [100] | [TriMIM][MeSO$_4$] | $n$ = 18 | 400EO | > 99/0,5 | 10 |
| 8 | Pt-92 [1000] | [TriMIM][MeSO$_4$] | $n$ = 18 | 400EO | > 99/0,1 | 22 |

Remarkably, even at very high catalyst loads of the ionic liquid no leaching of the catalyst into the product phase was observed which demonstrates the excellent solubility of hexachloroplatinic acid in *N*-butyl-3-methylpyridinium tetrafluoroborate. After the reaction the catalyst was completely recovered. A different observation was made for the the reaction of the same α,ω-SiH-functional polydimethylsiloxane ($n$ = 18) with the hydrophilic polyether 400EO using a Pt-92/1,2,3-trimethylimidazolium methylsulfate solution as catalyst phase which in our screening experiments turned out to the best combination for the hydrosilylation of hydrophilic polyethers (entries 5-8, Table 2). Here, the increase of the catalyst concentration leads to drastically reduced reaction times, but also to a significant leaching of the catalyst into the product phase. Probably, the shorter reaction times are due to the leaching of the catalyst into the substrates/products mixture. The hydrophilic polyether 400EO and the corresponding polyethersiloxane compete with the ionic liquid as solvent for the dissolution of the catalyst. Once, the catalyst has leached in significant amounts into the substrates/products mixture a conventional homogenous catalysis takes place, whereas in case of the reaction of the $\alpha,\omega$-SiH-functional polydimethylsiloxane ($n$ = 18) with the hydrophobic polyether 400PO the ionic liquid ($N$-butyl-3-methylpyridinium tetrafluoroborate) is the much better solvent for the catalyst (hexachloroplatinic acid), and the hydrosilylation reaction takes place only at the interface between the two phases catalyst/ionic liquid and substrates/products. In the latter system the reaction rate is mainly determined by the size of the interface between the two phases, i.e. the droplet size of the catalyst/ionic liquid phase and not so much by the catalyst concentration of the ionic liquid.

The hydrosilylation reaction can also be conventionally conducted by reaction of an olefin and a SiH-functional polydimethylsiloxane in the presence of a standard transition metal catalyst, and after the reaction the catalyst can be extracted with an ionic liquid. In some cases the use of an ionic liquid in the hydrosilylation process even improves the quality of the polyethersiloxanes regarding color compared to the standard process. An explanation might be the avoidance of catalyst reduction leading to the formation of colloidal metal particles which tend to color the product slightly brownish. In other words the ionic liquid seems to have a stabilizing effect on the catalyst.

Hydrosilylation reactions with ionic liquids derived from 1-alkylimidazole, i.e. in 2-position unsubstituted 1,3-dialkylimidazolium salts, did not give the desired polyethersiloxanes. We assume that the proton in 2-position of the 1,3-dialkylimidazolium salts react with the SiH-group of the SiH-functional polydimethylsiloxanes under formation of the corresponding 1,3-dialkylimidazolylidene which deactivates the hydrosilylation catalyst by coordination to the metal center (*17, 18*).

## Conclusions

A novel transition metal catalyzed hydrosilylation process is described. The use of an ionic liquid in this process allows for the immobilization, respectively, heterogenization and recovery of the expensive precious metal catalyst, as well as its direct reuse in a subsequent hydrosilylation reaction. From an economic and ecological point of view this process perfectly fits in the concept of "Sustainable Chemistry". Future research activities will aim at the prolongation of the catalyst life-time. For this, it is necessary to gain a deeper understanding of the catalytically active species in the catalyst/ionic liquid solution.

# References

1. Marcniec, B. in *Applied Homogeneous Catalysis with Organometallic Compounds*; Cornils, B.; Herrmann, W. A., Eds.; VCH: Weinheim, 1996, Volume 1, p.487.
2. Cornils, B.; Herrmann, W. A. in *Applied Homogeneous Catalysis with Organometallic Compounds*; Cornils, B.; Herrmann, W. A., Eds.; VCH: Weinheim, 1996, Volume 2, p. 573.
3. Wasserscheid, P. *"Transition Metal Catalysis in Ionic Liquids"* in *Ionic Liquids in Synthesis*; Wasserscheid, P.; Welton, T., Eds.; VCH: Weinheim, 2002, p. 213-257.
4. Seddon, K. R. *J. Chem. Technol. Biotechnol.* **1997**, *68*, 351-356.
5. *Ionic Liquids in Synthesis*; Wasserscheid, P.; Welton, T., Eds.; VCH: Weinheim, 2002.
6. Holbrey, J. D.; Seddon, K. R. *Clean Products and Processes* **1999**, *1*, 223-226.
7. Welton, T. *Chem. Rev.* **1999**, *99*, 2071-2083.
8. Wasserscheid, P.; Keim, W. *Angew. Chem. Int. Ed.* **2000**, *39*, 3772-3789.
9. http://www.solvent-innovation.de
10. Parshall, G. W. *J. Am. Chem. Soc.* **1972**, *94*, 8716-8719.
11. Wilkes, J. S.; Zaworotko, M. J. *J. Chem. Soc., Chem. Commun.* **1992**, 965-967.
12. Suarez, P. A. Z.; Dullius, J. E. L.; Einloft, S.; de Souza, R. F. *Polyhedron* **1996**, *15*, 1217-1219.
13. Chauvin, Y.; Mußmann, L.; Olivier, H. *Angew. Chem. Int. Ed.* **1995**, *34*, 1149-1155.
14. Davis, Jr., J. H. *"Synthesis of Task-specific Ionic Liquids"* in *Ionic Liquids in Synthesis;* Wasserscheid, P.; Welton, T., Eds.; VCH: Weinheim, 2002, p. 33-40.
15. http://www.degussa.com/en/structure/performance_materials/-oligomers_silicones.html
16. Aubin, S.; Le Floch, F.; Carrie, D.; Guegan, J. P.; Vaultier, M. *ACS Symposium Series 818* (Ionic Liquids), **2002**, 334-346.
17. Öfele, K.; Herrmann, W. A.; Mihalios, D.; Elison, M.; Herdtweck, E.; Scherer, W.; Mink, J. *J. Organomet. Chem.*, **1993**, *459*, 177-184.
18. Herrmann, W. A.; Goossen, L. J.; Köcher, C.; Artus, G. J. *Angew. Chem. Int. Ed.* **1996**, *35*, 2805-2807.

# Chapter 12

# Nonlinear Optical Ionic Liquids

Rico E. Del Sesto[1], Doug S. Dudis[2], Fassil Ghebremichael[3],
Norman E. Heimer[1], Tammy K. C. Low[1], John S. Wilkes[1,*],
and A. Todd Yeates[2]

Departments of [1]Chemistry and [3]Physics, U.S. Air Force Academy,
Colorado Springs, CO 80840
[2]Air Force Research Laboratory, Wright-Patterson Air Force
Base, OH 45433

Molecular orbital calculations indicate that molecules with a
high electron density diffused over a large volume will have
third order nonlinear optical activity. Anions often have
higher second hyperpolarizability values (gamma) than
similar neutral molecules. Also, molecules or ions containing
higher row elements have higher gammas. Salts with cations
that have their positive charge only weakly interacting with
the anion also enhance the third order nonlinear optical
activity. That looks like the recipe for ionic liquids. A
number of sulfur-containing mono- and dianion salts were
synthesized and characterized through Z-scan measurements.
Most were ionic liquids, and some showed significant third
order nonlinear optical behavior. The general features of
ionic liquids such as wide liquidus range, good thermal
stability, and low vapor pressure are particularly
advantageous for applications of nonlinear optical materials.
Potential applications are in optical limiting and other all-
optical devices.

## Introduction

The terms "nonlinear optics" and "ionic liquids" are not often mentioned in the same sentence. Here we describe some new materials that combine some of the favorable physical properties of ionic liquids with the potentially useful optical properties of third order nonlinear optical materials. A complete understanding of nonlinear optics (NLO) is not necessary to appreciate the interesting applications for NLO materials. A relatively superficial treatment of the appropriate optical properties, parameters and phenomena will be presented below. The niche for ionic liquids in the NLO materials field will be identified, and the preparation and properties of some remarkable ionic liquids that fill the niche will be described.

Others have shown that some organic salts exhibit nonlinear optical behavior. Some are second order (also called quadratic) nonlinear optical materials, such as the dihydrogen phosphate–organic cation salts described by Seddon and coworkers (1), and the stilbazolium $BF_4^-$, triflate, and tosylate salts of Seth Marder and coworkers (2). These are all necessarily solids. Reports of third order (also called cubic) nonlinear optical organic salts are much rarer. Some recent examples are silver phenylacetylides, where the NLO effect comes from the anion (3), and aryliodonium triflate salts, where cations are NLO active (4). Some of these latter salts actually meet the "boiling water" definition of an ionic liquid, but all are solids at room temperature.

Compounds with second order NLO activity must be non-centrosymmetric at both the molecular and materials level. Odd order (in practice just first and third) NLO materials have no symmetry restrictions at either the molecular or materials level. So it would seem easier to design and prepare third order compounds, and they could be liquids or solids. Our approach was to predict some anion structures that might have third order NLO activity, attempt to prepare some ionic liquids with those anions, and characterize the resulting compounds chemically, physically, and optically.

## Nonlinear Optics of Molecules and Materials

A rigorous treatment of nonlinear optics is outside the scope of this chemistry paper. See one of the many monographs in this area for background reading (5). However, a somewhat superficial introduction to nonlinear optics is needed to describe the measured optical properties, and also to appreciate some useful applications of these nonlinear optical ionic liquids. The polarization of a material may be described at the molecular (or ion) level by Equation 1, or at the macroscopic (or materials) level by Equation 2.

$$\mu = \mu_0 + \alpha \cdot E + \vec{\vec{\beta}} : EE + \vec{\vec{\gamma}} \vdots EEE + \cdots \tag{1}$$

$$P = P_0 + \chi^{(1)} \cdot E + \vec{\vec{\chi}}^{(2)} : EE + \vec{\vec{\chi}}^{(3)} \vdots EEE + \cdots \tag{2}$$

For the molecular polarization, Equation 1, $\alpha$ is the polarizatibility, $\beta$ is the first hyperpolarizability, and $\gamma$ is the second hyperpolarizability. The polarizatibilities are properties of individual molecules or ions, so they can in principle be calculated from knowledge of the structure. For the macroscopic polarization, Equation 2, $\chi^{(n)}$ is the n-th order susceptibility tensor. The values for the susceptibilities are properties of a material that can be measured experimentally, and are directly related to the molecular polarizabilities.

Two measurable third-order NLO properties are $n_2$, the nonlinear index coefficient and $\alpha_2$, the two-photon absorption (TPA) coefficient. These coefficients are related to the complex third-order NLO susceptibility through its real and imaginary parts, as shown in Equation 3(6, 7).

$$n_2 = \frac{3\pi}{2n_0} \mathrm{Re}\left\{\chi^{(3)}\right\},$$

$$\alpha_2 = \frac{3\omega}{2\varepsilon_0 n_0^2 c^2} \mathrm{Im}\left\{\chi^{(3)}\right\}. \tag{3}$$

$n_0$ is the index of refraction, $\varepsilon_0$ is the free space permittivity, $c$ is the speed of light in vacuum, and $\omega$ is the optical frequency. We used the z-scan method in order to determine experimentally the coefficients $n_2$ and $\alpha_2$. In this experiment, a laser beam was focused with a lens and the material under study moved across the focal plane in the direction of the beam propagation, say z-axis. The material experienced varying intensities, first increasing, as it got closer to the focal plane of the lens, and then decreasing past the focal plane. If $n_2$ is positive, the beam at a plane further down the propagation axis will expand making the intensity on that plane smaller at the "closed detector." As the material is moved past the focal plane, the beam will be focused tighter making the intensity at the plane larger. If $n_2$ is negative the reverse intensity behavior is observed. Similarly, if the whole beam at the observation plane is measured at the "open detector," as the material is scanned across the focal plane there will be a dip in intensity output as the absorption increases as the material traverses the focal plane. The nonlinear absorption coefficient $\alpha_2$ can then be determined from the size of the intensity dip. It can easily be imagined that the dip in the intensity is located at the inflection of the previous description, where the intensity went from minimum to a maximum as a function of $z$ before flattening out again. The two coefficients can be related through the Kramers-Kronig relations, or dispersion relations.

We are interested in the third-order optical properties of our materials. Specifically, we wish to use third-order NLO ionic liquids for light limiting applications, where the relevant third-order NLO effect is TPA. These materials show TPA effects and may be used as light limiting because at low intensities ("low" means everyday normal light intensities) the material absorbs light linearly according to Beer's Law. However, at high intensities (e.g. laser light) the absorption increases with intensity. A material that is an effective two photon absorber will be effective for eye and sensor protection from laser light.

To meet our goals the optimization of the optical limiting properties of a material should correspond with maximizing the third-order NLO properties. For both second-order and third-order NLO behavior, materials are typically built starting at the molecular level, which are engineered with strong electron donating groups tethered to a strong electron acceptor group via an extended $\pi$-conjugated system. There have been numerous reports of optimizing the polarizability of these molecules though increasing the length of the conjugation, as well as increasing the electron-directing effect of the functional groups attached to the chain (8).

## Prediction of Nonlinear Optical Activity

Molecular orbital computational methods predict properties of molecules (or ions) quite well. It is more difficult to predict materials properties, so we calculated $\gamma$, the second hyperpolarizability, to serve as an indicator of the relative third order responses of candidate materials for light limiting.

In an attempt to develop a diffuse Gaussian basis set usable for the *ab initio* calculation of the second hyperpolarizability ($\gamma$) of molecules of technologically important size, we noted that $\gamma$ was quite sensitive to the inclusion of diffuse functions in the basis set, but insensitive to the quality of the valence portion of the basis set. This observation indicated that the region of the electron density farthest from the nuclei, the "diffuse" region, generated the bulk of the second hyperpolarizability. The diffuse region contains those electrons least strongly bound to the molecule. It would seem reasonable then that an increase of the electron density in this region would lead to an increase in $\gamma$. Three methods to achieve this present themselves. The first is to use molecules in their excited states. Such molecules have been studied previously (9,10) and do lead to large enhancements in $\gamma$. A second approach involves including higher row elements in the molecule. It is well known that the ionization energies of the elements decrease as one moves down the rows. Thus the electrons are more loosely bound to the atoms and should increase the electron density in the diffuse region. A third approach involves adding electrons directly to a molecule producing a negatively charged ion. In this study we used *ab initio* Hartree-Fock based calculations to explore the consequences of these last two molecular design methods. Our preliminary results of such calculations were previously reported (11)

Some examples of the calculated second hyperpolarizabilities for some anions are in Table I. The $\gamma$ values calculated with the Restricted Hartree-Fock (RHF) protocol are all less than those where the MP2 correction was used. However, the relative order and second hyperpolarizabilities are the same for both methods, and appear to scale by a factor of 2-3. MP2 is much more computationally intensive, so RHF is a more convenient method. One can see that our intuitive sense of the effects of higher row elements and increased negative charge on the $\gamma$ values is borne out by the calculations. Monoanions with second row elements ($NO_2^-$ and $HCO_2^-$) have the lowest gammas, monoanions containing the third row element sulfur ($HOCS_2^-$ and $H_2NCS_2^-$) have higher gammas, and dianions containing sulfur have the highest predicted gammas.

**Table I. Predicted Second Hyperpolarizability of Some Anions**

| Anion | $\gamma x\ 10^{-3}$ a.u. (RHF) | $\gamma x\ 10^{-3}$ a.u. (MP2) |
|-------|-------|-------|
| $NO_2^-$ | 9 | 36 |
| $HCO_2^-$ | 12 | 31 |
| $HOCS_2^-$ | 66 | 117 |
| $H_2NCS_2^-$ | 84 | 156 |
| $O_2NCHCS_2^{2-}$ | 268 | 664 |
| $(NCCS)_2^{2-}$ | 435 | 1048 |

The structures and names of the anions in Table I are

nitrite      formate      xanthate

$NO_2^-$       $HCO_2^-$       $HOCS_2^-$

| dithiocarbamate | K-salt | dithiomaleonitrile, dtmn |
|:---:|:---:|:---:|
| $H_2NCS_2^-$ | $O_2NCHCS_2^{2-}$ | $(NCCS)_2^{2-}$ |

The calculations on these structures point to general types of anions that may have good third order nonlinearity, but of course real materials must also contain cations. The high charge density of small cations can decrease the polarizability of the diffuse anions, as demonstrated in Table II. The obvious approach to avoiding this effect is to employ cations that have the positive charge delocalized or screened from the anion by molecular foliage. Salts with large organic cations are often ionic liquids.

**Table II. Effect of Cations on Diffuse Anion Second Hyperpolarizability**

| Cation | $\gamma x\ 10^{-3}$ a.u. (RHF) |
|:---:|:---:|
| None | 268 |
| $Li^+$ | 42 |
| $Na^+$ | 71 |
| $K^+$ | 119 |

## Ionic Liquids

The use of an ionic liquid may have some major advantages as light limiting materials. The low volatility, high thermal stability, and relatively easy preparation are all favorable for sensor protection. However, the fact that ionic liquids are viscous liquids is the most advantageous property. When two photon absorption functions as it should during irradiation of high intensity laser light, the material necessarily absorbs considerable energy. We saw the consequences when we used laser pulses that were too long during the Z-scan measurements of optical response (*vide infra*). When a lot of energy is deposited in a small volume the ionic liquid does thermally decompose, producing a black char. The advantage of using a liquid film is that the decomposition products can diffuse away or the charred area can be removed by convective flow.

The calculations guide us to compounds that have heavy anions with delocalized electrons combined with large non-polarizing cations. Our experience tells us that many salts with such structures are liquid at or near room temperature. 1-Ethyl-3-methylimidazolium was the initial cation chosen for this study, but had to be abandoned due to the acidity of that cation and the basicity of the sulfur-containing anions. The cation used in all of the ionic liquids reported below is tetradecyl(trihexyl)phosphonium, which we will abbreviate as [PR$_4$].

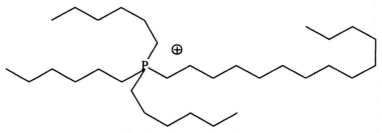

tetradecyl(trihexyl)phosphonium, [PR$_4$]

Some physical properties, the linear absorption, and the third order nonlinear optical response were measured. The method used for the nonlinear optical characterization was Z-scan, which detects the focusing of a laser beam by the intensity dependent refractive index of the material. The Z-scan experiments are the most difficult to do, and not all of the ionic liquids that we prepared have been optically characterized yet. The density, viscosity, melting or glass transition temperature, maximum linear absorption wavelength, and linear molar absorptivity are given in Table III. Some of the ionic liquids are not expected to be NLO active, but were prepared and characterized for comparison purposes.

Some of the anions in Table III are not exactly the same as the models we used to predict the gammas in Table I. We did not calculate two of the anions ([Co(NCS)$_4$]$^{2-}$ and [CoCB$^-$]), because they we synthezed faster than they were calculated. The structures of anions in Table II that are not shown below Table I are:

### Table III. Physical and Optical Properties

| [anion] | $T_g$ °C | $T_{dec}$[a] °C | Density,[b] g/cm³ | Viscosity[b] $\eta$, cP | $\lambda_{max}$,[c] nm |
|---------|------|------|------|------|------|
| [dtmn]²⁻ | -71 | 360 | 0.942 | 4780 | 394 |
| [K-salt]²⁻ | -79 | 350 | 0.960 | 560 | 396 |
| [ddtc]⁻ | -77 | 255 | 0.942 | 1470 | 305 |
| [xan]⁻ | -70[e] | 290 | 0.920 | 1480 | 345 |
| [N(CN)₂]⁻ | -67 | 395 | 0.904 | 490 | d |
| [Tf₂N]⁻ | -76 | 400 | 1.080 | 450 | d |
| [Co(NCS)₄]²⁻ | -72[e] | 405 | 0.963 | 2436 | 628 |
| [CoCB]⁻ | -71 | 440 | 1.00 | 3702 | 290 |

a. TGA onset temperature

b. At 20° C

c. Neat liquid in 10μm pathlength cell

d. Colorless

e. Melting temperature; not glass transition

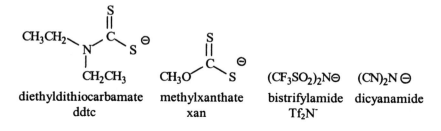

diethyldithiocarbamate    methylxanthate    bistrifylamide    dicyanamide
ddtc                      xan               Tf₂N⁻

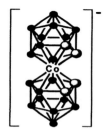

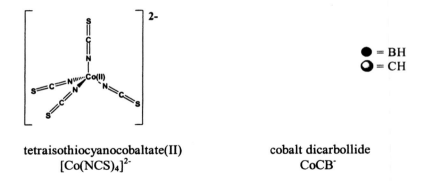

● = BH
◯ = CH

tetraisothiocyanocobaltate(II)
[Co(NCS)$_4$]$^{2-}$

cobalt dicarbollide
CoCB$^-$

The metal-containing anions were of interest for NLO applications in that they contain diffuse d-orbitals, they can be easily functionalized and tuned based on the ligand systems, and their total negative charge can be built up rapidly with appropriate ligands. Additionally, the cobalt bis(dicarbollide), CoCB, has boron-carbon cages which are σ-aromatic, thus providing the delocalization desired for high NLO behavior. One potential flaw with these metal-ligand systems is that the central metal cations are typically in the +2 or greater oxidation state, potentially concentrating the charge of the ligands adjacent to the metal. However, at the molecular level, this may not be the case as the molecular orbitals will differ significantly from the formal charges designated to the atomic components. There have been previous reports (12,13) of solutions or solids of the metal complexes similar to the metal-containing salts displaying NLO behavior, which provided motivation to choose these anions to develop into RTILs.

The preparation of the somewhat complicated looking ionic liquids is actually quite easy. The chloride salt of the tetradecyl(trihexyl)phosphonium cation is commercially available as CYPHOS IL 101. The sodium salts of the thiolate anions are all prepared by variations on the general addition reaction in Equation 4

$$Na^+ \; A\!-\!X^- \;+\; CS_2 \;\longrightarrow\; A\!-\!X\!-\!\overset{\displaystyle S}{\underset{\displaystyle S}{C}}{}^- \; Na^+$$

(4)

where X is a nucleophilic atom and A is the rest of the molecule. The sodium salts are then used to create the desired salt by a metathesis reaction with the tetradecyl(trihexyl)phosphonium chloride. Other large tetraalkyl phosphonium halides are also available, and they also combine with the large anions to form ionic liquids. See the experimental section for the details of the syntheses. The easy synthesis from readily available materials is a favorable feature of the ionic liquids listed in Table III.

## Nonlinear Optical Properties

Some details of this analysis are provided in an earlier paper (14). An example of the Z-scan results for the $[PR_4]_2[Co(NCS)_4]$ ionic liquid is demonstrated in Figures 1 and 2, where data from two experiments show the effect of incident laser beam power. The optical responses are dependent on the light intensity, as expected for the nonlinear optical phenomena. The experiments were done with very short laser pulses (picosecond or shorter) to avoid thermal effects that can obscure the nonlinear optical effects. The data from Figures 1 and 2, and similar data for other ionic liquids were used to obtain the $n_2$ and $\alpha_2$ values listed in Table IV. $CS_2$ and ZnSe are commonly used standards for NLO measurements, and we include them here for comparison purposes. The NLO characterization is incomplete at this time, but some structure-property correlations are already apparent. The monoanions have lower $\chi^{(3)}$ than the dianions, as seen by the $n_2$ values.

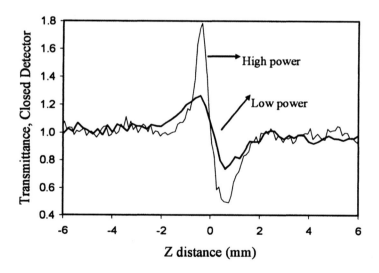

*Figure 1. Z-scan experiment on $[PR_4]_2[Co(NCS)_4]$ with the transmittance recorded at the open detector.*

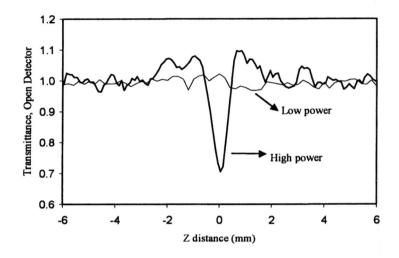

*Figure 2. Z-scan experiment on [PR₄]₂[Co(NCS)₄] with the transmittance recorded at the open detector.*

**Table IV. Nonlinear Optical Properties**

| Anion | Wavelength nm | $n_2$ measured $X10^{14}\ m^2\ W^1$ | $\alpha_2$ measured $X10^{11}\ m\ W^1$ |
|---|---|---|---|
| [xan]⁻ | 910 | -0.09 | 0[a] |
| [CoCB]⁻ | 910 | -0.22 | [b] |
| [ddtc]⁻ | 910 | -0.45 | 0[a] |
| [Co(NCS)₄]²⁻ | 910 | -1.9 | [b] |
| [dtmn]²⁻ | 910 | -6.1 | 10.8 |
| CS₂ | 1064 | 0.0003 | 0[a] |
| ZnSe | 532 | 0.0007 | 5.8 |

a. Low laser power used

b. Not yet determined

# Experimental

## Synthesis of Ionic Liquids

Alkali metal salts of the desired anions were either commercially available or easily synthesized through known reactions. Sodium diethyldithiocarbamate, Na[ddtc] (Fischer), sodium dicyanamide, Na[N(CN)$_2$] (Fluka), lithium bis(trifluoromethane)sulfonamide, Li[Tf$_2$N] (Aldrich), and cesium bis(dicarbollyl)cobalt(III), Cs[CoCB] (Katchem, Czech.) were used as received. Sodium dithiomaleonitrile, Na$_2$[dtmn], and potassium nitrodithioacetate, K$_2$[K-salt], were prepared as previously reported (15,16). Sodium methylxanthate, Na[xan], was prepared through the reaction of 1.2 eq. methanolic NaOH with 1 eq. carbon disulfide in methanol. Ammonium tetrathiocyanocobaltate(II), [NH$_4$]$_2$[Co(NCS)$_4$], was prepared from 4.2 eq. of [NH$_4$][SCN] and 1 eq. of CoCl$_2$ in H$_2$O, and the product was not isolated before proceeding to the following methathesis reaction to make the ionic liquid.

To prepare the ionic liquid materials, the above alkali metal salts were used in metathesis reactions with an appropriate number of equivalents of the tetradecyl(trihexyl)phosphonium chloride, CYPHOS 101 (Cytec), in organic solvents. Typically the alkali salts were water soluble, including the [ddtc], [xan], [N(CN)$_2$], [Tf$_2$N], and [Co(NCS)$_4$] salts. These salts were stirred with CHCl$_3$ solutions of the CYPHOS for 24 hours, separated and the organic layer washed several times with water and then evaporated off. Exceptions to this are the [CoCB] salt, which was reacted with CYPHOS in THF; and the [dtmn] and [K-salt] were reacted with CYPHOS in EtOH. In these cases, the alkali chloride by-product could be filtered out, the organic solvent evaporated, and the resulting ionic liquid purified through redissolving in CHCl$_3$ and washed with water several times. All of the resulting ionic liquids were dried under vacuum at 75-85°C.

## Properties of Ionic Liquids

Thermal gravimetric analyses were run at 10°/min on a TA Instruments SDT 2960, and differential scanning calorimetry (DSC) was performed at 5°C/min on a TA instruments DSC 2910. The TGA instrument was temperature calibrated from room temperature to 800° at 10°/min. using indium and aluminum. It was weight calibrated with ceramic standards provided with the instrument. The DSC instrument was calibrated using mercury, water, and indium at 5°C/min. Liquid densities were measured with a Mettler Toledo DE40 Density Meter-40. Viscosities were measured with a Cambridge Applied Systems Viscolab 4100 moving piston viscometer.

## Prediction of Second Hyperpolarizability

All calculations reported here were done with the General Atomic and Molecular Electronic Structure System (GAMESS) program (17). The orientation averaged values of the gammas were calculated at zero frequency using a finite-field technique with the GAMESS computational package. The geometries of the anions were optimized using the Restricted Hartree-Fock (RHF) level of theory with the Dunning-Hay basis set (18) enhanced with polarization and diffuse sp functions (19,20). The second hyperpolarizabilities of the anions were computed with the finite-field method using the same basis set except the diffuse sp functions were replaced with a set of diffuse p and d functions designed for NLO computations. This basis set will be designated DH(d)+pd. Both the RHF and MP2 levels of theory were used for the NLO computations of the anions. Because of its relative stability compared to the trithiocarbonate anion, we chose to study the effects of counter-ions by computing the NLO response of the salts of *aci*-nitro-dithioacetate (K-salt). The geometries of the salt molecules were found by first doing a short simulated annealing computation with AM1 followed by an *ab initio* optimization with a DH(d) basis enhanced with diffuse sp functions on the anion only. The NLO computations were done with the DH(d)+pd on the anion and DH(d) on the cation.

## Nonlinear Optical Characterization

The NLO properties of the materials were measured using a typical Z-scan setup. The samples were loaded into 0.01, 0.1 and 1mm quartz cuvettes depending on how strongly they absorbed. A Coherent mode-locked titanium:sapphire laser operating at wavelength 700-1000nm with 3ps pulses at 76MHz was used. When operating, only one pulse was used with the help of a peak picker and computer software. The spatial filtered beam with 3mm waist radius was focused with a 75mm lens. The sample was placed after the lens and could be translated ±45mm from the focal plane. A beam splitter was used after the sample, where one beam was directed to a detector with no aperture ("open" detector) and one towards a detector with aperture of $1mm^2$ ("closed" detector) at the far field of the lens. Peak powers of the pulses were measured to be 1.5-2.0 GW/m$^2$. All of the measurements were performed at room temperature.

## Conclusions

Ionic liquids with good third order nonlinear optical properties may be designed and prepared. Measurements show that the $\chi^{(3)}$ of the actual materials have a similar relative ordering to the molecular $\gamma$ predicted by theory.

## References

1. Aakeroy, Christer B.; Hitchcock, Peter B.; Moyle, Brian D.; Seddon, Kenneth R.. *J.. Chem. Soc. Chem. Commun.* **1989**, *23*, 1856-1859.
2. Marder, Seth R.; Perry, Joseph W.; Schaefer, William P. *Science* **1989**, *245*, 626-628.
3. Teo, Boon K.; Xu, Yi Hui; Zhong, Bing Yuan; He, Yuan Kang; Chen, Hui Ying; Qian, Wei; Deng, Yu Jun; Zou, Ying Hua *Inorg. Chem.* **2001**, *40*, 6794-6801.
4. Bykowski, Darren; McDonald, Robert; Hinkle, Robert J.; Tykwinski, Rik R. *J. Org. Chem.*, **2002**, *67*, 2798-2804.
5. Sauter, E. G. *Nonlinear Optics*; John Wiley & Sons: New York, NY, 1996.
6. Yang, L.; Dorsinville, R.; Wang, Q. Z.;Ye, P. X.; Alfano, R. R.; Zamboni, R.; Taliani, C. *Optics Lett.*, **1992**, *17*, 323-325.
7. Sheik-bahae, M.; Said, A. A.; Van Stryland, E. E. *IEEE J. Quantum Electron*, **1990**, *26*, 760.
8. Gubler, U.; Bosshard, C. *Adv. Polymer.Sci.*, **2002**, *158*, 123-191.
9. Zhou, Q. L.; Heflin, J. R.; Wong, K. Y.; Zamani-Khamiri, O.; Garito, A. F. *Org. Mol. Nonlinear Opt. Photonics*, NATO ASI Ser., Ser. E, 1991; Vol 194, pp 239-262.
10. Heflin, J. R.; Zhou, Q. L.; Wong, K. Y.; Zamani-Khamiri, O.; Garito, A. F. *Proc. SPIE-Int. Soc. Opt. Eng.*, 1337 (Nonlinear Opt. Prop. Org. Mater. 3), 1990; pp186-194.
11. Yeates, A. T.; Dudis, D. S.; Wilkes, J. S. *Proc. SPIE-Int. Soc. Opt. Eng,*, 4106 (Linear, Nonlinear, and Power-Limiting Organics), 2000; pp. 334-337.
12. Zhan, C.; Xu, W.; Zhang, D; Li, D; Lu, Z; Nie, Y; Zhu, D. *J. Mater. Chem.*, **2002**, *12*, 2945-2948. ().

13. García-Frutos, E. M.; O'Flaherty, S. M.; Maya, E. M.; de la Torre, G.; Blau, W.; Vázquez, P.; Torres, T. *J. Mater. Chem.*, **2003,** *13*, 749-753..
14. Del Sesto, Rico E.; Dudis, Douglas S.; Ghebremichael, Fassil; Heimer, Norman E.; Low, Tammy K.; Wilkes, John S.; Yeates A. T. In *Proceedings of the SPIE,* 2003; Vol. 5212, pp 292-298.
15. Davidson, A.; Holm, R. H. *Inorg. Synth.*, **1967,** *10*, 8.
16. Jensen, Kai Arne; Buchardt, Ole; Lohse, Christian. *Acta Chem. Scand.,* **1967,** *21*, 2797.
17. Schmidt, M. W.; Baldridge, K. K.; Boatz, J. A.; Elbert, S. T.; Gordon, M. S.; Jensen, J. H.; Koseki, S. Matsunaga, N.; Nguyen, K. A.; Su, S. J.; Windus, T. L.; Dupuis, M.; Montgomery, J. A. *J. Comput. Chem.,* , *14*, 1347- 1363.
18. Dunning, T. H. Jr.; Hay, P. J. In *Methods of Electronic Structure Theory*; Shaefer H. F. III Ed., Plenum Press: New York, NY, 1977; pp 1-27.
19. Clark, T.; Chandrasekhar, J.; Spitznagel, G. W.; Schleyer, P. von R. *J. Comput. Chem.,* **1983,** *4*, 294-301.
20. G. W. Spitznagel, Diplomarbeit, Erlangen, 1982

# Chapter 13

# Preparation and Properties of Polymerized Ionic Liquids as Film Electrolytes

## Hiroyuki Ohno and Masahiro Yoshizawa

Department of Biotechnology, Tokyo University of Agriculture and Technology, Koganei, Tokyo 184–8588, Japan

Ionic liquids are quite interesting materials for not only solvents but also electrolytes. In spite of non-volatility, liquid state restricts some applications especially for the industrial use as electrolytes. Polymerized ionic liquids should broaden the application field of this interesting material. Introduction of vinyl group on some imidazolium cations induced no inferior effect on the ionic liquid formation with proper anions. Similarly, anions containing vinyl groups could form liquid salts with vinyl imidazolium cations. Copolymerizations of these monomeric ions as well as homopolymerization of either polymerizable cations or anions have been examined. Introduction of spacer groups in between charged site and vinyl group was revealed to be effective to lower the glass transition temperature and accordingly to improve the ionic conductivity. Their electrochemical characteristics and physico-chemical properties are discussed with their structure.

Ionic liquids (ILs) are collecting much attention as a new ion conductive matrix (1). Since ILs are liquid which consist of only ions, these ILs have two brilliant features of very high ion concentration and high mobility of component ions at room temperature. Since lots of them show the ionic conductivity of over $10^{-2}$ S cm$^{-1}$ at room temperature (2,3), there are lots of applications as electrolyte materials for rechargeable lithium-ion battery (4-8), fuel cell (9-12), and solar cell (13-17) or capacitor (18-23).

However, when applying ILs as an electrolyte material, ILs cannot transport only target ions alternatively because ILs, which are used as solvent, can also migrate along with potential gradient. In order to improve this drawback, we have proposed various structures for the transport of only target ions such as zwitterionic liquids (ZILs) have both cation and anion, which form ILs in the same molecule (24-26), triple ion-type imidazolium salts have two anions and cation (27), polymerized ILs (28-34), etc.

## Introduction of ILs into Polymers

Although various methods are proposed to introduce ILs to polymer matrices, they are divided roughly into the following three; polymer gel electrolytes containing simple ILs as the medium (ILs *in* polymer), the formation of ILs generates polymer (ILs *for* polymer), and the polymerization of ILs themselves (ILs *as* polymer). Figure 1 shows the schematic concepts. Among these models, development of polymer gel electrolytes using ILs has been widely investigated due to facility of design and synthesis. There are a lot of host polymers reported up to present from synthetic polymers such as PVdF-HFP, (25,35,36) Nafion, (9,10b) and poly(hydroxyethyl methacrylate) (37,38) to biomaterial such as DNA (39). These gel systems have not only the ability of film formation but also quite high ionic conductivity of over $10^{-3}$ S cm$^{-1}$ at room temperature. Its high ionic conductivity should be based on high ion mobility of IL itself, that is, it is possible to introduce ILs into polymer matrices without damage of the excellent properties of ILs. However, when ILs are used as the electrolyte material, substantial problems such as the improvement of target ion transference number are newly found.

Since PEO/salt hybrids having salt structure on the chain ends of poly(ethylene oxide) (PEO) are liquid at room temperature when PEO chain is short, they can be positioned as a new category of ILs (40, 41). In addition, they show relative high ionic conductivity of about $10^{-5}$ S cm$^{-1}$ at room temperature because they form amorphous state in a wide temperature range. By mixing PEO/salt hybrids which fixed the opposite electric charges on the chain end, it was expected that pseudo-cross-linking of polymer (42). They showed the ionic conductivity of over $10^{-5}$ S cm$^{-1}$ at room temperature, and it was confirmed that the formation of IL on the chain end of PEO worked as a highly ion conductive path. However, their interaction was very weak to behave as polymers. Therefore, the third method, ILs *as* polymer, has attracted attention in recent years. In this report, we tried to introduce the characteristics of polymerized ILs and the ability of target ion transport of them.

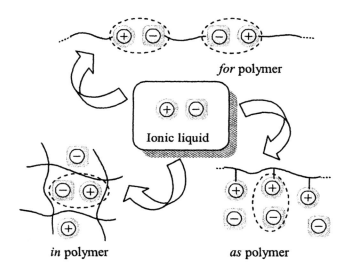

*for* polymer

Ionic liquid

*in* polymer          *as* polymer

*Figure 1. Introduction of ILs into polymers.*

## Design of Polyether-alternative Ion Conductive Matrices

The research for solid polymer electrolytes has been mainly deployed by using polyether matrices because these polyethers can dissolve salts and can transport the dissociated ions along with intra- and inter-molecular motion of these polyether chains. The polymer matrices, that satisfy both high polarity and low glass transition temperature, are not so common. However, the ionic conductivity of the reported polyether systems is about $10^{-4}$ S cm$^{-1}$ at room temperature (43), and it is difficult to improve cationic conductivity. This is attributable to the ion-dipole interaction of polyethers to target ions so called strong solvation of polyethers to cations. Therefore, alternative matrices have been looking for to solve this drawback. We prepared some polyampholytes, in which both cation and anion were fixed in the polymer main chain, as a polyether-alternative ion conductive matrix. These copolymers can provide ion conductive path for target ions prohibiting the migration of IL itself. These are mainly two categories of the polyampholyte; copolymers and poly(zwitterion)s.

## IL Copolymer

Copolymers were obtained by the polymerization of IL monomers (Figure 2) synthesized by the neutralization of acid and base both having vinyl group (34). In order to compare this simple system **I**, the couple of monomers **II** having alkyl spacer between vinyl group and sulfonate group was also synthesized and polymerized. The effect of alkyl spacer on the properties of IL copolymers is summarized in Table 1. Both couples showed only glass transition temperature (Tg), and were transparent and colorless liquids at room temperature. The ionic conductivity of the monomers was $10^{-4}$ - $10^{-3}$ S cm$^{-1}$ at room temperature. This should be based on quite low Tg of about -80 °C. On the other hand, **I** did not show Tg after polymerization, whereas **II** showed Tg at -31 °C even after polymerization, indicating that it maintains high mobility of charged units. However, both copolymers showed quite low ionic conductivity. It was confirmed that they contain no carrier ions which can support long-distance migration.

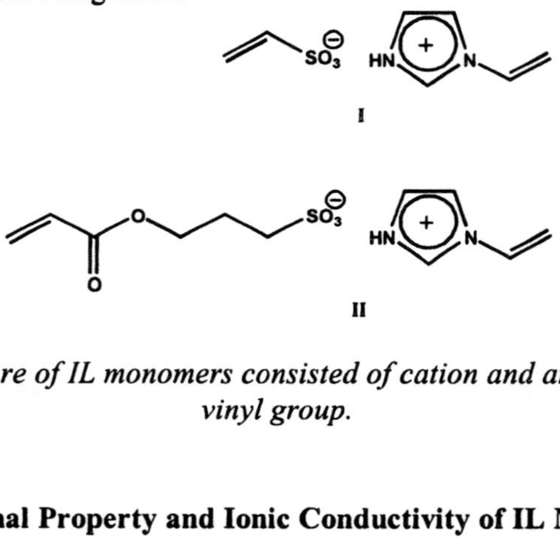

**I**

**II**

*Figure 2. Structure of IL monomers consisted of cation and anion both having vinyl group.*

**Table I. Thermal Property and Ionic Conductivity of IL Monomers and Their Polymers**

|  | Tg / °C | | $\sigma_i^a$ / S cm$^{-1}$ | |
|---|---|---|---|---|
|  | *Monomer* | *Polymer* | *Monomer* | *Polymer* |
| I | -83 | - | $3.5 \times 10^{-3}$ | $< 10^{-9}$ |
| II | -73 | -31 | $6.5 \times 10^{-4}$ | $< 10^{-9}$ |

[a] at 30 °C

An equimolar amount of LiTFSI to imidazolium cation unit was then added to those copolymers. Figure 3 shows the Arrhenius plots of the ionic conductivity for the copolymers after salt addition. PII (P implies polymer) having alkyl spacer showed one order higher ionic conductivity than PI without spacer. Although there was no Tg for both copolymers after adding the salt, it was considered that PII having alkyl spacer may form successive ion conductive path easily for carrier ions due to flexibility of the IL units.

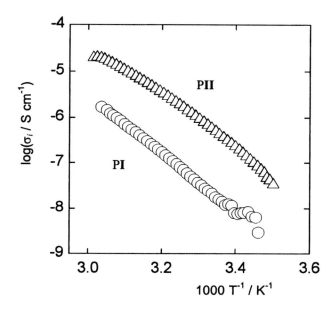

*Figure 3. Arrhenius plots of the ionic conductivity of IL coploymers containing an equimolar amount of LiTFSI.*

**Poly(zwitterionic liquid (ZIL))**

Zwitterionic monomers, which have both cation and anion in the same unit, were prepared. Figure 4 shows the structure of those monomers (24). Four different monomers were synthesized to examine the effect of the freedom of the imidazolium cation on the properties such as ionic conductivity and Tg.

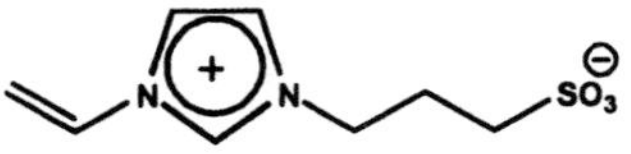

**III**

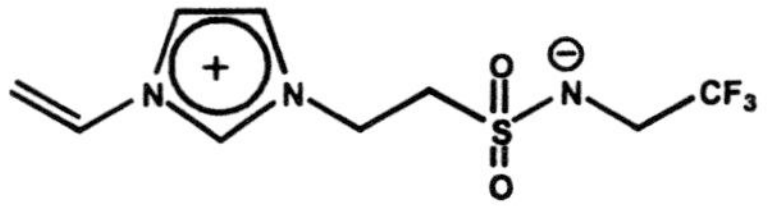

**IV**

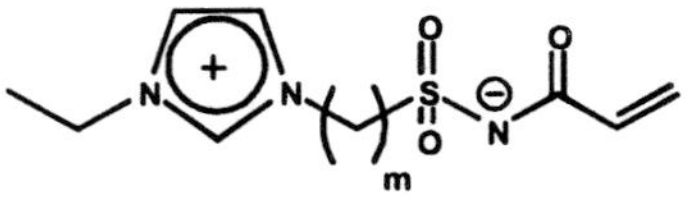

**V; m = 2**

**VI; m = 3**

*Figure 4. Structure of ZIL monomers.*

The ionic conductivity of thus prepared poly(ZIL)s was below $10^{-9}$ S cm$^{-1}$ as well as IL copolymers, suggesting that they contain no carrier ions. Then, the ionic conductivity of them was investigated after adding an equimolar amount of LiTFSI to monomer unit. Figure 5 shows the temperature dependence of the ionic conductivity for poly(ZIL)s containing LiTFSI. Their ionic conductivity strongly depended on the fixed position of imidazolium cation. The ionic conductivity of the polymers having an imidazolium cation on the main chain, **PIII** and **PIV**, containing LiTFSI equimolar to imidazolium unit, was about $10^{-8}$ - $10^{-9}$ S cm$^{-1}$. On the other hand, poly(ZIL)s having an immobilized counter-anion on the main chain (**PV** and **PVI** containing LiTFSI) showed relatively higher ionic conductivity of about $10^{-5}$ S cm$^{-1}$ at 50 °C in spite of their rubber-like properties. **PVI** having long alkyl spacer (m=3) showed slightly better ionic conductivity.

Galin and co-workers have also reported the ionic conductivity of poly(ZIL)s consisted of ammonium cation (44). They suggested that the dipole moment of poly(ZIL)s increase with expanding the distance between cation and anion fixed on the side chain. Accordingly, there is a possibility that the solubility of the added salts is different for each poly(ZIL). However, the ionic conductivity of **PIV** and **PV**, which may have almost the same polarity, were quite different. In order to obtain higher ionic conductivity for IL polymers, it was strongly suggested that the freedom of imidazolium cation is quite important factor. Although it is difficult to compare our results with Galin's results due to different anion structure, ionic conductivity of **PVI** was 1 or 2 orders higher than that of Galin's systems. This should be based on the difference of the fixed position for onium cation and the use of suitable added salt.

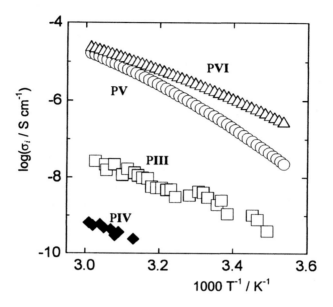

*Figure 5. Temperature dependence of the ionic conductivity of poly(ZIL)s containing an equimolar amount of LiTFSI.*

IL copolymers and poly(ZIL)s contained no carrier ions, but they could provide the ion conductive path and showed the ionic conductivity of about $10^{-5}$ S cm$^{-1}$ at room temperature after adding LiTFSI. Highly ion conductive polymers, which do not depend on polyether structure, were obtained by the polymerization of ILs.

## Single Ion Conductive IL Polymers

In order to realize selective ion transport in IL polymers, cation or anion, which form ILs, are fixed on the side chain of IL polymers. Since the fixed ions cannot migrate along with the potential gradient, it is expected that such IL polymers work as a novel single ion conductive matrix.

**Polycations**

Monomers were prepared from vinylimidazole, and their ionic conductivity was measured before and after polymerization (28,30). The monomer having TFSI anion was liquid at room temperature. It showed quite high ionic conductivity of about $10^{-2}$ S cm$^{-1}$ at room temperature, but the ionic conductivity of corresponding polymer system was below $10^{-6}$ S cm$^{-1}$, which is more than 4 orders lower than that of monomer. This big difference should be based on the decrease of carrier ions and of the mobility of IL itself because all of cations was fixed on the polymer chains.

To minimize the decrease of ionic conductivity for IL polymers, ionic liquid-type polymer brushes (ILPBs), which introduced flexible spacer between vinyl group and cation structure were synthesized (29,32,33). Figure 6 shows the structures of IL monomers having spacer. They have different spacer structure and length to investigate the effect of their conditions on the properties. ILPBs having TFSI anion showed the ionic conductivity of over $10^{-4}$ S cm$^{-1}$ at room temperature, which is equal to that of corresponding monomer. It was revealed that ILPBs were effective to maintain high ionic conductivity of IL monomers even after polymerization. In addition, alkyl spacer showed higher ionic conductivity than oligoether spacer. This difference should be based on the hindrance of ion transport by the ion-dipole interaction between oligoether and carrier ion.

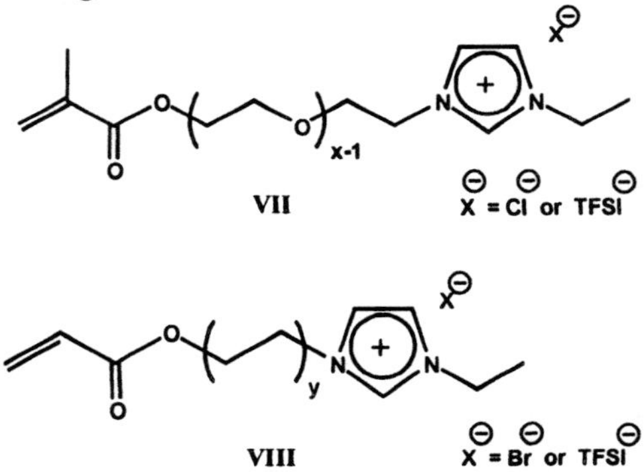

*Figure 6. Structure of IL monomers having spacer.*

Although ILPBs were found to maintain high ionic conductivity of corresponding monomers, their mechanical property was poor. Actually, most of them were obtained as sticky rubber, not film. We tried to improve it by using cross-linkers. Figure 7 shows the photograph of novel film electrolyte composed of ILPB with small amount of cross-linker. We succeeded to obtain transparent and flexible films without considerable decrease in the ionic conductivity as compared with that of corresponding monomers. These properties are the function of the added amount of cross-linker.

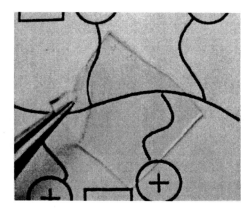

*Figure 7. Photograph of novel film electrolyte composed of ILPB with 1.3 mol% cross-linker.*

**Polyanions**

It was suggested that high mobility of imidazolium cation was needed to maintain high ionic conductivity of IL polymers in a previous section. Therefore, IL polymers, anions were fixed on the main chain, were synthesized. To investigate the effect of anion structure on the properties, some IL monomers having different anion structure were prepared by neutralization of ethylimidazole and acrylic acid (**AC**), styrene sulfonic acid (**SS**), vinylsulfonic acid (**VS**), or vinylphospholic acid (**VP**) as shown in Figure 8 (45). Table 2 summarizes thermal property and ionic conductivity of these IL monomers and their polymers. All monomers were liquid state at room temperature. Among these, **VS** monomer showed the highest ionic conductivity of $9.0 \times 10^{-3}$ S cm$^{-1}$ at 30 °C. This result reflected the lowest Tg of -95 °C. After polymerization, PVS also showed the highest ionic conductivity according to the lowest Tg as well as the monomer. Higher ionic conductivity of IL polymers may be realized by maintaining high mobility of IL unit structure in the polymer.

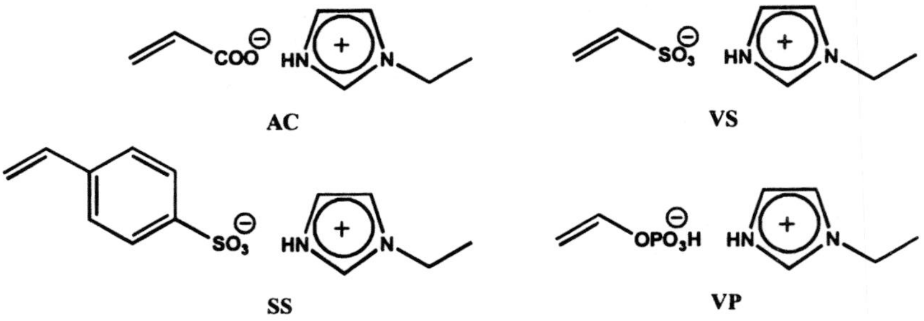

*Figure 8. Structure of IL monomers consisted of anion having polymerizable group.*

**Table II. Properties of IL monomers consisted of anion having polymerizable group and their polymers**

| | $Tg \, / \, ^{\circ}C$ | | $\sigma_i^{\,a} \, / \, S \, cm^{-1}$ | |
|---|---|---|---|---|
| | *Monomer* | *Polymer* | *Monomer* | *Polmer* |
| AC | -61 | -32 | $1.4 \times 10^{-4}$ | $1.2 \times 10^{-6}$ |
| SS | -77 | -[b] | $8.7 \times 10^{-5}$ | $1.1 \times 10^{-8}$ |
| VS | -95 | -63 | $9.0 \times 10^{-3}$ | $1.1 \times 10^{-4}$ |
| VP | -78 | -56 | $1.5 \times 10^{-4}$ | $2.9 \times 10^{-5}$ |

[a] at 30 °C, [b] not detected

## Conclusion

Highly ion conductive polymers, which do not depend on the polyether structure, were obtained by the polymerization of IL monomers. Thus prepared polymerized ILs worked as a novel single ion conductive matrix. In addition, it was revealed that ILPB having flexible spacer between polymerizable group and IL structure can maintain high ionic conductivity, which is almost equal to that of corresponding monomer, even after polymerization. Recently, we succeeded to obtain ILs consisted of lithium cation as a component ion for the first time (46). It will be possible to develop lithium cation conductive polymers by the combination of ILPB and lithium ILs. Functional design will be realized soon through the design of corresponding unit structure.

## Acknowledgements

The experimental contribution of Mr. Wataru Ogihara and Ms. Satoko Washiro were greatly acknowledged. The present study was supported by the Grant-in-Aid for Scientific Research from the Ministry of Education, Science, Sports and Culture, Japan (#14205136). The present study was carried out under the 21$^{st}$ Century COE program, Future Nano Materials.

## References

1.  *Ionic Liquids-The Front and Future of Material Development-*; Ohno, H., Ed.; CMC: Tokyo, 2003 (in Japanese). (English version will be published in September, 2004)
2.  Bonhôte, P.; Dias, A-P.; Armand, M.; Papageorgiou, N.; Kalyanasundaram, K.; Grätzel, M. *Inorg. Chem.* **1996**, *35*, 1168.
3.  Hagiwara, R.; Ito, Y. *J. Fluorine Chem.* **2000**, *105*, 221.
4.  Koch, V. R.; Nanjundiah, C.; Appetecchi, G. B.; Scrosati, B. *J. Electrochem. Soc.* **1995**, *142*, L116.
5.  MacFarlane, D. R.; Huang, J.; Forsyth, M. *Nature* **1999**, *402*, 792.
6.  Nakagawa, H.; Izuchi, S.; Kuwana, K.; Nukuda, T.; Aihara, Y. *J. Electrochem. Soc.* **2003**, *150*, A695.
7.  Hayashi, A.; Yoshizawa, M.; Angell, C. A.; Mizuno, F.; Minami, T.; Tatsumisago, M. *Electrochem. Solid-State Lett.* **2003**, *6*, E19.
8.  Sakaebe, H.; Matsumoto, H. *Electrochem. Commun.* **2003**, *5*, 594.
9.  Doyle, M.; Choi, S. K.; Proulx, G. *J. Electrochem. Soc.* **2000**, *147*, 34.
10. (a) Sun, J.; MacFarlane, D. R.; Forsyth, M. *Electrochim. Acta* **2001**, *46*, 1673. (b) Sun, J.; Jordan, L. R.; Forsyth, M.; MacFarlane, D. R. *Electrochim. Acta* **2001**, *46*, 1703.
11. (a) Noda, A.; Susan, M. A. B. H.; Kudo, K.; Mitsushima, S.; Hayamizu, K.; Watanabe, M. *J. Phys. Chem. B* **2003**, *107*, 4024. (b) Susan, M. A. B. H.; Noda, A.; Mitsushima, S.; Watanabe, M. *Chem. Commun.* **2003**, 938.
12. Yoshizawa, M.; Xu, W.; Angell, C. A. *J. Am. Chem. Soc.* in press.
13. Papageogiou, N.; Athanassov, Y.; Armand, M.; Bonhôte, P.; Pettersson, H.; Azam, A.; Grätzel, M. *J. Electrochem. Soc.* **1996**, *143*, 3099.
14. Kawano, R.; Watanabe, M. *Chem. Commun.* **2002**, 330.
15. Kubo, W.; Kitamura, T.; Hanabusa, K.; Wada, Y.; Yanagida, S. *Chem. Commun.* **2002**, 374.
16. Wang, P.; Zakeeruddin, S. M.; Exnar, I.; Grätzel, M. *Chem. Commun.* **2002**, 2972.
17. Matsumoto, H.; Matsuda, T.; Tsuda, T.; Hagiwara, R.; Ito, Y.; Miyazaki, Y. *Chem. Lett.* **2001**, 26.

170

18. Lewandowski, A.; Swiderska, A. *Solid State Ionics* **2003**, *161*, 243.
19. Ue, M.; Takeda, M.; Toriumi, A.; Kominato, A.; Hagiwara, R.; Ito, Y. *J. Electrochem. Soc.* **2003**, *150*, A499.
20. Stenger-Smith, J. D.; Webber, C. K.; Anderson, N.; Chafin, A. P.; Zong, K.; Reynolds, R. *J. Electrochem. Soc.* **2002**, *149*, A973.
21. McEwen, A. B.; Ngo, H. L.; LeCompte, K.; Goldman, J. L. *J. Electrochem. Soc.* **1999**, *146*, 1687.
22. McEwen, A. B.; McDevitt, S. F.; Koch, V. R. *J. Electrochem. Soc.* **1997**, *144*, L84.
23. Nanjundiah, C.; McDevitt, S. F.; Koch, V. R. *J. Electrochem. Soc.* **1997**, *144*, 3392.
24. Yoshizawa, M.; Hirao, M.; Ito-Akita, K.; Ohno, H. *J. Mater. Chem.* **2001**, *11*, 1057.
25. Ohno, H.; Yoshizawa, M.; Ogihara, W. *Electrochim. Acta* **2003**, *48*, 2079.
26. Yoshizawa, M.; Narita, A.; Ohno, H. *Aust. J. Chem.* in press.
27. Yoshizawa, M; Ohno, H. *Ionics* **2002**, *8*, 267.
28. Ohno, H.; Ito, K. *Chem. Lett.* **1998**, 751.
29. Yoshizawa, M.; Ohno, H. *Chem. Lett.* **1999**, 889.
30. Hirao, M.; Ito, K.; Ohno, H. *Electrochim. Acta* **2000**, *45*, 1291.
31. Hirao, M.; Ito-Akita, K.; Ohno, H. *Polym. Adv. Technol.* **2000**, *11*, 534.
32. Ohno, H. *Electrochim. Acta* **2001**, *46*, 1407.
33. Yoshizawa, M.; Ohno, H. *Electrochim. Acta* **2001**, *46*, 1723.
34. Yoshizawa, M.; Ogihara, W.; Ohno, H. *Polym. Adv. Technol.* **2002**, *13*, 589.
35. Fuller, J.; Breda, A. C.; Carlin, R. T. *J. Electrochem. Soc.* **1997**, *144*, L67.
36. Fuller, J.; Breda, A. C.; Carlin, R. T. *J. Electroanal. Chem.* **1998**, *459*, 29.
37. Noda, A.; Watanabe, M. *Electrochim. Acta* **2000**, *45*, 1265.
38. Tsuda, T.; Nohira, T.; Nakamori, Y.; Matsumoto, K.; Hagiwara, R.; Ito, Y. *Solid State Ionics* **2002**, *149*, 295.
39. (a) Ohno, H.; Nishimura, N. *J. Electrochem. Soc.* **2001**, *148*, E168. (b) Nishimura, N.; Ohno, H. *J. Mater. Chem.* **2002**, *12*, 2299.
40. (a) Ohno, H.; Ito, K. *Polymer* **1995**, 36, 891. (b) Ito, K.; Tominaga, Y.; Ohno, H. *Electrochim. Acta* **1997**, *42*, 1561. (c) Ito, K.; Nishina, N.; Ohno, H. *J. Mater. Chem.* **1997**, *7*, 1357. (d) Tominaga, Y.; Ito, K.; Ohno, H. *Polymer* **1997**, *38*, 1949.
41. Yoshizawa, M.; Ito-Akita, K.; Ohno, H. *Electrochim. Acta* **2000**, *45*, 1617.
42. Nakai, Y.; Ito, K.; Ohno, H. *Solid State Ionics* **1998**, *113-115*, 199.
43. Hooper, R.; Lyons, L. J.; Mapes, M. K.; Schumacher, D.; Moline, D. A.; West, R. *Macromolecules*, **2001**, *34*, 931.
44. Galin, M.; Mathis, A.; Galin, J-C. *Polym. Adv. Technol.* **2001**, *12*, 574.
45. Ohno, H.; Yoshizawa, M.; Ogihara, W. *Electrochim. Acta* in press.
46. Ogihara, W.; Yoshizawa, M.; Ohno, H. *Chem. Lett.* **2002**, 880.

# Chapter 14

# An Ionic Liquid-Based Optical Thermometer

## Sheila N. Baker[1], T. Mark McCleskey[1], and Gary A. Baker[2,*]

[1]Chemistry Division, MS J514 and [2]Bioscience Division, MS J586,
Los Alamos National Laboratory, Los Alamos, NM 87545

We report on a high performance ratiometric luminescent thermometer based on the reversible temperature-dependent excited-state self-association of the probe 1,3-*bis*(1-pyrenyl)propane dissolved at a few parts per million in the ionic liquid [$C_4$mpy][$Tf_2$N]. The thermometer performs well over the 25–140 °C range, is rapid and completely reversible, and exhibits an average imprecision of less than ± 0.20 °C over the truncated temperature range 25 °C≤T≤ 120 °C. The precision near physiological temperature is slightly better (± 0.12 °C) suggesting possible applications on the operating table and in the battlefield.

While it is frequently important to know the temperature for a given application, it is often problematic to measure in nanospaces or along surfaces. Compared with analytes such as $H^+$ and metal ions, the development of molecular sensors to follow changes in environmental properties, particularly temperature, has lagged behind, despite fundamental and technological importance. Real time temperature monitoring can, for example, lead to process optimization, waste minimization and energy conservation in an industrial setting. Precise and real time in vivo monitoring of temperature is also of paramount significance in biomedical and cancer diagnosis and during hypothermia therapy or surgery where temperature fluctuations of even a few degrees can be telling or even life-threatening. While conventional contact techniques such as the use of liquid-in-glass thermometers, thermistors, thermocouple taps and resistance temperature detectors (RTDs) certainly have their place (1), using light as the information carrier has several benefits. For example, optical temperature sensors (so-called opt(r)odes) may be deployed where it is undesirable or impossible to connect a wire such as where electromagnetic noise is strong, in explosive or corrosive environments and near/at high-speed moving parts such as turbine blades. Non-contact optical approaches also offer utility for obtaining temperature with high spatial resolution. For example, this is useful for mapping temperature fluctuations at the cellular level, in microfluidic chips and microelectromechanical systems (MEMS) and for locating heat "bottlenecks" in an integrated circuit

While remote two-dimensional infrared thermography offers some of these advantages, few objects truly behave like blackbodies. There are also several limitations to this approach such as strong absorption of radiation by water vapor and ordinary glass materials and the fact that radiation from a solid object seldom exhibits distinctive thermal signatures. On the other hand, luminescent signals are inherently multidimensional offering sensitive, selective and rapid feedback. For this reason luminescence is often the observable of choice in designing chemosensors and in the operation of molecular-level devices. In the past, luminescent thermometers have been designed around temperature-dependent excited-state decay times, intensities, intensity ratios, and excitation or emission wavelength maxima (2-4).

Now receiving serious consideration with the promise of significant environmental benefits, ionic liquids (ILs) have gained worldwide notoriety. Driven by both scientific curiosity and the potential for deriving useful new materials, ILs are attracting attention as media for and/or constituents of composite and hybrid materials (5). In the literature to date, however, we found just one example where an ionic liquid was an integral component in the fabrication of a diagnostic or sensory device. In this report, Dai and co-workers showed that exposure of an IL-coated resonant crystal of a quartz crystal microbalance (QCM) to organic vapors resulted in sorption/plasticization which was readily detected from the resulting frequency shift (6). Spurred by these findings and based on recent results from our group (7) and those from the Pandey lab (8), we were led to the possibility of using an IL as a thermofluid for developing an effective broad range optical thermometer based on the viscosity dependence of intramolecular excimer formation for 1,3-bis(1-pyrenyl)propane, BPP.

## Experimental

The ionic liquid 1-butyl-1-methylpyrrolidinium bis(trifluoromethylsulfonyl) imide, [$C_4$mpy][$Tf_2N$], was synthesized and purified using recently disclosed methods (9). Even traces of water can significantly reduce the bulk viscosity and accelerate solute diffusion within ionic liquids (7c). Achieving a reproducibly low water level is imperative if one wishes to create a stable and robust sensor free of calibration and drift issues. For this reason, drying was carried out under identical conditions for all IL samples reported herein. Specifically, 10 g samples of IL in 50 mL Falcon tubes fitted with plastic screw caps drilled with an identical hole pattern were flash frozen in liquid nitrogen for 5 minutes. Lyophilization was then carried out for a full week (Labconco, Freeze-dry/Freezone 4.5) at a condenser temperature of −50 °C and a pressure of 99 μbar.

## Results and Discussion

At this point it is important to consider the basics of excimer formation. Although more complex photophysical treatments are sometimes used, the simplest and a frequently adequate description of excimer formation kinetics is that of the Birks model (10). The ratio of excimer to monomer fluorescence intensities ($R = I_E/I_M$) is, according to Figure 1, given by

$$\frac{I_E}{I_M} = \frac{k_a k_E}{(k_E + k_{IE} + k_d)k_M} \tag{1}$$

where $k_M$ and $k_E$ are the monomer and excimer radiative rate constants, $k_{IM}$ and $k_{IE}$ are the corresponding rates for nonradiative decay, and $k_a$ and $k_d$ represent the monomolecular rates for excimer association and dissociation. Note that according to the Birks model the excimer is entirely dissociative in the ground state. Under conditions where $k_E \gg k_d$, this intensity ratio essentially reflects the variation in the cyclization rate constant $k_a$ which depends on both solvent viscosity and overall solvent quality (10). That is, $k_a$ is small for good solvents and large for poor solvents. Significantly, this expression also suggests that the temperature range and sensitivity for our optical thermometer considered here might be tuned by variation in the ionic liquid employed.

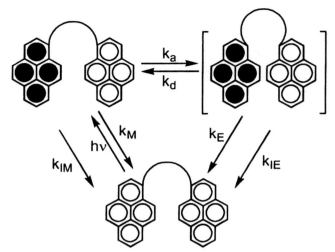

*Figure 1. Birks kinetic scheme for intramolecular excimer formation.*

In our experience, due to instability, particularly for prolonged exposure to elevated temperatures, hexafluorophosphate containing ionic liquids were unsuitable for fabricating robust optical temperature sensors. Following investigation of a number of materials, however, we identified [C$_4$mpy][Tf$_2$N], as an excellent matrix for demonstrating this novel class of molecular temperature sensor. Panel A of Figure 2 illustrates that the spectral response of the BPP in [C$_4$mpy][Tf$_2$N] system is strongly governed by temperature. Upon heating, the excited-state equilibrium is shifted in favor of excimer formation resulting from a reduced bulk viscosity evidenced by the appearance of a prominent band near 475 nm. Upon cooling there is an almost perfect superposition of the entire emission profile demonstrating the full reversibility of the system. We also note that the raw emission spectra exhibit an isoemissive point at 446 nm at 90 °C and below (data not shown) validating the simple two-state model assumed above. Parenthetically, we also note that excitation spectra (Figure 2B) monitored at the monomer (M, 375 nm) and excimer (E, 475 nm) emission colors were identical, ruling out inter- or intramolecular ground-state pyrene interaction or pre-association.

In order to generate an analytical working curve, temperature was controlled using a ±0.2 °C temperature control stage similar to one described by us earlier (*11*). In practice, a working curve was obtained by integrating 10-nm slices centered at 376 ($I_M$) and 476 nm ($I_E$). Photoluminescence measurements were made following slow heating or cooling (1-2 °C/min) plus a 10 min thermal

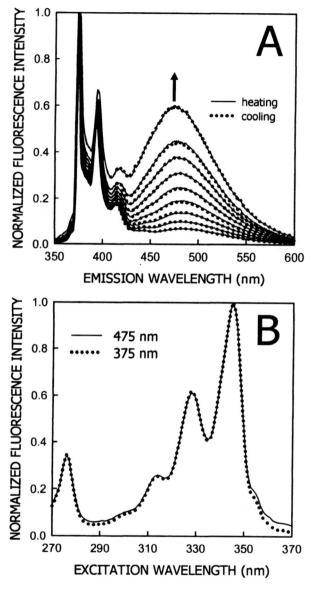

*Figure 2. BPP in [C₄mpy][Tf₂N]. (A) Normalized emission spectra during a heating (solid lines) and cooling (dotted profile) cycle and (B) normalized excitation spectra monitored at the monomer and excimer emission bands.*

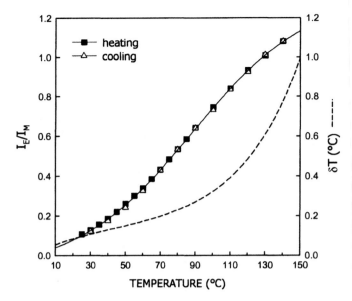

*Figure 3. Temperature response of the BPP in [C₄mpy][Tf₂N] optical
thermometer to heating and cooling portions of a cycle. The dashed line
represents the uncertainty in temperature determination predicted from eq (2).*

equilibration at each temperature (Figure 3). The uncertainty in the temperature
($\delta T$) as provided in Figure 3 was computed based on the imprecision in the
excimer to monomer ratio ($\delta R$) according to

$$\delta T = (\partial T / \partial R)\delta R \tag{2}$$

where $\partial T/\partial R$ was obtained from differentiation of a sigmoidal fit to the $R$ vs $T$
curve. Importantly, the seamless overlap of $I_E/I_M$ values for heating and cooling
segments of a thermal cycle indicate that the sensor response to temperature is
completely reversible. Moreover, since the response is two-color, that is
ratiometric, the sensor response is independent of the illumination source
intensity, the optical configuration, or the luminophore concentration. This is
illustrated by results for a photobleaching study given in Figure 4. Under
continuous ultraviolet irradiation for 24 h at a constant temperature (60 °C)
there was significant and similar photobleaching at the monomer and excimer
bands (~15%). This drift is canceled out in their ratio ($I_E/I_M$), however, which

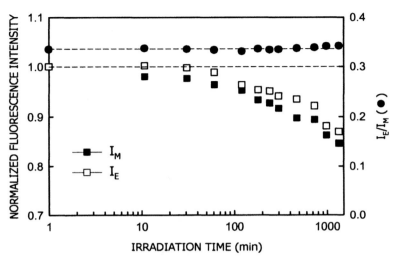

Figure 4. Photobleaching study of BPP in [C$_4$mpy][Tf$_2$N] (T=60 °C).

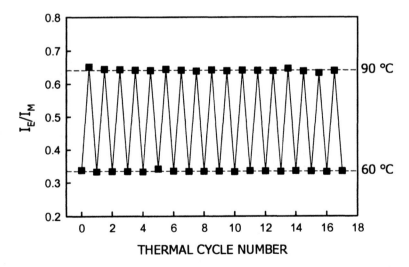

Figure 5. Thermocycling study of BPP in [C$_4$mpy][Tf$_2$N] demonstrating its reversibility and robustness.

178

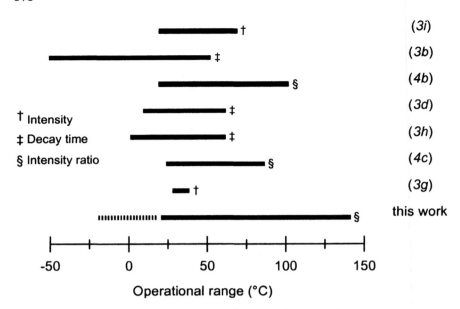

*Figure 6. A dynamic range comparison between the optical temperature sensor discussed in this Chapter and representative ones from the literature.*

changed by well less than 1.0%, corresponding to an error in temperature estimation of better than 0.35 °C. The full reversibility and robustness of the optical thermometer is illustrated in Figure 5 for repeated cycling between 60 and 90 °C over an 8 h period. Average deviations from the analytical working curve
resulted in errors in temperature determination of only 0.16 and 0.30 °C at 60 and 90 °C, respectively. Figure 5 also shows that the IL-based temperature sensor is completely reversible for 17 such cycles. In practice, the sensor did not reach failure for 100 cycles, the maximum number tried.

In contrast to excimer or exciplex based molecular thermometers reported in the past (2,4), the dynamic range of our luminescent thermometer is not limited by the boiling point of the otherwise most suitable solvents (aliphatic hydrocarbons and paraffin oils) but is instead ultimately dictated by the thermal stability of the fluorophore itself. As a result, the useful range for our thermometer extends well beyond those reported, spanning the 25 to 140 °C range. Beyond 150 °C we noted minor hysteresis which became pronounced at T ≥ 170 °C. If held past 200 °C for several minutes the sensor became completely unreliable. As might be expected, recovery from more transient exposure to elevated temperatures is improved. A comparison between the BPP in [C$_4$mpy][Tf$_2$N] optical thermometer reported here and other optically-based molecular thermometers from the open literature is provided in Figure 6. It should be noted that at the time of these experiments active cooling was not operational for our temperature controller based on modifications to a commercially available microscope stage (*11*). Therefore, our lower temperature bound was limited to room temperature. Given that [C$_4$mpy][Tf$_2$N] remains a liquid down to −18 °C (*12*), it is likely that the dynamic range of our thermometer could be further expanded (this prospect is displayed by the broken segment for our thermometer range in Figure 6).

We have also discovered that under isothermal conditions this material has potential for use as an optical oxygen sensor. For example, when a solution of BPP in [C$_4$mpy][Tf$_2$N] is spin-coated (2,000 r.p.m.) onto a coverslip and held at 70 °C by contacting a copper plate using thermally conductive grease, $R=0.431\pm0.002$ under ambient atmosphere. Under strong air flow the fluorescence intensity is relatively unchanged and $R$ is 0.467. In contrast, under nitrogen the intensity is much higher and $R$ becomes 0.634. Presumably, this behavior is due to differential collisional quenching of the monomer and excimer forms of BPP by oxygen.

## Summary

We have shown that the diffusion-controlled intramolecular cyclization for a bis-pyrenyl luminophore within an IL can be switched off and on simply by changing the temperature. This effect forms the basis for a highly effective optical thermometer which enjoys the following performance features: (a) inexpensive and simple, both in construction and operation; no specialized equipment is required beyond a standard spectrofluorimeter. Further, since the associated wavelengths are well resolved, dispersive optics are unnecessary opening the possibility for compact, fieldable devices (*e.g.*, handheld Palm Pilot sensors); (b) modular architecture, *i.e.*, the thermometer may be constructed in many formats including thin films; (c) reversible and rapid reponse on the order of one second; (d) broad dynamic range (120+ °C); (e) self-referencing; (f) precise. In fact, variations in temperature can be measured with an imprecision of ca. 0.35 °C over the entire range (25–140 °C) and with minimal averaging. Areas of potential utility for this technology include surgical, security and defense, industrial, lab-on-a-chip and aerodynamic applications. Efforts in these areas and others are currently underway in our laboratory.

## Acknowledgements

We are indebted to Professor Siddharth Pandey (Indian Institute of Technology, Delhi) for many helpful discussions and preprints which were the genesis of this research, as well as for his kindness and generosity in and out of science. This work was supported in part by a Frederick Reines Fellowship to G.A.B.

## References

1. Childs, P. R. N.; Greenwood, J. R.; Long, C. A. *Rev. Sci. Instrum.* **2000**, *71*, 2959.
2. Lou, J. F.; Finegan, T. M.; Mohsen, P.; Hatton, T. A.; Laibinis, P. E. *Rev. Anal. Chem.* **1999**, *18*, 235.
3. (a) Stump, N. A.; Burns, J. B.; Dai, S.; Mamantov, G.; Young, J. P.; Peterson, J. R. *Spectrosc. Lett.* **1993**, *26*, 1073. (b) Fister III, J. C.; Rank, D.; Harris, J. M. *Anal. Chem.* **1995**, *67*, 4269. (c) Sun, T.; Zhang, Z. Y.; Grattan, K. T. V.; Palmer, A. W. *Rev. Sci. Instrum.* **1998**, *69*, 4179. (d) Liebsch, G.; Klimant, I.; Wolfbeis, O. S. *Adv. Mater.* **1999**, *11*, 1296. (e) Engeser, M.; Fabbrizzi, L.; Licchelli, M.; Sacchi, D. *Chem. Commun.* **1999**, *13*, 1191. (f) Barabas, L.; Gorguinpour, C.; Leung, A. F.; McCoy, J. *Appl. Spectrosc.* **2001**, *55*, 1382. (g) Uchiyama, S.; Matsumura, Y.; de Silva, A. P.; Iwai, K. *Anal. Chem.* **2003**, *75*, 5926. (h) Brewster, E. R.; Kidd, M. J.; Schuh, M. D. *Chem. Commun.* **2001**, *12*, 1134. (i) Schrum, K. F.; Williams, A. M.; Haerther, S. A.; Ben-Amotz, D. *Anal. Chem.* **1994**, *66*, 2788.
4. (a) Murray, A. M.; Melton, L. A. *Appl. Opt.* **1985**, *24*, 2783. (b) Lou, J.; Hatton, T. A.; Laibinis, P. E. *Anal. Chem.* **1997**, *69*, 1262. (c) Chandrasekharan, N.; Kelly, L. A. *J. Am. Chem. Soc.* **2001**, *123*, 9898. (d) Buckner, S. W.; Forlines, R. A.; Gord, J. R. *Appl. Spectrosc.* **1999**, *53*, 115.
5. (a) Holbrey, J. D.; Seddon, K. R. *J. Chem. Soc., Dalton Trans.* **1999**, *13*, 2133. (b) Nishimura, N.; Ohno, H. *J. Mater. Chem.* **2002**, *12*, 2299. (c) Snedden, P.; Cooper, A. I.; Scott, K.; Winterton, N. *Macromolecules* **2003**, *36*, 4549.
6. Liang, C.; Yuan, C.-Y.; Warmack, R. J.; Barnes, C. E.; Dai, S. *Anal. Chem.* **2002**, *74*, 2172.
7. (a) Baker, S. N.; Baker, G. A.; Kane, M. A.; Bright, F. V. *J. Phys. Chem. B* **2001**, *105*, 9663. (b) Baker, S. N.; Baker, G. A.; Bright, F. V. *Green Chem.* **2002**, *4*, 165. (c) Werner, J. H.; Baker, S. N.; Baker, G. A. *Analyst*

2003, *128*, 786. (d) Baker, S. N.; Baker, G. A.; Munson, C. A.; Chen, F.; Bukowski, E. J.; Cartwright, A. N.; Bright, F. V. *Ind. Eng. Chem. Res.* 2003, *42*, 6457. (e) Baker, S. N.; McCleskey, T. M.; Pandey, S.; Baker, G. A. *Chem. Commun.* 2004, 940.

8. (a) Fletcher, K. A.; Pandey, S. *Appl. Spectrosc.* 2002, *56*, 266. (b) Fletcher, K. A.; Pandey, S. *Appl. Spectrosc.* 2002, *56*, 1498. (c) Fletcher, K. A.; Baker, S. N.; Baker, G. A.; Pandey, S. *New J. Chem.* 2003, *12*, 1706.

9. Warner, B.; McCleskey, T. M.; Burrell, A. K.; Hall, S. "Reversible Electro-optic Device Employing Aprotic Molten Salts," U. S. Patent Application, 2003.

10. (a) Zachariasse, K. A.; Kühnle, W.; Weller, A. *Chem. Phys. Lett.* 1978, *59*, 375. (b) Goldenberg, M.; Emert, J.; Morawetz, H. *J. Am. Chem. Soc.* 1978, *100*, 7171. (c) Zachariasse, K. A.; Kühnle, W.; Weller, A. *Chem. Phys. Lett.* 1980, *73*, 6. (d) Redpath, A. E. C.; Winnik, M. A. *J. Am. Chem. Soc.* 1982, *104*, 5604. (e) Snare, M. J.; Thistlethwaite, P. J.; Ghiggino, K. P. *J. Am. Chem. Soc.* 1983, *105*, 3328. (f) Zachariasse, K. A.; Duveneck, G.; Busse, R. *J. Am. Chem. Soc.* 1984, *106*, 1045. (g) Martinho, J. M. G.; Reis e Sousa, A. T.; Winnik, M. A. *Macromolecules* 1993, *26*, 4484. (h) Winnik, F. M. *Chem. Rev.* 1993, *93*, 587. (i) Kane, M. A.; Baker, G. A.; Pandey, S.; Maziarz, E. P.; Hoth, D. C.; Bright, F. V. *J. Phys. Chem. B* 2000, *104*, 8585. (j) Pandey, S.; Kane, M. A.; Baker, G. A.; Bright, F. V.; Fürstner, A.; Seidel, G.; Leitner, W. *J. Phys. Chem. B* 2002, *106*, 1820.

11. (a) Baker, S. N.; Baker, G. A.; Munson, C. A.; Bright, F. V. *Appl. Spectrosc.* 2001, *55*, 1273. (b) Munson, C. A.; Baker, G. A.; Baker, S. N.; Bright, F. V. *Langmuir* 2004, *20*, 1551.

12. MacFarlane, D. R.; Meakin, P.; Sun, J.; Amini, N.; Forsyth, M. *J. Phys. Chem. B* 1999, *103*, 4164.

Chapter 15

# Room Temperature Ionic Liquids as Solvent Media for the Photolytic Degradation of Environmentally Important Organic Contaminants

Qiaolin Yang and Dionysios D. Dionysiou[*]

Department of Civil and Environmental Engineering, University of Cincinnati, Cincinnati, OH, 45221–0071
[*]Corresponding author: dionysios.d.dionysiou@uc.edu

The photolytic degradation of an organic contaminant, naphthalene, in 1-butyl-3-methylimidazolium hexafluorophosphate has been investigated. The original parent compound could be degraded in this ionic liquid using 253.7 nm UV radiation. Oxygen and the purity of ionic liquid with respect to UV light absorbing compounds had a significant effect on the transformation rate of naphthalene. The addition of oxygen to the naphthalene ring and the rupture of the ring due to direct photolytic effect are discussed as possible reaction pathways that contributed to the transformation of the compound. Intermediate product identification using Electrospray TOF MS, and GC-MS revealed the formation of bicycle[4,2,0]octa-1,3,5-triene, 2'hydroxyacetophenone, and phenol among the stable intermediate products. With the use of non-toxic ionic liquids and further optimization and modification, this process has the potential for further development into a two-step process for extraction of organic pollutants from solid matrices, such as contaminated soils or dredged sediments, using ionic liquids followed by *in-situ* photodegradation of the organic contaminants in the ionic liquid extractant phase with simultaneous regeneration of the ionic liquids.

## Introduction

As a new generation of solvents, room temperature ionic liquids (RTILs) have attracted increasing research interest among chemists and engineers during the last few years. A distinct advantage of RTILs is their lack of detectable vapor pressure. As a result, they are currently considered as a promising replacement to volatile organic compounds (VOCs) which are a source of a number of environmental pollution problems.

Development of ionic liquids has experienced two different chronological periods. Chloroaluminate-based ionic liquids, frequently referred to as the *first generation* of RTILs, are a mixture of organic chlorides and aluminum chlorides. These ionic liquids are generally reactive with water and their handling requires an environment with exclusion of moisture. Therefore, the use of such ionic liquids will not be practical in open to the atmosphere environmental and other applications. Since these ionic liquids exhibit high electrical conductivity, enlarged electrochemical window and enhanced thermal stability, they were initially used as electrolytes in high-energy batteries that were sealed and had no contact with the atmosphere. (*1*) Successful synthesis of the second generation of RTILs in the early 1990s substantially expanded the applications of RTILs. These second generation RTILs share the advantages of their predecessors but also remain stable when exposed to water and air. (*1-3*) In addition, by fine-tuning their structure, they can be designed to satisfy specific task-specific applications. (*4, 5*) Since the development of these water and air stable RTILs, the number and diversity of applications of RTILs increased dramatically. Currently they are studied extensively in various applications dealing with electrochemistry, chemical synthesis, catalysis, and liquid-liquid separations. (*1-11*)

However, in the field of photochemistry, ionic liquids received relatively less attention. Only a few photochemical reactions in ionic liquids have been studied to date. Photooxidation of iron(II) diimine complexes to iron(III) diimine complexes in an ionic liquid consisting of aluminum chloride and ethylpyridinium bromide (2:1 mole ratio) is the first of such published processes. (*12*) This reaction gave a ferric yield of approximately 100%. The proposed reaction mechanism was that ethylpyridinium cations could accept electrons to form the corresponding radicals and subsequent dimerization of these species resulted in the formation of a colored compound.

Later, photolysis of anthracene (An) (*13*) and 9-methylanthracene (9-CH$_3$An) (*14*) conducted in 1-ethyl-3-methylimidazolium chloride

(EMIC)/AlCl$_3$ ionic liquids indicated that reactions in basic (55 mol % EMIC) and acidic (45 mol % EMIC) solvents generated different products. Irradiation of An in the basic ionic liquid yielded the dimer formed via 4+4 cycloaddition as in conventional solvents. In the acidic ionic liquid, the result was substantially altered due to the participation of HCl into the reaction. HCl was present as impurity in the solvent. Irradiation of 9-CH$_3$An in the basic ionic liquid yielded 4+4 dimer as major product and six minor products not previously detected in conventional solvents. Investigation of reaction mechanism indicated that the six minor products were formed via electron transfer from the excited state 9-CH$_3$An to the solvent cations followed by a series of subsequent reactions. N-Butylpyridinium chloride/AlCl$_3$, an ionic liquid containing better oxidizing cation than EMIC, showed again that the solvent cations could act as electron acceptors, which supported the proposed reaction mechanism.

In addition to behaving as electron acceptors, ionic liquids can participate in reactions as hydrogen donors in the presence of highly reactive species. It was found that the triplet excited state of benzophenone ($^3$Bp*) in imidazolium-based ionic liquids was able to abstract an alkyl chain H atom from the cations of the solvents and produced benzophenone ketyl radical. (15) The activation energy required to initiate the reaction was significantly high, and the reaction rate of H-abstraction was one order of magnitude lower than that observed in conventional solvents. The reaction might have involved a large change in geometry, which altered the electrostatic interaction between the solvents ions and resulted in raising the activation energy. Although ionic liquids are generally more stable than common solvents, this result indicates that they are not inert under certain extreme conditions.

Another study concerning amine mediated photoreduction of benzophenones in imidazolium-based ionic liquids showed that benzhydrol was the only detectable product. (16) This result was in contrast to those observed in non-imidazolium-based ionic liquids and conventional solvents where benzpinacol was the only product. The requirement of imidazolium cation for the formation of benzhydrol suggested that the solvent cation was involved in the reaction mechanism. (16)

Besides participating directly in the chemical reactions, ionic liquids can influence the reactions physically. It has been reported that the high viscosity of the ionic liquids could decrease the reaction rates of the bimolecular processes in the photolysis of 9-CH$_3$An. (14) In other diffusion controlled photochemical processes of molecular couples such as the quenching of the phosphorescence of

[3]Bp* by naphthalene as well as photoinitiated Diels-Alder cyclization reaction between [1]O$_2$ and diphenylbenzofuran (DPBF), the high viscosity of the ionic liquids resulted in higher activation energies. (*11*)

The effect of viscosity on the reactions was also illustrated in the electron transfer process between the ruthenium tris(4,4'-bipyridyl)/methylviologen ([Ru(bpy)$_3$]$^{2+}$/MV$^{2+}$) couple in 1-butyl-3-methylimidazolium hexafluorophosphate ([bmim][PF$_6$]). (*17*) Although the solvent anions were able to reduce the electrostatic repulsion between the two reaction cations via a charge screening effect and thus to facilitate the reaction, the formation of the encounter complex was diffusion controlled and was significantly affected by the viscous ionic solvent. As a result, the overall reaction rate of the forward electron transfer from [Ru(bpy)$_3$]$^{2+}$ to MV$^{2+}$ was not enhanced compared to those in water and acetonitrile. On the formation of the final products ([Ru(bpy)$_3$]$^{3+}$ and MV$^{+}$), the electrostatic repulsion between [Ru(bpy)$_3$]$^{3+}$ and MV$^{+}$ within the encounter complex could be counteracted by the anions of the solvent, and the diffusion of the products could not compete with the back electron transfer from MV$^{+}$ to [Ru(bpy)$_3$]$^{3+}$. Consequently, this should give low cage escape efficiency. However, the latter process involved significant molecular reorientation that was limited by the viscous and significantly ordered environment and was thus the rate-determining step. It was suggested that the influence of the ionic liquid on the back electron transfer and final product formation was of similar extent. Hence, the alteration in the cage escape yield was not significantly different than those in less viscous solvents. The increase of entropy of the system implied that the formation of the encounter complex was accompanied with structure-breaking involving solvent ion freeing.

More recently, (i) the energy transfer between xanthone triplet and naphthalene, (ii) hydrogen transfer between xanthone triplet and diphenylmethane, (iii) quenching of singlet excited state 2,4,6-triphenylthiopyrylium ion (TPTP$^+$) by neutral, negative and positive quenchers, namely, biphenyl, I$^-$ and Co$^{2+}$, and (v) quenching of anthracene triplet by methylviologen (MV$^{2+}$) in [bmim]PF$_6$ illustrated that these diffusion controlled processes were about two orders of magnitude slower than those in conventional solvents due to the higher viscosity of ionic liquids. It was also found that the lifetime of triplet excited state of TPTP$^+$ was one order of magnitude longer than that in the conventional solvents. (*18*)

These studies provided critical information on the photochemical properties of ionic liquids and helped to better understand other chemical reactions conducted in these solvents. To explain all these results, it is essential to fully understand the nature of ionic liquids. It is also important, however, to further explore the potential of RTILs in new applications. Ionic liquids are capable of solubilizing a number of organic compounds to high concentrations.

(*4*) In some applications, this may be very beneficial. For example, in photodegradation reactions following first or higher order kinetics, increasing the initial concentration of reactants over a certain range will result in higher reaction rates. This is impractical in some cases when water is used as solvent because the solubility of some substances in water is very low. A good example is that of polycyclic aromatic hydrocarbons (PAHs), which are sparingly soluble in water. In general, the solubility of the PAHs in water decreases as the number of aromatic rings increases. Another concern is the environmental implications of VOCs. The use of VOCs has created major environmental problems due to their volatility. The search for replacements to these solvents is critical for the sustainable development. To be a candidate, ionic liquids need to be fully assessed from all aspects, particularly for their toxicity. While more studies are required to fully examine the environmental implications of ionic liquids, it is worthwhile to investigate such solvents in applications that utilize some of their key advantages.

In this study, a representative ionic liquid, [bmim][PF$_6$], is examined as an alternative reaction medium for the photodegradation of naphthalene, a relevant environmental pollutant. This ionic liquid was selected because it has been involved in a number of published research studies concerning RTILs. PAHs are environmental contaminants of great health concern due to their carcinogenic potency. They can be released into the environment as byproducts of the incomplete combustion of fossil fuels and by industrial waste discharge. Natural processes, such as volcanic eruptions, are another source for the formation of PAHs. U.S. EPA is regulating certain PAHs (i.e., benzo(a)-pyrene) and has listed some others (i.e., naphthalene, anthracene and pyrene) as priority pollutants. (*19*) Among PAHs, naphthalene has the simplest structure. We thus selected this compound as a probe to investigate the potential of photodegrading PAHs in ionic liquids. This experimental approach can be further developed into a two-step process for the remediation of soils or dredged sediments contaminated by organic compounds such as PAHs, polychlorinated biphenyls (PCBs), and pesticides. The first step is extraction of the organic pollutants from the solid matrix using ionic liquids as extractants. The subsequent step is the *in-situ* photodegradation of the pollutants in the ionic liquid phase with simultaneous regeneration of the extractant phase. Here, we report the photodegradation of naphthalene in [bmim][PF$_6$]. The influence of the impurities present in the ionic liquid and the role of oxygen on the reaction rates were also investigated.

## Experimental

Naphthalene (99+%) was obtained from Aldrich and used as received. [bmim][PF$_6$] (97%, w/w) was purchased from Sachem, Inc. (Austin, TX). This ionic liquid was either used as received or purified with activated carbon. The purification was achieved using FILTRASORB 400 activated carbon obtained from Calgon Carbon Corp. (Pittsburgh, PA). The activated carbon was washed with water and then dried at 105 °C overnight before use. It has been reported that ionic liquids can absorb moisture from atmosphere, and that equilibrium can be reached given enough time. (20, 21) Aqueous and ionic liquid solutions of naphthalene were prepared by dissolving the compound in double deionized water (18 MΩ) and [bmim][PF$_6$], respectively. Photodegradation was conducted in an approximately 20 mL cylindrical quartz photoreactor with reaction space of 5 mL. Two opposite UV-C sources generated predominantly 253.7 nm UV radiation. The solutions were mixed with a magnetic stirrer during the photodegradation process. To investigate the effect of oxygen on the reactions, the ionic liquid containing naphthalene was first bubbled with pure oxygen (dry; water content 1 ppmv) or nitrogen (dry; water content 3 ppmv) for 1 hour. The solution was subsequently irradiated with UV-C radiation while oxygen or nitrogen bubbling was continued during the irradiation. It should be noted that the 2 hours of bubbling could result in less than 1% loss of naphthalene concentration in the ionic liquid when the initial concentration was 1.11 mM. In water, 1 hour bubbling could reduce the naphthalene concentration to zero when the initial concentration was 0.16 mM.

The concentrations of naphthalene in both water and ionic liquid were quantified using a Series 1100 HPLC (Agilent) equipped with a reverse phase amide column (RP-16 Discovery Supelco) and a UV-Vis Diode Array Detector. Water samples were injected directly. Ionic liquid samples were dissolved in acetonitrile prior to injection into the HPLC. The mobile phase used was a mixture of acidic water (0.01N sulfuric acid) and acetonitrile. The flow rate of the mobile phase was 1.5 mL/min. Identification of intermediates was conducted using both Q-TOF II MS (Waters) and HP 6890 series GC-MS (Hewlett-Packard) equipped with a Zebron ZB-5 column (15m×0.25mm×0.25μm) and a single quadruple mass analyzer. Electrospray source was used for Q-TOF II MS. Since compounds like naphthalene are difficult to be detected in Q-TOF II MS because they are nonpolar, coupling of GC-MS and Q-TOF II MS can avoid the overlooking of possible products. Helium was used as the carrier gas for GC-MS. The oven temperature was increased from the initial value of 40 °C to 170 °C at a rate of 10 °C/min and was then hold for 1 min. The temperature was then increased to 300 °C at a rate of 20 °C/min. Samples were dissolved in acetonitrile. The results of purification of the ionic liquid were obtained by comparing the absorbance of

[bmim][PF$_6$] before and after the treatment recorded with Diode Array UV-Vis Spectroscopy (Hewlett-Packard 8452A). The samples were prepared by dissolving the ionic liquid in acetonitrile and the path length of the cuvette in which the sample was contained was 1 cm. The ionic liquid treated by activated carbon was first filtered using 0.1 μm filter membranes prior to UV-Vis analysis. For the examination of the stability of the ionic liquid, 250 MHz $^1$H NMR (Bruker AC-250) was employed. The samples were prepared by dissolving the ionic liquid in CDCl$_3$. Tetramethylsilane (TMS) was used as the internal standard.

**Results and Discussion**

As illustrated in the HPLC chromatograms in Figure 1, the height of the peak of naphthalene decreased with extending treatment time. Figure 2 shows the reduction in the concentration of naphthalene in the ionic liquid as a function of irradiation time. The results show that naphthalene is degraded under UV-C radiation while no change in naphthalene concentration was observed for the control experiment in the absence of UV radiation. When the initial concentration was 0.16 mM, the transformation rate of naphthalene in the first hour of the reaction was $6.8 \times 10^{-4}$ mM min$^{-1}$. This rate increased to $1.9 \times 10^{-3}$ and $2.5 \times 10^{-3}$ mM min$^{-1}$ at initial concentrations of naphthalene of 0.41 and 1.11 mM, respectively. The removal efficiency (i.e., % reduction of the initial amount) of naphthalene over six hours of irradiation decreased with an increase in the initial concentration. The corresponding values at these three initial concentrations of naphthalene (0.16, 0.41 and 1.11 mM) were 90%, 86% and 76%, respectively. This is expected since the input photon flux was the same in all three cases. After six hours of irradiation, the final concentration of naphthalene in the three cases was 0.016, 0.057, and 0.27 mM, respectively.

Oxygen has been observed to significantly influence the photo-transformation of naphthalene. In the presence of oxygen, the color of the solution did not change after the six-hour irradiation process. On the other hand, in the experiment involving purging with nitrogen (i.e., anoxic conditions), the color of the solution turned to brownish after UV-C radiation. In this series of experiments, reaching anoxic or almost anoxic conditions when purging the ionic liquid with nitrogen for prolonged time is an assumption since dissolved oxygen measurements were not performed. Figure 3 shows the initial reaction rates of naphthalene degradation in the presence and absence of oxygen. It should be noted that the small loss of naphthalene due to volatilization was taken into account when calculating the initial reaction rates. When the solution was exposed to air (equilibration with air), the transformation rate of naphthalene was approximately two times higher than that obtained under anoxic conditions. Purging the solution with pure oxygen resulted in only a small enhancement of the reaction rate. This suggests that the

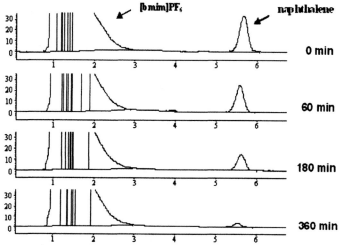

**Figure 1** *Degradation of naphthalene with initial concentration of 0.41 mM in [bmim][PF$_6$] under UV-C radiation based on the HPLC chromatograms*

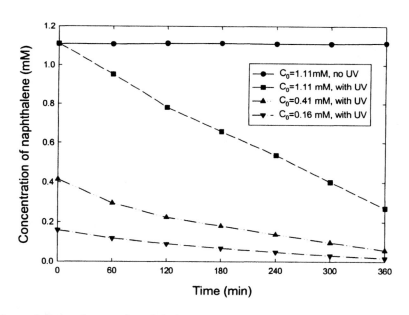

**Figure 2** *Degradation of naphthalene with different initial concentrations (C$_0$) in [bmim][PF$_6$] under UV-C radiation.*

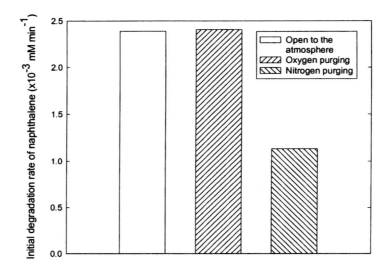

**Figure 3** *Effect of oxygen on the initial photodegradation rates of naphthalene in [bmim][PF₆] under UV-C radiation (Note: The concentration reduction due to volatilization of naphthalene during purging was taken into account).*

oxygen concentration in the ionic liquid is high enough or the transfer rate of oxygen from the atmosphere into the ionic liquid phase is sufficiently fast to compensate the consumption of oxygen during the photochemical reactions.

GC-MS analysis indicated that the possible major reaction intermediate of the reaction in the absence of oxygen was bicycle[4,2,0]octa-1,3,5-triene, as shown in Figure 4. This compound had a mass to charge (m/z) value of 104, and the fragmentation of 78 was observed. The compound was not observed in Electrospray TOF MS spectrum, which suggests its nonpolar nature. The same product was also found in GC-MS analysis when oxygen was present. Electrospray TOF MS analysis revealed that 2'-hydroxyacetophenone and phenol were also possible intermediates in the presence of oxygen, as shown in Figures 5 and 6. These two compounds showed m/z values of 137.1 and 95.1, respectively, as a consequence of addition of a proton to each molecule. The m/z values of 136 and 94 were observed in GC-MS, and fragmentation of 121 and 66 were also found for two compounds, respectively. It should be noted that bicycle[4,2,0]octa-1,3,5-triene, 2'-hydroxyacetophenone and phenol have also been reported as intermediate degradation products of naphthalene photodegradation in water. *(22)* In the present study, other low molecular weight compounds were detected in all the experiments but could not be identified.

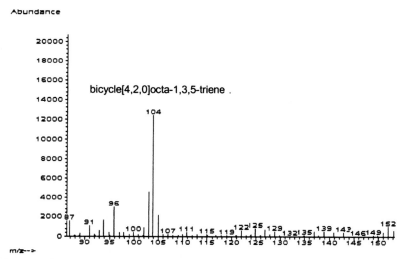

**Figure 4** *Formation of bicycle[4,2,0]octa-1,3,5-triene in [bmim][PF₆] under UV-C radiation in the absence of oxygen based on the GC-MS chromatogram.*

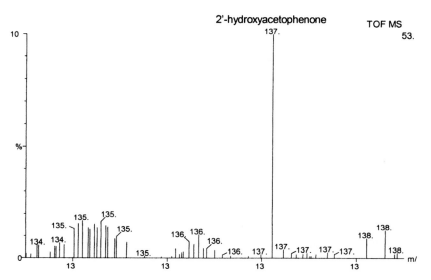

**Figure 5** *Formation of 2'-hydroxyacetophenone in [bmim][PF₆] under UV-C radiation in the presence of oxygen based on the electrospray TOF MS chromatogram.*

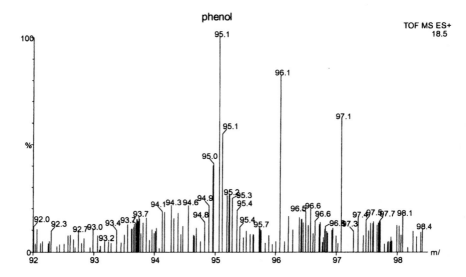

**Figure 6** *Formation of phenol in [bmim][PF₆] under UV-C radiation in the presence of oxygen based on the electrospray TOF MS chromatogram.*

The possible reaction mechanism is that in the presence of oxygen, 1-naphthol is first formed. (22, 23) However, it may also react fast to yield other byproducts and may not accumulate at detectable concentrations. As a result, it could not be detected in the MS spectra obtained in this study. 1-naphthol has been reported to be able to abstract hydrogen from H-donors. (24) Chemical interaction of 1-naphthol with the ionic liquid to generate other byproducts was not confirmed in this work but it is not excluded. Two reaction pathways may take place in the presence of oxygen: (i) addition of oxygen to the ring, and (ii) direct photolytic destruction of the ring. These two pathways are illustrated in Scheme 1.

In the presence of nitrogen, the rupture of the ring due to absorption of high energy UV radiation was predominant, and this resulted in the formation of bicycle[4,2,0]octa-1,3,5-triene and other low molecular weigh compounds as illustrated in Scheme 2. The color change of the solution indicates some reactive species might be formed and their reactions resulted in the formation of brownish compounds.

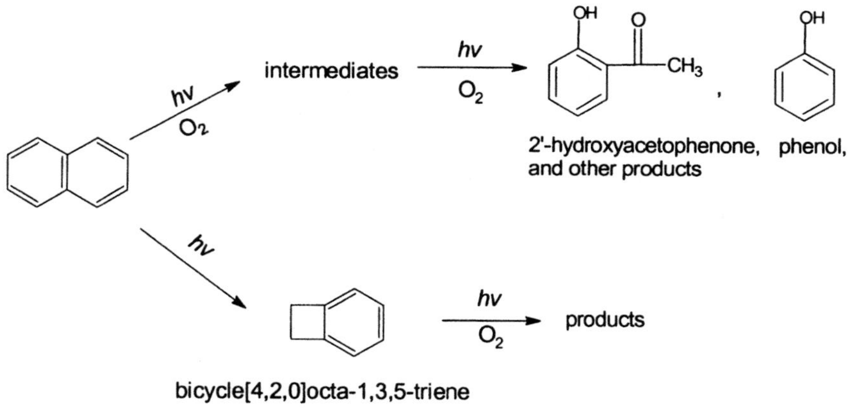
hv
O2
intermediates
hv
O2
OH
O
CH3
2'-hydroxyacetophenone, and other products
OH
phenol,
hv
hv
O2
products
bicycle[4,2,0]octa-1,3,5-triene

*Scheme 1*

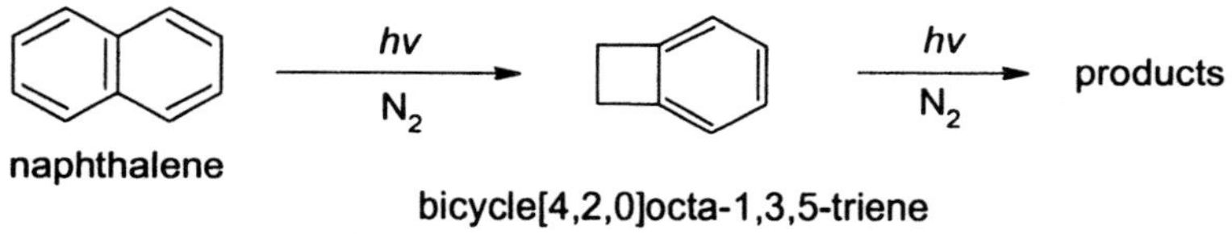
naphthalene
hv
N2
bicycle[4,2,0]octa-1,3,5-triene
hv
N2
products

*Scheme 2*

Under the same experimental conditions, the photodegradation of naphthalene in water was performed and compared to that in the ionic liquid. As shown in Figure 7, at similar initial concentrations (0.16 mM), the photo-transformation rate of naphthalene was significantly higher in water. The rate in the first hour of the reaction was $2.4\times10^{-3}$ mM $min^{-1}$, compared to $6.8\times10^{-4}$ mM $min^{-1}$ in the ionic liquid. This rate can be reached in the ionic liquid by using a higher initial concentration of 1.11 mM, as indicated above. The reduction of the reaction rate in the ionic liquid may be due to the presence of impurities that dye the solvent yellowish. It has been reported that high quality ionic liquids based on $[bmim]^+$ cation and anions including $[PF_6]^-$, $[BF_4]^-$, $[CF_3SO_3]^-$, $[CF_3CO_2]^-$ and $[(CF_3SO_2)_2N]^-$ are colorless, although they are not 100% pure. (*25*) Less pure ionic liquids exhibit color ranging from yellowish to orange. The formation of the color can be attributed to excessive heating during the synthesis of imidazolium salt. (25) Impurities have a considerable effect on the physical properties of ionic liquids. The presence of water can decrease the viscosity of the ionic liquids. (20, 26) On the other hand, chloride impurity makes the solvents more viscous. In photochemistry, impurities in solvents can alter the reaction mechanism and make the results unpredictable. (13) Another major concern for the impurities is their color. Depending on the wavelength of the incident light, the color can interfere with the penetration of light in the solvents and attenuate the effective photon flux used in the reaction. This will result in a decrease in the reaction rate. For this reason, the availability of spectroscopic grade ionic liquids in studies dealing with photochemistry in ionic liquids is of great importance.

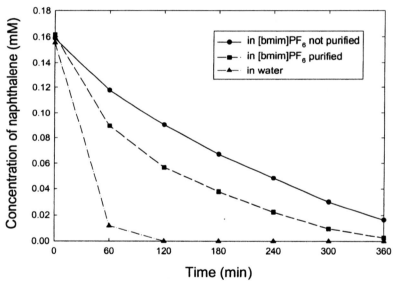

**Figure 7** Degradation of naphthalene at the same initial concentration in [bmim][PF$_6$] and in water under UV-C (253.7 nm) radiation.

A number of precautions have been described for the synthesis of colorless ionic liquids, (11) and a purification procedure using acidic alumina and activated charcoal has been developed for removing the color from impure ionic liquids. (25) In this work, we also used activated carbon to purify the ionic liquid. The yellowish color of the ionic liquid faded after prolonged treatment with activated carbon. Figure 8 presents results from UV-Vis spectroscopic analysis of the ionic liquid before and after purification. It is clearly seen that the absorbance of the ionic liquid in the region between 245 nm and 300 nm was reduced after purification.

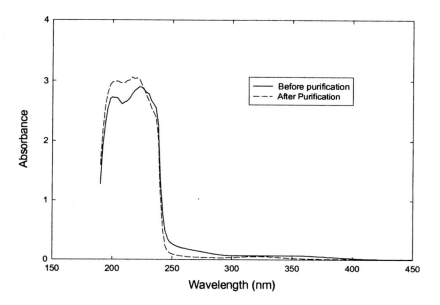

***Figure 8*** *Absorbance of non-purified and purified [bmim][PF₆] (Note: samples were dissolved in acetonitrile).*

Since the UV sources used in our study generate radiation (253.7 nm) mainly within this region, it is reasonable to expect a higher reaction rate by using purified ionic liquid. The degradation of naphthalene in the purified ionic liquid is also shown in Figure 7. Compared to that in the non-pure ionic liquid, the degradation rate increased. However, this rate is still lower than that in water. These results suggest that some factors other than the impurities also affect the photodegradation process. The high viscosity of [bmim][PF₆] may play a role in the reaction. It has been reported that the viscosity of [bmim][PF₆] (397 cP, water equilibrated at 25 °C) and some other ionic liquids are at least two orders of magnitude higher than that of water (0.8904 cP at 25 °C). (*20*) These highly viscous solvents could lower the reaction rates of bimolecular processes (*11, 14, 15, 17, 18*) and the second order reaction rate constants of the reactions involving $CCl_3O_2^{\bullet}$ radicals and chlorpromazine as well as $CCl_3O_2^{\bullet}$ radicals and Trolox compared to those observed in water. (27, 28) This suggests that in addition to the direct photolytic destruction of the parent or other intermediate compounds, other diffusion-controlled reactions also contribute to the degradation process.

The stability of the ionic liquid under UV-C radiation was also investigated. Examined by $^1$H NMR (250 MHz), the molecular structure and bonding of [bmim][PF$_6$] before and after irradiation with UV-C radiation for 360 min did not show any detectable destruction of alteration. However, since NMR is sensitive only when the conversion of ionic liquids is above 1%, (29) further verification is required. Such studies are underway and will be reported in the future.

## Conclusions

This study showed that degradation of naphthalene in [bmim][PF$_6$] using UV-C radiation can take place. Two possible pathways during photolytic transformation of naphthalene were (i) ring rupture due to direct photolysis, and (ii) oxygen involving photodegradation reactions. Compared to water, the photodegradation of naphthalene in [bmim][PF$_6$] was slower most probably due to diffusion-controlled photodegradation reactions as it was also expected from the much higher viscosity of [bmim][PF$_6$]. The photodegradation reactions were much faster in the presence of oxygen compared to anoxic conditions, which also supports the existence of the second pathway reported above. The presence of UV-C-absorbing organic impurities in [bmim][PF$_6$] had a dramatic negative effect on the photodegradation rates most probably due to light attenuation effects and additional competition between naphthalene and impurities for photon energy absorption. The ionic liquid was found to be stable to some extent under the experimental conditions applied in this work. This resistance to photodegradation is not expected from conventional volatile organic solvents, which are known to degrade under such conditions.

In preparation of this manuscript, the toxic potent of [bmim][PF$_6$] has been reported. The purification process in the synthesis of this ionic liquid can result in the decomposition of [bmim][PF$_6$] to form 1-butyl-3-methylimidazolium fluoride hydrate and a possible toxic product, HF. (30) Although the photodegradation of naphthalene in [bmim][PF$_6$] has been demonstrated to be feasible, the possible toxicity of this ionic liquid may cause problems when it is used for environmental remediation. For that purpose, ionic liquids which give no or acceptable impact on the environment are required.

Based on this work, with the use of environmentally friendly ionic liquids that are resistant to UV photolysis or other harsh oxidation conditions, this method, after further study and optimization, may have the potential for further development into a two-step remediation process. The first step concerns extraction of organic pollutants from a contaminated solid matrix using ionic liquids. In a second step, the extracted organic pollutants are destroyed *in-situ* in the ionic liquid solvent with the simultaneous regeneration of the solvent. However, further fundamental studies are required to fully understand the implications of the ionic liquid solvent in this photodegradation process.

## Acknowledgements

We are thankful to Dr. Koka Jayasimhulu and Dr. Stephen Macha (Department of Chemistry, University of Cincinnati) for identification of intermediates using ElectroSpray-TOF and Dr. Elwood Brooks (Department of Chemistry, University of Cincinnati) for his assistance with the NMR analysis. We are also grateful to the National Science Foundation (Grant CTS-0086725) and the NOAA/UNH Cooperative Institute for Coastal and Estuarine Environmental Technology (CICEET) for providing financial support to this project. Qiaolin Yang is also grateful to Sigma Xi Scientific Society for a Grant-in-Aid Research Scholarship.

## References

1. Wilkes, J. S. *Green Chem.*, **2002**, 4, 73.
2. Seddon, K. R. *J. Chem. Tech. Biotechnol.*, **1997**, 68, 351.
3. Earle, M. J. and Seddon, K. R. *Pure Appl. Chem.*, **2000**, Vol. 72, No. 7, 1391.
4. Huddleston, J. G. Willauer, H. D. Swatloski, R. P. Visser, A. E. and Rogers, R. D. *Chem. Commun.*, **1998**, 1765.
5. Visser, A. E Swatloski, R. P. Reichert, W. M. Mayton, R. Sheff, S. Wierzbicki, A. Davis, J. H. and Rogers, R. D. *Chem. Commun.*, **2001**, 135.
6. Bonhôte, P. Dias, A.-P. Papageorgiou, N. Kalyanasundaram, K. and Grätzel, M. *Inorg. Chem.*, **1996**, 35, 1168.
7. Fuller, J. Breda, A. C. and Carlin, R. T. *J. Electroanal. Chem.*, **1998**, 459, 29.
8. Welton, T. *Chem. Rev.*, **1999**, 2071.
9. Holbrey, J. D. and Seddon, K. R. *Clean Products and Processes*, **1999**, 1, 223.

10. Fadeev, A. G. and Meagher, M. M. *Chem. Commun.*, **2001**, 295.
11. Gordon, C. M. McLean A. J. Muldoon M. J. and Dunkin I. R. in *Ionic Liquids- Industrial Applications for Green Chemistry*; Editors, Rogers, R. D. and Seddon, K. R.; ACS Symposium Series 818; American Chemical Society, **2002**; 428.
12. Chum, H. L. Koran, D. and Osteryoung, R. A. *J. Am. Chem. Soc.*, **1978**, 100, 310.
13. Hondrogiannis, G. Lee, C. W. Pagni, R. M. and Mamantov, G. *J. Am. Chem. Soc.*, **1993**, 115, 9828.
14. Lee, C. Winston, T. Unni, A. Pagni, R. M. and Mamantov, G. *J. Am. Chem. Soc.*, **1996**, 118, 4919.
15. Muldoon, M. McLean, A. J. Gordon, C. M. and Dunkin, I. R. *Chem. Commun.*, **2001**, 2364.
16. Reynolds, J. L. Erdner, K. R. and Jones, P. B. *Org. Lett.*, **2002**, Vol. 4, No. 6, 917.
17. Gordon, C. M. and McLean, A. J. *Chem. Commun.*, **2000**, 1395.
18. Álvaro, M. Ferrer, B. García, H. and Narayana, M. *Chemical Physics Letters*, **2002**, 362, 435.
19. http://oaspub.epa.gov/wqsdatabase/wqsi_epa_criteria.rep_parameter
20. Huddleston, J. G. Visser, A. E. Reichert, W. M. Willauer, H. D. Broker, G. A. and Rogers, R. D. *Green Chem.*, **2001**, 3, 156.
21. Anthony, J. L. Maginn, E. J. and Brennecke J. F. *J. Phys. Chem. B*, **2001**, 105, 10942.
22. Tuhkanen, T. A. and Beltrán, F. J. *Chemosphere*, **1995**, Vol. 30, No. 8, 1463.
23. McConkey, B. J. Hewitt, L. M. Dixon, D. G. and Greenberg, B. M. *Water, Air and Soil Pollution*, **2002**, 136, 347.
24. Payne, J. R. and Phillips, C. R. *Environ. Sci. Technol.*, **1985**, Vol. 19, No. 7, 569.
25. Farmer, V. and Welton, T. *Green Chem.* **2002**, 4, 97.
26. Seddon, K. R. Stark, A. and Torres M.-J. *Pure Appl. Chem.*, **2000**, Vol. 72, No. 12, 2275.
27. Behar, D. Gonzalez, C. and Neta, P. *J. Phys. Chem. A*, **2001**, 105, 7607.
28. Behar, D. Neta, P. and Schultheisz, C. *J. Phys. Chem. A*, **2002**, 106, 3139.
29. Allen, D. Baston, G. Bradley, A. E. Gorman, T. Haile, A. Hamblett, I. Hatter, J. E. Healey, M. J. F. Hodgson, B. Lewin, R. Lovell, K. V. Newton, B. Pitner, W. R. Rooney, D. W. Sanders, D. Seddon, K. R. Sims, H. E. and Thied, R. C. *Green Chem.*, **2002**, 4, 152.
30. Swatloski, R. P. Holbrey, J. D. and Rogers, R.D. *Green Chem.*, **2003**, 5, 361.

Chapter 16

# Brønsted Acid-Base Ionic Liquids as Fuel Cell Electrolytes under Nonhumidifying Conditions

Md. Abu Bin Hasan Susan[1,2], Akihiro Noda[1], and Masayoshi Watanabe[1,*]

[1]Department of Chemistry and Biotechnology, Yokohama National University, 79–5 Tokiwadai, Hodogaya-ku, Yokohama 240–8501, Japan
[2]Permanent address: Department of Chemistry, University of Dhaka, Dhaka 1000, Bangladesh

Brønsted acid–base ionic liquids, derived from a simple combination of a wide variety of organic amines with bis(trifluoromethane sulfonyl)imide are proton-conducting and show electroactivity for $H_2$ oxidation and $O_2$ reduction at a Pt electrode under nonhumidifying conditions. Protic ionic liquids of this variety can be used as electrolytes for fuel cells operating at elevated temperatures.

## Fuel Cells — A Brief Introduction

The recent trend towards environmentally benign power sources for vehicles in urban environment (*1*) and on-site-use electricity/heat cogeneration systems has shifted the bias from conventional fuels and heat engine technology such as the internal combustion engines or the gas turbine and resulted in surge of interest in fuel cells. Fuel cells are electrochemical devices in which the free energy of a chemical reaction is converted directly into electrochemical energy without messy combustion. The efficiency achieved by the fuel cells is high since, unlike the heat engines, the devices utilize the electrochemical reaction that usually occurs when hydrogen combines with oxygen to generate water and the process is not governed by Carnot's law. The high thermodynamic efficiency and low or 'zero' tail pipe emission make fuel cells viable power sources for many applications, including ground transport, distributed power generation and portable electronics.

While there are many different configurations of such fuel cells, the polymer electrolyte fuel cell (PEFC) seems to be the most suitable for terrestrial transportation applications. Figure 1 shows the schematic diagram of a PEFC. A key part of the cell is the polymer electrolyte membrane, which in most cases is a perfluoro sulfonate ionomer, typically Nafion®. While the Nafion® based PEFCs offer several advantages like high ionic conductivity, good chemical stability, and excellent mechanical strength, their large-scale commercialization suffers from some major technological drawbacks as well as the high cost. The protons formed at the anode migrate towards the cathode under the influence of the electric field. Water molecules act as the proton carrier and proton is hopped from the protonated form ($H_3O^+$) to the neutral molecule ($H_2O$). This necessitat-

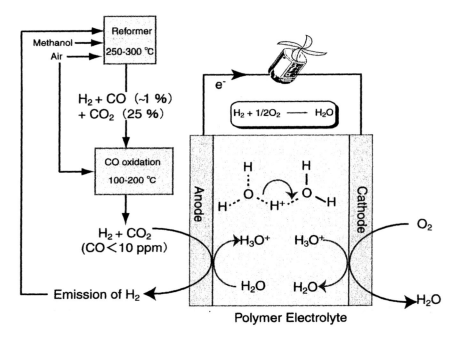

*Figure 1. Schematic diagram of a polymer electrolyte fuel cell.*

es humidification of the membrane. Insufficient water causes a loss of conductivity and can also result in localized heating and failure of the membrane. At the other extreme too much water causes flooding at the cathode, hindering gas transport to the membrane/electrode interface. The water profile inside the membrane is determined by the extent of humidification, the amount of water accompanying the protons as they migrate from anode to cathode (the electro-osmotic drag), the amount of water produced at the cathode, and the extent of back-diffusion of water through the membrane as the result of the concentration gradient. Water management is, therefore, rather difficult.

Fuel is another major hurdle. Hydrogen obtained from a fuel reformer tends to contain trace amounts of CO, which function as a catalyst poison to reduce the catalytic activity of Pt for the anode reaction. Hydrogen gas with a very high grade of purity should therefore be used, although the onboard storage of hydrogen is problematic. An alternative is the elevated temperature operation (intermediate temperature range of 120–200 °C) (*2, 3*), which is limited to ca. 80 °C in the conventional PEFCs, above which the evaporation of water causes significant decrease in ionic conductivity.

## Ionic Liquids: Unique Properties for Fuel Cell Applications

The realization that the constraints, as discussed in the preceding section, are primarily associated with the limited temperature range of operation due to the use of water led to focus on investigation of a system where the water dependence will be less marked. Anhydrous, highly proton-conducting polymer electrolytes with high operational temperature are, therefore, critically sought. By operating above the boiling point of water (100 °C), the product water is produced as an easily manageable gas, and less expensive catalysts with superior performance at elevated temperatures can be used. It is, therefore, not surprising that significant attention has been paid to the investigation of anhydrous proton-conducting polymer electrolytes using phosphoric acid, imidazole, benzimidazole and pyrazole-based proton-conducting polymers, and even electrolytes where proton transport would proceed within hydrogen bonds fixed to a polymer backbone: a "polymer-bound proton solvent" (*4–6*). However, despite the phenomenological detail, there had been no report on the demonstration of positive fuel cell tests using these electrolytes. Although high ionic conductivity could be achieved in some cases, concerns regarding the relatively poorer thermal and electrochemical stability imply that we still have far to go to realize a "dream membrane" (*7*) for fuel cell applications.

Quite reasonably, attention has been focused on the ubiquitous ionic liquids, which have the potential of maintaining high ionic conductivity over a wide temperature range. Ionic liquids, due to, *inter alia*, their immeasurably low vapor pressure, and greater thermal and electrochemical stability are very promising as fuel cell electrolytes. Fuller and Carlin (8) recognized the unique physicochemical properties of the ionic liquids to be ideal for fuel cell applications and incorporated aprotic ionic liquids into Nafion® membranes. The ionic liquid–Nafion® polymer composite system appears to be quite complex and is dependent on a number of variables including the initial hydration of the Nafion®, soaking times, and temperature cycling. However, they have successfully demonstrated that the ionic liquids do not degrade the physical properties of the membrane at elevated temperatures and the ionic conductivity is retained as the composite is heated and cooled. Both the cation and anion of the ionic liquid can be chosen to impart hydrophilic or hydrophobic character to the liquid. This tuning of the miscibility behavior of ionic liquids is particularly important since it enables the optimization of performance through proper matching of the ionic liquid to the hydrophilic and hydrophobic regions inherent to the Nafion® polymeric structure. Recently, Doyle et al. (9) have shown that high proton conductivity is possible at elevated temperatures for Nafion® membranes swelled with aprotic ionic liquid solvents even under almost anhydrous condition however, the mechanism of proton conduction in these ionic liquid-based systems is not clear. MacFarlane and co-workers (10) have also identified a number of aprotic ionic liquids as candidates for impregnation into polymer membranes having hydrophilic groups (Nafion®) or a high degree of porosity to accommodate the electrolyte (porous polytetrafluoroethylene, Teflon®). These polymers also have high temperature stability however, the considerably lower values of conductivity achieved at high temperatures suggests that the proton conduction is unlikely to be facilitated in such cases.

From the viewpoint that the enhancement of conductivity is caused merely by ionic liquids and protons are immobile, it will be impracticable to use these systems for fuel cell applications. The ionic liquids either need to act as proton solvents or themselves be capable of conducting protons for their true use in proton conductors. This chapter addresses the perspective of the use of Brønsted acid–base ionic liquids, which constitute a very important subgroup of ionic liquids as fuel cell electrolytes (11). We discuss the proton acid–base chemistry and proton transport behavior and aim at evaluating the protic ionic liquids as electrolytes under nonhumidifying and elevated temperature conditions for a new mesothermal fuel cell (12–15).

## Ionic Liquids: Protic vs. Aprotic

Ionic liquids constitute two major classes: aprotic and protic. Most of the ionic liquids of interest are aprotic with the cation formed by the addition of a group other than a proton, usually an alkyl group to a base site on the parent base molecule (*16, 17*). Figure 2 shows some of the combinations of cations and anions for typical aprotic ionic liquids. The rupture of the carbon–nitrogen bond once formed is very difficult and the iconicity is guaranteed. Protic ionic liquids, on the other hand, are formed by the transfer of a proton from a Brønsted acid to a Brønsted base. Angell and co-workers (*18, 19*) have established that the drive to transfer a proton can be measured by the difference in the $pK_a$ values of the acid and the base. When acids and bases used are strong with large difference in $pK_a$, the transferred proton locates very strongly on the base and the reverse process does not occur with any significant probability before the decomposition temperature is reached and the protic ionic liquids under such condition are as true a salt as the aprotic ionic liquids.

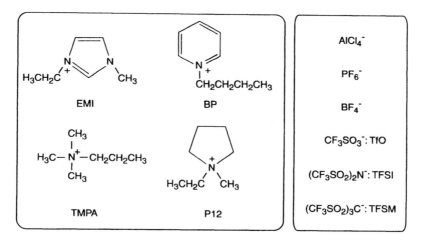

*Figure 2. Typical organic cations and counter anions for aprotic ionic liquids.*

The intriguing phenomenon of ionic liquid has been first reported for a protic ionic liquid, ethyl ammonium nitrate in 1914 (*20*). However, in reality the scientific community has experienced rapid progress in the development of aprotic ionic liquids. Despite the less attention to the protic systems, it is natural that imidazole and its derivatives, and some other amine compounds neutralized by a strong mono-protonic acid like bis(trifluoromethane sulfonyl)imide (*21*), HBF$_4$ (*22–24*), HPF$_6$, (*25*), trifluoroacetic acid (10), and trifluoromethane sulfo-

**Table I. Thermal Properties and Ionic Conductivities of the Neutral Salts of Organic Amines with HTFSI**

| Organic Amine (mp/ °C)[a] | Molar Ratio, [Amine]:[HTFSI] | $T_m$/°C[b] | $T_d$/°C[c] | σ at 130 °C/ $10^{-2}$ Scm$^{-1}$[d] |
|---|---|---|---|---|
| Pyrrolidine (-58) | 1:1 | 35.0 | 373 | 3.96 |
| Pyridine (-42) | 1:1 | 60.3 | 314 | 3.04 |
| Piperidine (-13) | 1:1 | 37.9 | 363 | 2.35 |
| Acridine (108) | 1:1 | 116.1 | 353 | —[e] |
| Butylamine (-50) | 1:1 | 16.2 | 352 | 1.04 |
| Dibutylamine (-62) | 1:1 | 42.6 | 325 | 1.26 |
| Triethylamine (-115) | 1:1 | 3.5 | 350 | 3.23 |
| Diphenylamine (52) | 1:1 | 51.5 | 193 | 0.85 |
| Imidazole (89) | 1:1 | 73.0 | 379 | 2.71 |
| Pyrazole (67) | 1:1 | 58.9 | 265 | 2.65 |
| Pyrazine (54) | 1:1 | 53.6 | 229 | 3.38 |
| Piperazine (108) | 1:1 | 172.7 | 358 | —[e] |
| Benzimidazole (173) | 1:1 | 101.9 | 368 | 1.31 |
| Morpholine (-7) | 1:1 | 58.5 | 349 | 1.08 |
| Quinoxaline (29) | 1:1 | 74.1 | 244 | 1.65 |
| 4,4'- Trimethylenedipyridine (57) | 1:2 | 62.0 | 386 | 1.05 |
| 4,4'- Trimethylenedipiperidine (66) | 1:2 | 167.3 | 403 | —[e] |
| 1,2,4- Triazole (119) | 1:1 | 22.8 | 287 | 2.20 |
| 1,2,3- Benzotriazole (96) | 1:1 | 136.6 | 230 | —[e] |

[a] Data in parentheses are melting points of organic amines. [b] Onset of an endotherm peak (melting point, Tm) during heating scans from −150 °C using differential scanning calorimetry (DSC). [c] Temperature of 10% weight loss during heating scans from room temperature using TG. [d] Ionic conductivity, σ has been determined by complex impedance method in the frequency range of 5 Hz to 13 MHz at AC amplitude of 10 mV. [e] Not measured.

nic acid (*10*) form neutral salts. Analyses of the melting point, viscosity and ionic conduction behavior of such neutral salts, some of which are liquid at ambient or sub-ambient temperatures, have been reported but no information on the proton conduction is available. It is our recognition (*12–15*) for the first time that the protic ionic liquids have the ability to conduct proton under anhydrous conditions and can serve as fuel cell electrolytes.

## Brønsted Acid–Base Systems as Ionic Liquids

A simple combination of a wide variety of organic amines (Table I) as Brønsted bases with a super-strong acid, bis(trifluoromethane sulfonyl)imide (HTFSI) under solvent-free conditions yields the protic ionic liquids of our interest. The preparation is rather simple: just to mix stoichiometric amounts of

HTFSI and amines without the aid of any solvent and to heat above the respective melting points in an inert atmosphere. Since the acids do not essentially have to be a strong one, a wide variety of acids may also be combined with different organic amines.

Table I (*12*) shows the thermal properties and ionic conductivities of a series of neutral salts. The neutral salts exhibit high thermal stability as an ionic liquid, for instance, the commencement of decomposition for the salt of 4,4'-trimethylene dipiperidine with HTFSI can be observed at temperatures >400 °C. Notably, the neutral salts of 1,2,4-triazole, triethylamine and butylamine are liquid at room temperature. Reflecting the characteristics of ionic liquids, the equimolar salts exhibit high ionic conductivity values.

The thermal behavior of the neutral salts with an added base or acid shows interesting changes. Figure 3 depicts a phase diagram of the imidazole (Im)–

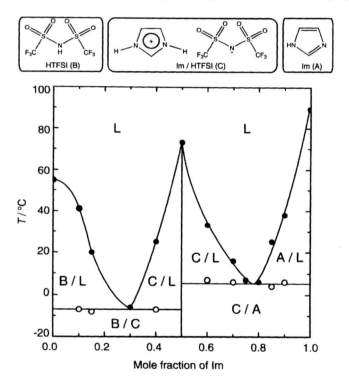

*Figure 3. Phase diagram of the system Im–HTFSI. A, B, C, and L refer to Im, HTFSI, Im/HTFSI equimolar salt, and liquid, respectively. (Reproduced with permission from Reference 13. Copyright 2003.)*

HTFSI system. Im, HTFSI and the 1:1 composition of Im–HTFSI are all solid at room temperature with melting point, $T_m$ as 89, 55, and 73 °C, respectively. Resembling the thermal behavior of typical Lewis acid–base ionic liquids (26), $T_m$ changes with change in the mole fraction in the Im–HTFSI system. There is a eutectic at ca. −6 °C for compositions between pure HTFSI and the equimolar salt and also between the equimolar salt and Im at ca. 6 °C. This is a typical phase diagram of binary mixtures (HTFSI and salt, salt and Im), where they form homogeneous liquids but are phase-separated in the solid state. Some of the compositions with certain molar ratios of Im and HTFSI are liquid at room temperature. At some HTFSI-rich compositions ([Im]/[HTFSI] = 25/75 and 2/8), the mixtures do not crystallize and form homogeneous glasses.

## Proton Transport in Brønsted Acid–Base Ionic Liquids

The composition dependence of ionic conductivity is rather interesting and informative for proton transport in such protic ionic liquid systems. Figure 4a (13) shows the composition dependence of ionic conductivity at certain temperatures for the Im–HTFSI system. The ionic conductivity above the melting point increases with increasing Im mole fraction with a sharp decrease for neat Im. This is in good agreement with Kreuer et al. in excess Im, but sharply contrasts with the protonic acid excess compositions (4). The ionic conductivity as can be seen in Figure 4b (13) for [Im]/[HTFSI] = 9/1 attains the value of ca. 0.1 Scm$^{-1}$ at 130 °C. Since the number of ion carriers, HIm$^+$ and TFSI$^-$, are maximum at the equimolar composition, the increase in ionic conductivity with increasing Im mole fraction indicates the enhanced ionic mobility. The conductivity is due not only to ionic species like HIm$^+$ and TFSI$^-$, but also to intermolecular proton transfer. The number of proton signals in the $^1$H NMR spectrum of a base-rich composition of Im/HTFSI using D$_2$O containing 1 wt % 2,2'-dimethyl-2-silapentane-5-sulphonic acid (DSS) as an external standard is three, although four different $^1$H nuclei exist in the system and the peaks corresponding to the proton bonded to the N–atom of Im and HIm$^+$ cannot be separated (13). This is due to fast proton exchange between Im and the HIm$^+$ ($>10^9$ s$^{-1}$) in Im excess compositions, which transcends NMR resolution.

The self-diffusion coefficients (Table II) determined by pulsed-gradient spin–echo (PGSE)-NMR measurements of some of the Im/HTFSI compositions also help to evidence proton transport in the system and envisage the proton conduction mechanism (13). The measurements of the self-diffusion coefficients for the proton attached to the nitrogen atom (NH Proton) and bonded to the carbon atom of Im (CH proton), and the fluorine atom of TFSI, have been made by using the $^1$H and $^{19}$F nuclei, respectively. If a fast intermolecular proton transfer exists in the system, the self-diffusion coefficient of the NH proton of Im (HIm) should be larger than that of the CH proton (Im). In fact, the diffusion coefficient at 30 °C of HIm ($^1$H) has been found to be larger than the matrix diffusion coefficient of Im ($^1$H) for Im-rich compositions. This implies that

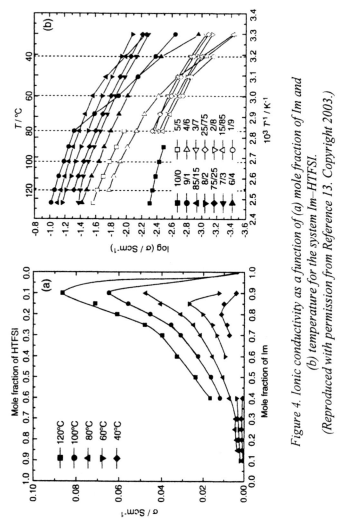

*Figure 4. Ionic conductivity as a function of (a) mole fraction of Im and (b) temperature for the system Im–HTFSI. (Reproduced with permission from Reference 13. Copyright 2003.)*

**Table II. Diffusion Coefficient for Im/HTFSI Compositions at 30 °C**

| [Im]/ [HTFSI] | $D / 10^{-7}\ cm^2s^{-1}$ | | | $D_{H^+}/$ $10^{-7}cm^2s^{-1}$ | $t_+$ | Grotthuss Contribution in proton transport (%) |
|---|---|---|---|---|---|---|
| | Im($^1$H)[a] | HIm($^1$H)[b] | TFSI($^{19}$F)[c] | | | |
| 8 / 2 | 5.7 | 6.3 | 3.7 | 8.1 | 0.69 | 30 |
| 7 / 3 | 3.8 | 4.1 | 2.5 | 4.5 | 0.64 | 16 |

[a] $^1$H bonded to carbon atoms of imidazole. [b] $^1$H attached to the nitrogen atom of imidazole. [c] $^{19}$F of TFSI.

excess Im in HIm$^+$ accelerates proton transfer dominantly via the Grotthuss mechanism in Im-rich compositions (4). The HIm$^+$ and TFSI$^-$ species also considerably influence the conduction behavior as the transport of the matrix following the vehicle mechanism.

The experimental diffusion coefficients of HIm ($^1$H) include both protonated Im (HIm$^+$) and neat Im. Thus, the H$^+$ diffusion coefficient is calculated by the experimental diffusion coefficients (Im($^1$H) and HIm ($^1$H)) with the component mole fraction (27). If the diffusion coefficient of Im ($^1$H) is assumed to be the same as the self-diffusion coefficient of the protonated imidazolium (vehicle mechanism), Grotthuss contribution to the proton transport for [Im]/[HTFSI] =8/2 and 7/3 can be calculated as 30 and 16 %, respectively. Furthermore, the apparent $^1$H transference number compared to TFSI($^{19}$F) is 0.69 and 0.64 for [Im]/[HTFSI] = 8/2 and 7/3, respectively (13). The increase in the Im mole fraction from the equimolar salt renders dominant conducting properties to change from the vehicle to the Grotthuss mechanism (4). The proton conduction, therefore, follows a combination of Grotthuss and vehicle-type mechanisms (28, 29) (Figure 5) with the contribution from the particular mechanism being a variable of the composition.

## Electrochemical Polarization and Fuel Cell Performance of Brønsted Acid–Base Ionic Liquids

The equimolar salts as well as the base-rich compositions are proton-conducting. This has been corroborated by conducting simple direct current polarization experiment using a U-shaped glass tube with two Pt-wire electrodes (proton pump cell) with the anode under H$_2$ or N$_2$ bubbling atmosphere for Im–HTFSI as well as some other systems. The current detected under N$_2$ atmosphere is quite low, whereas a noticeable change in every case is eminent upon change to H$_2$ gas atmosphere resulting in the observation of higher current (12–15). Furthermore, evolution of gas (H$_2$) is confirmed as bubbles at the cathode. A similar experiment conducted on an aprotic ionic liquid, 1-ethyl-3-

**Grotthuss mechanism (Proton hopping)**

**Vehicle mechanism (Matrix transport)**

*Figure 5. Illustration of proton and ion conduction mechanisms (28).*

methylimidazolium bis(trifluoromethane sulfonyl)imide, results in an imperceptible change in the current in $N_2$ to $H_2$ atmosphere, indicating that for proton conduction in neutral ionic liquids, the species should be a protic one.

The electrode reactions of the ionic liquid systems at the three-phase interface of the ionic liquid/Pt/$H_2$ or $O_2$, have been explored by conducting cyclic voltammetric (CV) measurements using a two-compartment glass cell under dry Ar, $H_2$ or $O_2$ gas bubbling atmosphere. Figure 6 (13) shows the CV behavior for the Im–HTFSI system under base-rich conditions. In these experiments, Pt and Pt-black electrodes immersed in ionic liquids have been used as a working electrode (W.E.) and a counter electrode (C.E.), respectively, and a Pt electrode with $H_2$ bubbling as a reference electrode (R.E.). When the W.E. is in Ar atmosphere, the voltammograms show remarkable reduction and oxidation currents at around 0 V and gas bubbles are observed on the W.E. in the reduction process. The potential of 0 V, therefore, corresponds to the hydrogen redox potential, and the R.E. can be considered as a reversible hydrogen electrode (RHE). Upon change in the working atmosphere from Ar to $H_2$ atmosphere, the results of CVs apparently change as shown in Figure 6. The steady-state $H_2$ oxidation on the Pt working electrode gave rise to the large oxidation currents. The $H_2$ bubbles could be visually observed on the counter electrode. These are comprehensible and decisive evidence of the proton conduction in the system. When the W.E. is under $O_2$ bubbling atmosphere, a radical change in the shapes of the CVs (Figure 6) is observed. The cathodic current is observed at a potential below 0.8 V $vs.$ RHE, and the waves, corresponding to the evolution and reoxidation of $H_2$, are overlapped below 0 V $vs.$ RHE. The cathodic current, judged from the potential, appears to be assigned to the 4-electron oxygen reduction reaction. The equimolar salts of Im (12) and some other amines with HTFSI also show similar behavior.

Figure 7 represents current $vs.$ potential ($vs.$ RHE) characteristics of $H_2/O_2$ fuel cells under nonhumidifying conditions for the neutral salt as well as base-rich compositions for the Im–HTFSI system. The measurements have been conducted at 130 °C for the neutral salt and at 80 °C for the Im-rich compositions with the working and counter electrodes under $O_2$ and $H_2$ bubbling atmosphere. The compositions have been found to be electroactive at a Pt electrode for $H_2$ oxidation and $O_2$ reduction. The electrode reactions (vide supra) can be summarized as:

Anode: $H_2 + 2Im \rightarrow 2HIm^+ + 2e^-$
Cathode: $\frac{1}{2}O_2 + 2ImH^+ + 2e^- \rightarrow H_2O + 2Im$
Net: $H_2 + \frac{1}{2} O_2 \rightarrow H_2O$

Although a potential drop with increasing current density is apparent, we have been able to demonstrate for the first time the performance of an $H_2/O_2$ fuel cell using a Brønsted acid–base ionic liquid as a proton-conducting nonaqueous electrolyte.

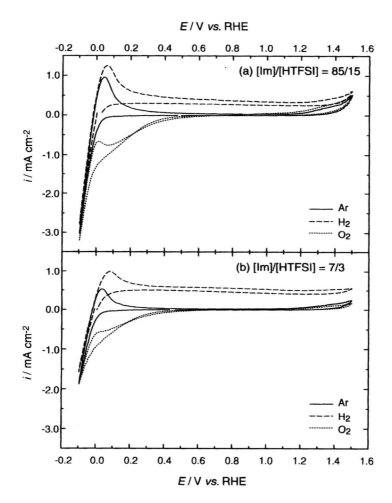

*Figure 6. Cyclic voltammograms for [Im]/[HTFSI] = (a) 85/15 and (b) 7/3 compositions at 80 °C. Scan rate is 50 mVs$^{-1}$. W.E. is a Pt wire in Ar, H$_2$ or O$_2$ atmosphere. C.E. is a Pt-black wire and R.E. is a PT wire in H$_2$ atmosphere. (Reproduced with permission from Reference 13. Copyright 2003.)*

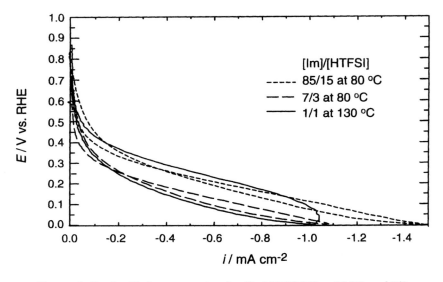

*Figure 7. Fuel cell characteristics for [Im]/[HTFSI] = 85/15 and 7/3 compositions at 80°C and for equimolar composition at 130 °C. W.E. is a Pt wire in O₂ atmosphere. C.E. is a Pt black wire in H₂ atmosphere and R.E. is a Pt wire in H₂ atmosphere. The cell potential is scanned from open circuit potential (OCP) to 0 V (vs. RHE) for the first scan followed by scanning to the OCP successively, at 10 mVs⁻¹. (Reproduced with permission from Reference 13. Copyright 2003.)*

## Structure Diffusion and "Good" Ionic Liquids: the Trade-off

In the present fuel cell system, the Im molecule functions not only as a proton carrier like water in acidic electrolytes but also as an $H^+$ donor and an acceptor for $O_2$ reduction and $H_2$ oxidation reactions, respectively. The major contribution in proton conductivity in the neutral salts comes from the translational dynamics of the protonated amines (vehicle mechanism) in the ionic liquids. For proton conduction in these systems by the Grotthuss mechanism, the protonated amines require free proton acceptor sites for proton exchange, as is evident for the systems with base-rich compositions. Interestingly, we have evidenced proton transport in some of the neutral salts without stoichiometric excess of free amines. As classified by Angell and co-workers (*18*), "poor" ionic liquids are formed by the combination of Brønsted acids and bases with a small difference in $pK_a$ values leading to an incomplete proton transfer, while for "good" ionic liquids the difference in $pK_a$ is very large and the amounts of free constituent acids and bases are virtually negligible for complete proton transfer. While "good" ionic liquids are desired for manifold applications, the hopping mechanism may be more pronounced for relatively "poorer" ionic liquids. Depending on the components of organic amines, the trace amount of the free base persisting from the equilibrium between neutral salt and the starting amine and HTFSI in our system can exhibit proton conductivity by the Grotthuss mechanism (14, 15). Whether or not the TFSI⁻ functions as a proton acceptor is not clear and investigation is underway.

To realize fast proton conductors, fast proton exchange between protonated and free amine by the Grotthuss mechanism should be ensured, which can be achieved by using base-rich compositions for protic ionic liquid systems. Proton conduction by structure diffusion involves intermolecular proton transfer and structural reorganization by hydrogen bond breaking and forming processes (4, 30) and naturally will be influenced by the geometry and difference in $pK_a$ values of the acid and base comprising an ionic liquid (15), rotation and local mobility of the Brønsted base in the system (5), temperature (4, 29) and composition (4, 13). Our current interest lies in the correlation of the factors governing proton transport in such Brønsted acid–base ionic liquid systems. If an optimum combination can be chosen retaining the unique properties of the ionic liquids, we will be able to realize fast and stable anhydrous proton conductors at elevated temperatures.

## Concluding Remarks

The concept of protic ionic liquid may be used with compatible polymers for the construction of solid-state proton conductors (31), and further, an acidic site and/or a basic site can be affixed to a polymer backbone. This finding heralds a new field of fuel cells under nonhumidifying conditions, which may be operated at temperatures above 100 °C and provide a firm underpinning for the development of solid-state anhydrous proton conductors. In addition, development of fundamental, molecular-based descriptions and models, will aid in tuning Brønsted acid–base ionic liquids to desirable chemistry for their true use as fuel cell electrolytes.

## Acknowledgments

This research was supported in part by Grant-in-Aid for Scientific Research (#14350452 and #16205024) from the Japanese Ministry of Education, Science, Sports, and Culture and by Technology Research Grant Program from the NEDO of Japan. M.A.B.H.S. also acknowledges a post-doctoral fellowship from JSPS. In addition, we thank Hirofumi Nakamoto for his support in conducting some experiments.

# References

1. Steele, B. C. H.; Heinzel, A. *Nature* **2001**, *414*, 345 and references therein.
2. Kreuer, K. D. *CHEMPHYSCHEM* **2002**, *3*, 771.
3. Yang, C; Costamagna, P; Srinivasan, S.; Benziger, J.; Bocarsly, A. B. *J. Power Sources* **2001**, *103*, 1.
4. Kreuer, K. D.; Fuchs, A.; Ise, M.; Spaeth, M.; Maier, J. *Electrochim. Acta* **1998**, *43*, 1281.
5. Schuster. M.; Meyer, M. H.; Wegner, G.; Herz, H. G.; Ise, M.; Schuster. M.; Kreuer, K. D.; Maier, J. *Solid State Ionics* **2001**, *145*, 85.
6. Savinell, R.; Yeager, D.; Tryk, D.; Landau, U.; Wainright, J.; Weng, D.; Lux, K.; Litt, M.; Rogers, C. *J. Electrochem. Soc.* **1994**, *141*, L46.
7. Wengenmayr, R. *MaxPlanckResearch* **2001**, *2*, 25.
8. Fuller, J.; Carlin, R. T. *Molten Salts XII*, Eds. Trulove, P. C; De Long, H. C.; Stafford, G. R.; Deki, S. Electrochem. Soc. Pennington, NJ, **1999**, *41*, 27.
9. Doyle, M.; Choi, S. K.; Proulx, J. *J. Electrochem. Soc.* **2000**, *147*, 34.
10. Sun, J.; Jordan, L. R.; Forsyth, M.; MacFarlane, D. R. *Electrochim. Acta* **2001**, *46*, 1703.
11. Freemantle, M. *C&E News* April 21, **2003**, *81 (16)* 48.
12. Susan, M. A. B. H.; Noda, A; Mitsushima, S.;Watanabe, M. *Chem. Commun.* **2003**, 938.
13. Noda, A; Susan, M. A. B. H.; Kudo, K.; Mitsushima, S.; Hayamizu, K.;Watanabe, M. *J. Phys. Chem. B* **2003**, *107*, 4024.
14. Susan, M. A. B. H.; Yoo, M.; Nakamoto, H.; Watanabe, M. *Chem. Lett.* **2003**, *32*, 836.
15. Susan, M. A. B. H.; Nakamoto, H.; Yoo, M.; Watanabe, M. *Trans. Mater. Res. Soc., Jpn* **2004**, *29*, in press.
16. Noda, A; Hayamizu, K.;Watanabe, M. *J. Phys. Chem. B* **2001**, *105*, 4603.
17. Xu, W.; Cooper, E. I.; Angell, C. A. *J. Phys. Chem. B* **2003**, *107*, 6170.
18. Angell, C.A.; Xu, W.; Yoshizawa, M.; Belieres, J. –P. in *Proceedings of the International symposium on ionic liquids in Honour of Marcelle Gaune-Escard* (Carry le Rouet, France, June 26–28, 2003); Eds. Oye, H. A.; Jagtoyen, A. 389–398.
19. Yoshizawa, M.; Xu, W.; Angell, C. A. *J. Am. Chem. Soc.* **2003**, *125*, 15411.
20. Walden, P. *Bull. Acad. Imper. Sci.* **1914**, 1800.
21. Yoshizawa, M.; Ogihara, W.; Ohno, H. *Electrochem. and Solid State Lett.* **2001**, *4*, E25.
22. Hirao, M.; Sugimoto, H.; Ohno, H. *J. Electrochem. Soc.* **2000**, *147*, 4168.
23. Christie, S.; Subramanian, S.; Wang, L.; Zaworotko, M. J. *Inorg. Chem.* **1993**, *32*, 5415.
24. Holbrey, J. D.; Seddon, K. R. *J. Chem. Soc., Dalton Trans.* **1999**, 2133.
25. Holbrey, J. D.; Seddon, K. R. *Clean Prod. Process.* **1999**, *1*, 223.
26. Wilkes, J. S.; Levinsky, J. A.; Wilson, R. A.; Hussey, C. L. *Inorg. Chem.* **1982**, *21*, 1263.

27. Dippel, Th.; Kreuer, K. D.; Lassègues, J. C.; Rodriguez, D. *Solid State Ionics* **1993**, *61*, 41.
28. Kreuer, K. D.; Rabenau, A.; Weppner, W. *Angew. Chem. Int. Ed. Engl.* **1982**, *21*, 208.
29. Kreuer, K. D. *Chem. Mater.* **1996**, *8*, 610.
30. Kreuer, K. D. *Solid State Ionics* **2000**, *136–137*, 149.
31. Noda, A.; Watanabe, M. *Electrochim. Acta* **2000**, *45*, 1265.

# Fundamentals

# Chapter 17

# Understanding Reactions in Ionic Liquids

Lorna Crowhurst, N. Llewellyn Lancaster, Juan M. Perez-Arlandis, and Tom Welton[*]

[1]Department of Chemistry, Imperial College of London, Exhibition Road, London SW7 2AY, United Kingdom
[*]Corresponding author: t.welton@ic.ac.uk

## Abstract

Since the discovery of organic salts that are liquid at room temperature and stable to both air and water; ambient-temperature ionic liquids have attracted growing interest as solvents for organic synthesis, using both stoichiometric and catalytic methodologies. Rather than taking a random approach to the selection of reactions for study in ionic liquids, we have been attempting to quantify ionic liquid effects, so that a rational approach to the selection of reactions can be used. In order to do this, we have needed to characterise the ionic liquids' abilities to interact with solute species and to investigate a number of reactions. We have chosen to select simple reactions, principally $S_N2$ processes, and to investigate the effect of the ionic liquids on their rates and to compare these results to those in molecular solvents. In this paper we summarise our findings.

## Introduction

Ionic liquids are being advanced as a class of "green" solvents that have the potential to be used in place of volatile organic solvents in synthesis [1]. Although estimates vary, there is no doubt that the number of combinations of anions and cations that will give rise to ionic liquids is vast. This possibility for synthetic variation has lead to ionic liquids being described as "Designer Solvents" [2]. However, in order to be able to realise this possibility it is necessary to know exactly what is being designed. Here we are interested in designing ionic liquids for increased rates of reaction. There is a vast range of

© 2005 American Chemical Society

potential ionic liquids; we have restricted ourselves to examples that are liquid at room temperature and give a variety of combinations of cation and anion. The ionic liquids used, together with abbreviations, are shown in Figure 1.

A number of organic reactions have been studied in ionic liquids, with varying degrees of success. This has most often involved taking a well-known reaction and performing it in an ionic liquid to see if it "goes". However, this form of study does little to explain how the use of ionic liquids can affect the reactions conducted in them, or how the ionic liquids might be more generally applied. This is in spite of the fact that the study of solvent effects on organic reactions is a well-established area of research [3]. Such studies are required to extrapolate the results to other solvents and to predict reaction outcomes (both rates and selectivities). That is, to select the right reactions to try in ionic liquids and the right ionic liquids to use for a given reaction. It is just as important to be able to identify reactions for which ionic liquids are likely to be disadvantageous and should be avoided.

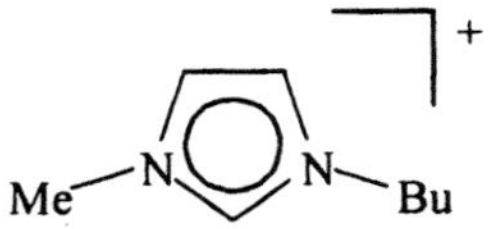

$[BF_4]^-$, $[PF_6]^-$, $[SbF_6]^-$, $[CF_3SO_3]^-$ ([TfO]$^-$), $[(CF_3SO_2)_2N]^-$ ([Tf$_2$N]$^-$)

[bmim]$^+$

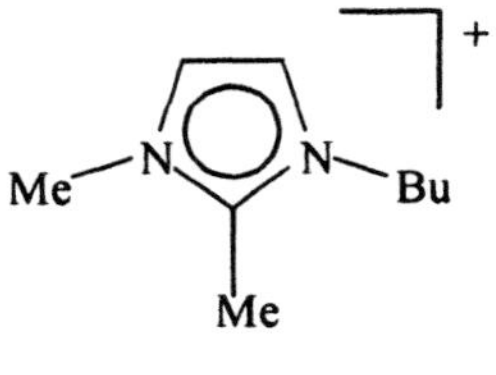

$[BF_4]^-$, $[(CF_3SO_2)_2N]^-$ ([Tf$_2$N]$^-$)

[bm$_2$im]$^+$

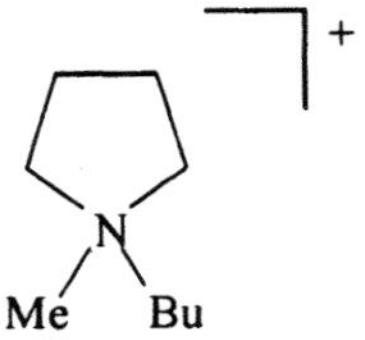

$[(CF_3SO_2)_2N]^-$ ([Tf$_2$N]$^-$)

[bmpy]$^+$

**Figure 1**: Ionic liquids used

It is well known that the outcome of a reaction is affected by the local microenvironment generated around a solute species by a solvent; changing the solvent can impact both on equilibria and rates [3]. Since ionic liquids have the potential to provide reaction media that are quite unlike any other available at room temperature, it is possible that they will have dramatic effects on reactions in them and there have been many claims of great improvements in reaction yields and rates when using ionic liquids [4]. In order to understand how such effects may arise we have chosen to study the rates of a range of relatively simple and well-understood reactions in a variety of ionic liquids and molecular solvents.

However, we must also consider how we are going to describe the ionic liquids. Unless we find some set of properties that can be measured, or predicted, against which the rates of the various reactions can be compared, we will generate a set of numbers that will ultimately fail us in our aims to achieve true design. These parameters must also be able to link the ionic liquids to the molecular solvents we are comparing them to.

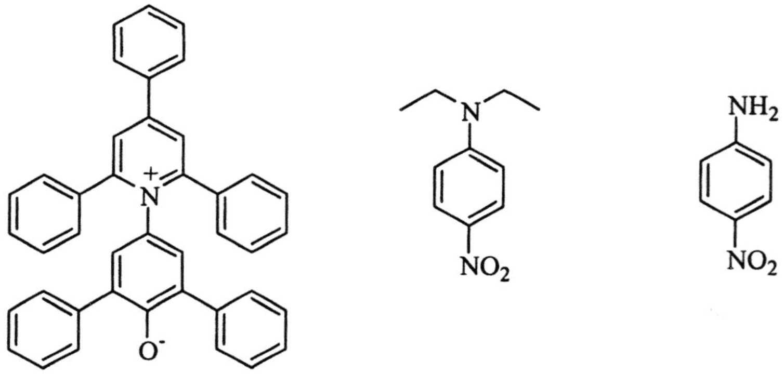

**Figure 2**: Reichardt's dye (A), *N*, *N*-diethyl-4-nitroaniline (B) and 4-nitroaniline (C)

The most important characteristics of a solvent are those that determine how it will interact with potential solutes. For molecular solvents, this is most commonly recorded as the polarity of the pure liquid, as expressed through its dielectric constant. A direct measurement of dielectric constant is not possible for conducting liquids and so is not available for the ionic liquids.

The Kamlet-Taft [4] system, based on the comparison of effects on the uv-vis spectra of sets of closely related dyes, was selected to probe particular solvent properties. These complimentary scales of hydrogen bond acidity ($\alpha$), hydrogen bond basicity ($\beta$) and dipolarity/polarizability effects ($\pi^*$) are the ones that we use here. It should be noted that these are empirically derived parameters and not fundamental molecular properties. The values can change according to the selection of dye sets or by the precise method used to calculate the values. We have chosen to use Reichardt's dye, 4-nitroaniline and $N$, $N$-diethyl-4-nitroaniline, which are shown in Figure 2.

The parameters $\pi^*$, $\alpha$ and $\beta$ were calculated using the following equations:

$$\pi^* = 0.314(27.52 - \nu_{N,\ N\text{-diethyl-4-nitroaniline}})$$

$$\alpha = -0.186(10.91 - \nu_{\text{Reichardt's dye}}) - 0.72\pi^*$$

$$\beta = (1.035\nu_{N,\ N\text{-diethyl-4-nitroaniline}} + 2.64 - \nu_{\text{4-nitroaniline}})/2.80$$

# Results and discussion

## Kamlet/Taft parameters

The, $\alpha$, $\beta$, and $\pi^*$ values are listed in Table 1. As far as possible, the literature values for molecular solvents included in the table (in parentheses) are only those obtained using the aforementioned dye set.

### $\pi^*$ values

All of the $\pi^*$ values (Table 1) for the ionic liquids are high in comparison with non-aqueous molecular solvents [5-7]. Although differences between the ionic liquids are small, both the cation and the anion can be seen to affect the value. All of the [Tf$_2$N]$^-$ based ionic liquids lie at the low end of the range of values observed. The [bmim]$^+$ ionic liquids have lower values than the [bm$_2$im]$^+$ ionic liquids with a common anion and the [bmpy][Tf$_2$N] ionic liquid has the lowest value of all.

In molecular solvents, $\pi^*$ reports the effects of the dipolarity and polarizability of the solvent. However, $\pi^*$ is actually a value derived from the change in the energy of the absorption maximum of $N$, $N$-diethyl-4-nitroaniline that is largely induced by the local electric field generated by the solvent. Therefore, it is no surprise that $\pi^*$ has been greatly affected by the ion-dye interactions now possible in the ionic liquid. It can be seen that as the charge on the anions becomes delocalised over more atoms the value of $\pi^*$ decreases, due to the decrease in the strength of these Coulombic interactions. In the cations, on the other hand, it appears that the decrease in Coulombic interactions caused by delocalising the charge around the imidazolium ring is

**Table 1**: Kamlet/Taft parameters for a selection of ionic liquids and molecular solvents

| Solvent | $\pi^*$ (Lit.) | $\alpha$ (Lit.) | $\beta$ (Lit.) |
|---|---|---|---|
| [EtNH$_3$][NO$_3$] [8] | (1.12) | (1.10[†]) | (0.46[§]) |
| [bmim][SbF$_6$] | 1.039 | 0.639 | 0.146 |
| [bmim][BF$_4$] | 1.047 | 0.627 | 0.376 |
| [bmim][PF$_6$] | 1.032 | 0.634 | 0.207 |
| [bmim][TfO] | 1.006 | 0.625 | 0.464 |
| [bmim][N(Tf)$_2$] | 0.984 | 0.617 | 0.243 |
| [bm$_2$im][BF$_4$] | 1.083 | 0.402 | 0.363 |
| [bmpy][N(Tf)$_2$] | 0.954 | 0.427 | 0.252 |
| [bm$_2$im][N(Tf)$_2$] | 1.010 | 0.381 | 0.239 |
| Water [6] | (1.33) | (1.12) | (0.41) |
| Methanol [6] | (0.73) | (1.05) | (0.61) |
| Acetonitrile | 0.799 | 0.350 | 0.370 |
| Acetone | 0.704 | 0.202 | 0.539 |
| Dichloromethane | 0.791 | 0.042 | -0.014 |
| Toluene | 0.532 | -0.213 | 0.077 |
| Hexane [6] | (-0.12) | (0.07) | (0.04) |

[†]recalculated, [§] average value from more than one dye set

more than compensated for by the increased polarizability of the delocalised system. The effect of delocalized $\pi$-electron systems can be seen in the relatively high $\pi^*$ value of toluene in comparison to hexane (Table 1).

It should be noted that the ionic liquids used in this study are all composed of relatively similar cations. Poole et al. have measured $\pi^*$ for a series of alkylammonium nitrates and thiocyanates (e.g. [EtNH$_3$][NO$_3$], Table 1), which have still higher values [8], albeit with a different set of dyes. This probably reflects the closeness of approach to the charge centres of these salts that is possible and hence greater ion-dye Coulombic interactions.

## $\alpha$ values

The $\alpha$ values for the ionic liquids are listed in Table 1. The values are largely determined by the nature of the cation, although there is a smaller anion effect. In general, the values for the [bmim]$^+$ are moderately high (c.f., $t$-BuOH) [5]. It has long been known that all three of the imidazolium ring protons are acidic [9, 10], and it was predicted that hydrogen bonding to solutes would be significant in the absence of hydrogen bond accepting anions [11]. The [bm$_2$im]$^+$ ionic liquids have the lowest $\alpha$ values, reflecting the loss of the proton on the 2-position of the ring. [bmpy][Tf$_2$N] has a slightly higher $\alpha$ value than [bm$_2$im][Tf$_2$N]. The higher $\alpha$ value of [EtNH$_3$][NO$_3$], which is an N-H hydrogen bond donor, shows that the opportunity arises to prepare strongly hydrogen bond donating ionic liquids. The hydrogen bond donor abilities of ammonium cations are known to be higher than that of their neutral amines [12], hence the appearance of a large Coulombic contribution to the $\alpha$ value.

Focusing on the [bmim]$^+$ salts, there is a clear anion effect seen. With the exception of the [N(Tf)$_2$]$^-$ ion, which appears not to fit the simple trend, as the anion becomes more basic (increasing $\beta$) the hydrogen bond donor ability of the ionic liquid decreases.

## $\beta$ values

The $\beta$ values for the ionic liquids studied here are moderate and dominated by the nature of the anion (Table 1). As the conjugate bases of strong acids, the anions of the ionic liquids might be expected to have low $\beta$-values in comparison to other solvents. However, although those found in this study are not as high as for acetone, they are comparable to acetonitrile, which is thought of as an electron pair donor solvent [5]. This shows again that there is an important Coulombic contribution to the hydrogen bonds formed between the ionic liquid and solute species.

The different β values obtained for the different [Tf$_2$N]$^-$ ionic liquids show that the cation has an influence, but with the limited data set here no trend can be clearly discerned at this stage.

## The rates of nucleophilic substitution reactions

We have investigated the reactions of a series of nucleophiles, both neutral and charged, with methyl p-nitrobenzenesulfonate (me-p-nbs) to form the p-nitrobenzenesulfonate ion (p-nbs, Figure 3). The substrate has a $\lambda_{max}$ at 253 nm whilst the p-nitrobenzenesulfonate ion product has a $\lambda_{max}$ at 275 nm. The progress of each reaction was monitored in situ by UV/vis spectroscopy, using an excess of the nucleophile. It was usually possible to record absorbance values at both 253 nm and 275 nm and to observe an isosbestic point.

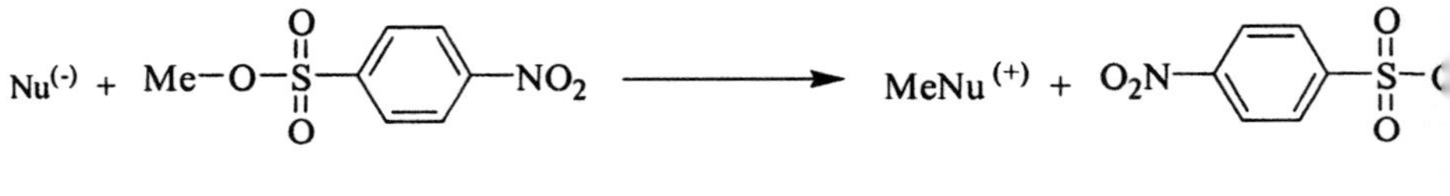

**Figure 3**: The reaction of nucleophiles with methyl p-nitrobenzenesulfonate to form the p-nitrobenzenesulfonate ion

*Neutral nucleophiles*

Initial studies of the reactions of $^n$butylamine (BuNH$_2$), di-$^n$butylamine (Bu$_2$NH) and tri-$^n$butylamine (Bu$_3$N) in [bmim][N(Tf)$_2$] ([bmim]$^+$ = 1-butyl-3-methylimidazolium) revealed that the amines were sufficiently basic to remove a proton from the imidazolium ring. Similar problems have been noted for this cation by others [13]. For this reason, the ionic liquid we used was [bmpy][N(Tf)$_2$] ([bmpy]$^+$ = 1-butyl-1-methylpyrrolidinium), whose saturated alkyl groups cannot be deprotonated by the amines.

The reactions in [bmpy][N(Tf)$_2$] are compared to the same reactions in acetonitrile and dichloromethane (Table 2). Immediately it becomes clear that, using the same substrate, all of the amines are very much more nucleophilic in the ionic liquid than in the molecular solvents examined here. The Hughes-Ingold concept is that solvent effects on the rates of reaction arise from the differential stabilization of the activated complex of the reaction with respect to the reagents in different solvents [14, 15]. The prediction for an S$_N$2 reaction of a neutral nucleophile with a neutral substrate is that, during activation, a complex will form in which charge separation will occur (Figure 4). This charge-separated complex has a greater dipole moment than the neutral starting materials and is preferentially stabilized in more polar environments. Hence, the effect of increased solvent polarity will be to increase the rate at which the charge separation occurs, *i.e.*, to increase the rate of reaction. Therefore, in respect of their effect on these nucleophilic substitutions, the ionic liquids can be regarded as highly polar solvents.

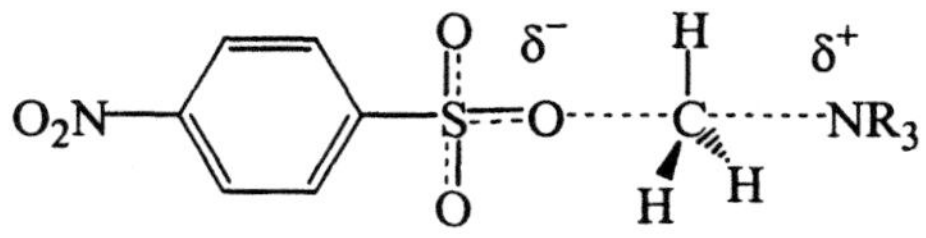

**Figure 4:** The activated complex for the reaction of tri-"butylamine with me-*p*-nbs.

Closer inspection of the results for the different amines, however, shows there are other contributions to the rate of the reactions. In all of the solvents used the secondary amine is the most nucleophilic. In dichloromethane there is little difference in the nucleophilicities of the amines with a ratio of 1:1.21:2.75 for $^nBuNH_2:^nBu_3N:^nBu_2NH$. In [bmpy][N(Tf)$_2$] the ratio is 1:3.48:4.79 for $^nBu_3N:^nBuNH_2:^nBu_2NH$, so while $^nBu_3N$ is now the least nucleophilic amine $^nBuNH_2$ and $^nBu_2NH$ have rather similar nucleophilicities. This is very pronounced in acetonitrile (1:6.03:7.08 for $^nBu_3N:^nBuNH_2:^nBu_2NH$). Clearly, in the ionic liquid and acetonitrile, the presence of a N-H protons is important in determining the relative nucleophilicities of the amines, with the first having a dramatic effect and the second having a slightly lesser influence.

**Table 2:** Second order rate constants for the reaction of amines with me-*p*-nbs in ionic liquids at 25 °C, relative rate constants and a comparison with polar molecular solvents.

| Solvents | $k_2/M^{-1}s^{-1}$ ($k_2$(solvent)/$k_2$(CH$_2$Cl$_2$)) | | | Kamlett-Taft parameters | | |
|---|---|---|---|---|---|---|
| | $^nBuNH_2$ | $^nBu_2NH$ | $^nBu_3N$ | $\pi^*$ | $\alpha$ | $\beta$ |
| [bmpy][N(Tf)$_2$] | 0.358 | 0.493 | 0.103 | 0.954 | 0.427 | 0.252 |
| | (21.70) | (10.90) | (5.15) | | | |
| CH$_3$CN | 0.155 | 0.182 | 0.0257 | 0.799 | 0.350 | 0.370 |
| | (9.39) | (4.01) | (1.29) | | | |
| CH$_2$Cl$_2$ | 0.0165 | 0.0454 | 0.0200 | 0.791 | 0.042 | -0.014 |
| | (1) | (1) | (1) | | | |

An alternative analysis of data is provided by comparing the increases in $k_2$ for the different amines as the reaction is transferred from dichloromethane to the ionic liquid. The relative rates $\{k_2(\text{solvent})/k_2(\text{CH}_2\text{Cl}_2)\}$ for the different amines are given in Table 2. The increases in the rates of reaction on transfer from dichloromethane to the ionic liquid are in the order $^n\text{Bu}_3\text{N} < {}^n\text{Bu}_2\text{NH} < {}^n\text{BuNH}_2$.

It can be seen that when considering $^n\text{Bu}_3\text{N}$ in the solvents dichloromethane, acetonitrile and [bmpy][N(Tf)$_2$], $k_2$ increases with increasing $\pi^*$, that is the reaction is very similar in dichloromethane and acetonitrile and considerably faster in the ionic liquid. This is entirely consistent with the Hughes-Ingold approach, with the caveat that $\pi^*$ best represents their generalized notion of polarity. All of the ionic liquids that we have studied have high $\pi^*$ values, with little variation between them. Hence, using all ionic liquids should lead to increased rates for this and similar reactions.

The rather more dramatic increases in the rates of the reactions of $^n\text{Bu}_2\text{NH}$ (x 10.9) and $^n\text{BuNH}_2$ (x 21.7) as the solvent is changed from dichloromethane to the ionic liquid require further explanation. Both acetonitrile and [bmpy][N(Tf)$_2$] can act as hydrogen bond acceptors (high $\beta$), whereas dichloromethane cannot. Analysis of the activated complexes for the reactions of $^n\text{Bu}_2\text{NH}$ and $^n\text{BuNH}_2$ shows that as the amine attacks the carbon of the substrate, the amine begins to develop positive charge. This will lead any protons bound to the nitrogen to become stronger hydrogen bond donors. Therefore a hydrogen bond accepting solvent will preferentially stabilize the activated complex with respect to the reagents and increase the rate of the reaction, by interacting with these protons (Figure 5). This effect is greater when more protons are bound to the nitrogen. While this effect is present in the ionic liquid reactions, it would seem to be the only significant difference in the reactions in acetonitrile and dichloromethane.

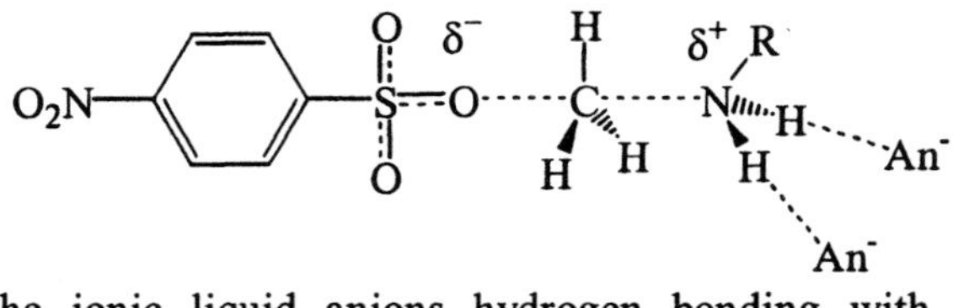

**Figure 5**. The ionic liquid anions hydrogen bonding with the emerging ammonium ion in the activated complex for the reaction of $^n\text{BuNH}_2$ with me-*p*-nbs.

*Charged nucleophiles*

We have also studied the reactions of the halides Cl⁻, Br⁻ and I⁻ with methyl p-nitrobenzenesulfonate. The relative nucleophilicities of the halides in the ionic liquids studied in this work are shown in Table 3. Comparison of data for reactions of the same substrate with different sources of halide in dichloromethane is also made.

The first point of note is that the difference between the slowest reaction in an ionic liquid and the fastest is only a factor of three. Whereas, the reactions in the ionic liquids are orders of magnitude decelerated in comparison to the reaction in dichloromethane, whether this is by the free ion or the ion-pair.

**Table 3:** Second order rate constants for the reaction of halides with me-*p*-nbs in ionic liquids at 25 °C, relative rate constants and a comparison with two polar molecular solvents.

| Solvents | $k_2/M^{-1} s^{-1}$ | | | Kamlett-Taft parameters | | |
|---|---|---|---|---|---|---|
| | Cl⁻ | Br⁻ | I⁻ | $\pi^*$ | $\alpha$ | $\beta$ |
| [bmim][N(Tf)₂] | 0.0124 | 0.0195 | 0.0232 | 0.984 | 0.617 | 0.243 |
| [bm₂im][N(Tf)₂] | 0.0296 | 0.0221 | 0.0238 | 1.010 | 0.381 | 0.239 |
| [bmpy][N(Tf)₂] | 0.0391 | 0.0226 | 0.0188 | 0.954 | 0.427 | 0.252 |
| CH₂Cl₂, ion-pair [16] | 0.51 | 0.42 | - | 0.791 | 0.042 | -0.014 |
| CH₂Cl₂, free ion [16] | 1.04 | 0.46 | - | 0.791 | 0.042 | -0.014 |

The reactions can once again be interpreted using a Hughes-Ingold approach [14, 15]. The nucleophiles used can be approximated as point charges which then combine with the substrate me-*p*-nbs to give an activated complex with a single negative charge distributed over several atoms. Transferring the reaction from a non-polar to a polar solvent is expected to stabilise the ground state with respect to the activated complex, thus slowing the reaction. This is, indeed, what is observed.

In the ionic liquids studied bromide nucleophilicity, $k_2$, was approximately constant, whilst iodide showed a little variation. However, the rate of reaction of the chloride shows a great deal more variation. For the reaction of Cl⁻ in [bmim][N(Tf)₂] the value of $k_2$ is less than half of the value in [bm₂im][N(Tf)₂] and less than a third of the value in [bmpy][N(Tf)₂]. Clearly there is some interaction, or interactions, between the ionic liquid and the Cl⁻ ion that is lowering its nucleophilicity in [bm₂im][N(Tf)₂] and yet more so in [bmim][N(Tf)₂]. Chloride is the best hydrogen bond acceptor of the halides (being hard, of high charge density and the most coordinating). The change in observed nucleophilicities might therefore be explained by the degree of stabilisation of the chloride ion *via* hydrogen bonding to the cation of the ionic liquid. The [bmim]$^+$ cation has the highest $\alpha$ value followed by [bmpy]$^+$ and [bm$_2$im]$^+$. This does not fit the trend in $k_2$ in a simple manner.

By studying the reaction at various temperatures we can evaluate the influence of activation enthalpy and entropy upon the reactions (Table 4) and hopefully get a more detailed picture of the reaction.

**Table 4**: Activation enthalpies, entropies and free energies for the reaction of chloride with me-*p*-nbs.

| Solvent | $\Delta H^{\ddagger}/$ kJ mol$^{-1}$ | $\Delta S^{\ddagger}/$ J K$^{-1}$ mol$^{-1}$ | $\Delta G^{\ddagger}_{298\,K}/$ kJ mol$^{-1}$ | Kamlett-Taft parameters | | |
|---|---|---|---|---|---|---|
| | | | | $\pi^*$ | $\alpha$ | $\beta$ |
| [bmim][N(Tf)$_2$] | 71.8 | -42.2 | 84.4 | 0.984 | 0.617 | 0.243 |
| | (1.6) | (5.4) | (3.2) | | | |
| [bm$_2$im][N(Tf)$_2$] | 71.8 | -37.3 | 82.9 | 1.010 | 0.381 | 0.239 |
| | (1.6) | (5.3) | (3.2) | | | |
| [bmpy][N(Tf)$_2$] | 68.5 | -43.9 | 81.6 | 0.954 | 0.427 | 0.252 |
| | (1.8) | (6.0) | (3.6) | | | |
| CH$_2$Cl$_2$, ion-pair [16] | 79.5 | 7.9 (10) | 77.2 | 0.791 | 0.042 | -0.014 |
| | (2.9) | | (5.9) | | | |
| CH$_2$Cl$_2$, free ion [16] | 54.4 | -58.6 | 71.9 | 0.791 | 0.042 | -0.014 |
| | (2.5) | (8.4) | (5.0) | | | |

Values of $\Delta H^{\ddagger}$ in the ionic liquids are similar to those for the ion-pair in dichloromethane, but very different to that for the free ion in dichloromethane. Therefore, it is not the change of solvent that is leading to the change in the activation enthalpy in going from the free ion in dichloromethane to the other systems, but rather the association of the chloride with a cation (ion-pair in dichloromethane) or cations (ionic liquids). This suggests that the same process is occurring in the rate-limiting step in both cases, i.e., as the chlorine-carbon bond is being formed in the activated complex, a cation-chloride association is being broken with an associated enthalpic cost. This is opposed to the "free ion" case where the chlorine-carbon bond is formed without the concomitant breaking of another interaction, giving a lower activation enthalpy.

In an ionic liquid, where there is no molecular solvent present to separate the ionic species, a free (molecular solvent solvated) ion cannot occur. An ionic species would always be expected to be coordinated by counter ions of the ionic liquid and we would expect that the chloride will always have its first coordination sphere dominated by cations. In fact recent structural studies in a related system confirm this expectation, and this is discussed later [17]. It is noteworthy that our data suggest $\Delta H^{\ddagger}$ more similar to that for the ion-pair in dichloromethane than for the free ion in dichloromethane. An ion-pair is a cation and anion that are coordinated to some extent and contained within a solvent shell; in our system that solvent shell is ionic, but is still more analogous to an ion-pair than a free anion.

The $\Delta S^{\ddagger}$ values, however, are more similar to those for the reaction by the free ion in dichloromethane, than the ion-pair in dichloromethane [16]. Since the activation enthalpies are so similar to the ion-pair situation in dichloromethane, this needs to be explained. This reaction follows an $S_N2$ reaction pathway and will form an activated complex as shown in Figure 3. In molecular solvents, this complex is coordinated by neutral solvent molecules. In ionic liquids it is likely that the cations will interact with the chloride and developing $p$-nbs.

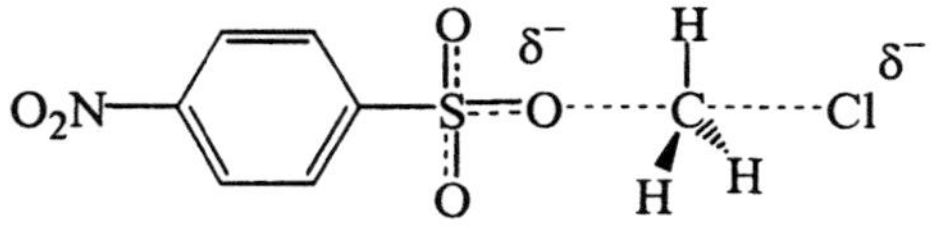

**Figure 6:** The activated complex for the reaction of me-$p$-nbs with chloride.

The activation step of an $S_N2$ reaction is an associative process and would therefore be expected to have a negative entropy. In dichloromethane the reaction by the ion-pair has a small positive activation entropy because, as the activated complex is formed, the cation of the ion-pair is liberated as a free solvated cation. This process counterbalances the loss of entropy associated with the formation of the activated complex. When performed in dichloromethane, the anionic leaving group $p$-nbs does not form an ion-pair in dichloromethane but becomes a free solvated anion. In the ionic liquid $p$-nbs does associate with the cations of the ionic liquid. It is proposed that the entropy gained by liberating a cation from its association with the chloride ion is cancelled out by the association of another cation with $p$-nbs. Hence, the activation entropy for the reaction was what might be expected for an $S_N2$ process, with a value similar to that for reaction by the free ion in dichloromethane.

Finally, the $\Delta G^{\ddagger}_{298\ K}$ values have been compared, and a slight trend emerges. It appears that the highest $\Delta G^{\ddagger}$ value is observed in the [bmim]$^+$ ionic liquid, and that it falls a little for [bm$_2$im]$^+$, and still more for [bmpy]$^+$. This trend is to be expected, given that these values should follow the same relationship as the $k_2$ values. These $\Delta G^{\ddagger}$ values are higher than those observed for reaction by either the free ion or the ion-pair in dichloromethane, suggesting, as one would expect, that in fact this reaction (with its charged reagent and its transition state with a delocalised charge) is slightly less favoured in ionic liquids than in dichloromethane. However, the activation parameters alone are not sufficient to explain the observed differences in the nucleophilicities of the halides in the different ionic liquids.

These parameters reveal the size of the entropy and enthalpy barriers on going from the ground state of the *available reactive forms* of the reagents and the activated complex. The ground state for the reaction has first to be achieved to give such an available chloride ion before the reaction can occur. The evidence shows that the chloride is not always available to react with the substrate when dissolved in the ionic liquids, although there is still a linear relationship between concentration of chloride and $k_{obs}$. So, it seems that there is an equilibrium (Figure 7) that must be accounted for.

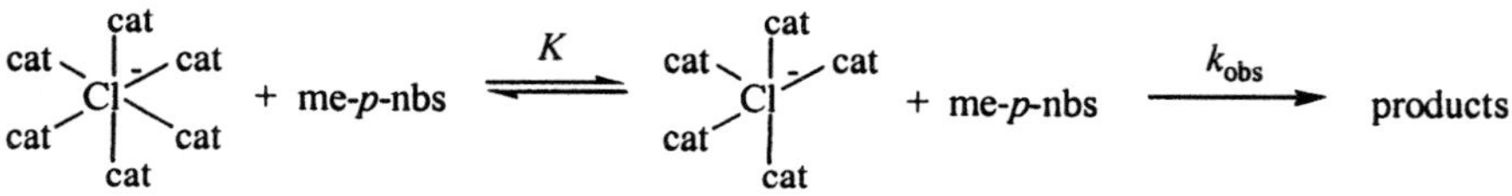

**Figure 7**: The formation of the "available" form of Cl⁻ in an ionic liquid.

The left hand side of the equilibrium represents a fully coordinated, "unavailable" chloride, whereas on the right hand side one face of the chloride ion is exposed to the substrate following the dissociation of one [bmim]⁺ cation, giving an "available" chloride. This loose association of available chloride with the substrate represents the ground state for the reaction in this system. It should be noted that UV/vis spectroscopy showed that there is no significant interaction between available chloride and the substrate to give rise to a formal intermediate. The extent to which the nucleophile is coordinated by the cation will affect $K$, and thus $k_{obs}$ and $k_2$. This is consistent with our results.

Although the nature of the complexes proposed above, and of the equilibrium constant, has not been determined in this work some initial proposals can be made. Hardacre et al.[17], have demonstrated by neutron diffraction that a Cl⁻ ion is fully coordinated by six cations within 6.5 Å of the halide in the related [mmim]Cl ([mmim]⁺ = 1,3-dimethylimidazolium cation). For the reaction to occur the Cl⁻ ion must first come into close proximity with the substrate me-p-nbs. To do this, the Cl⁻ ion must dissociate from at least one cation. At first glance, this appears to be simply a matter of breaking the cation-chloride hydrogen bond and so would be expected to correlate with α. However, our data show that the order of availability of chloride to react is [bmim][N(Tf)₂] < [bm₂im][N(Tf)₂] < [bmpy][N(Tf)₂]. Closer inspection of the process reveals that, as well as breaking the cation-chloride hydrogen bond, the formation of the "available" chloride also requires the separation of the cation from the anion and the insertion of the neutral substrate. The contribution of this charge separation to $\Delta G^{\ddagger}$ is best modelled by considering $\pi^*$. [bm$_2$im][N(Tf)$_2$] has the highest value of $\pi^*$ of all of the ionic liquids used here and we propose that the energetic cost of the separation of charges in this ionic liquid is sufficient to cause an inversion of the expected reactivities of Cl⁻ in [bm$_2$im][N(Tf)$_2$] and [bmpy][N(Tf)$_2$], when only hydrogen bonding is considered.

## Conclusions

This work shows that all ionic liquids are not the same. Therefore one cannot simply take a conventional organic reaction and replace the solvent with a single ionic liquid, then expect the result to be the same as would be achieved in all other ionic liquids. However, by the same token this work reveals that the claims made that ionic liquids can be tailor-made for a given reaction, are true [2]. It is possible to imagine that ionic liquids can be made to have the ideal combination of cation and anion for a given reaction.

We can generalise to say that, the Hughes-Ingold rules [14, 15] do seem to apply to ionic liquids and can be used to predict whether the use of ionic liquids is likely to be helpful (or unhelpful) for any given process. However, which ionic liquid should be used requires a deeper understanding of the solvent-solute interactions that are possible between all of the reacting species, any intermediate species and the activated complex for the reaction and the ionic liquids.

## Acknowledgements

We would like to thank the Leverhulme Trust for a fellowship (NLL) and the Kodak Foundation (JMPA) and GSK (LC) for studentships.

## References

1. T.Welton, *Chem. Rev.*, 1999, **99**, 2071; (b) P. Wasserscheid and W. Keim, *Angew. Chem. Int. Ed.*, 2000, **39**, 3772; (c) R. Sheldon, *Chem. Commun.*, 2001, 2399; (d) *Ionic Liquids in Synthesis*, P. Wasserscheid and T. Welton,

Eds., VCH Wiley, Weinheim, 2002.J.S. Smith, *Title of whatever (optional)*, Journal of the Best Sciences, **45**, 1989, p123

2. M. Freemantle, *Chem. Eng. News*, 1998, **76**, 32.

3. C. Reichardt, *Solvents and Solvent Effects in Organic Chemistry*, 2nd Edition, VCH (UK) Ltd, Cambridge, 1998.

4. (a) M. J. Kamlet and R. W. Taft, *J. Am. Chem. Soc.*, 1976, **98**, 377; (b) R. W. Taft and M. J. Kamlet, *J. Am. Chem. Soc.*, 1976, **98**, 2886; (c) T. Yokoyama, R. W. Taft and M. J. Kamlet, *J. Am. Chem. Soc.*, 1976, **98**, 3233; (d) M. J. Kamlet, J. L. Addoud and R. W. Taft, *J. Am. Chem. Soc.*, 1977, **99**, 6027.

5. M. J. Kamlet, J. L. Addoud, M. H. Abraham and R. W. Taft, *J. Org. Chem.*, 1983, **48**, 2877.

6. (a) R. M. C. Gonçalves, A. M. N. Simões, L. M. P. C. Albuquerque, M. Rosés, C. Ràfols and E. Bosch, *J. Chem. Res. (S)*, 1993, 214; (b) I. Person, *Pure & Appl. Chem.*, 1986, **58**, 1153.

7. P. Nicolet and C. Laurence, *J. Chem. Soc., Perkin Trans. 2*, 1986, 1071.

8. (a) P. H. Shetty, P. J. Youngberg, B. R. Kersten and C. F. Poole, *J. Chromatogr.*, 1987, **411**, 61; (b) S. K. Poole, P. H. Shetty, and C. F. Poole, *Anal. Chim. Acta* 1989, **218**, 241.

9. S. Tait and R. A. Osteryoung, *Inorg. Chem.*, 1984, **23**, 4352.

10. A. G. Avent, P. A. Chaloner, M. P. Day, K. R. Seddon and T. Welton, *J. Chem. Soc., Dalton Trans.*, 1994, 3405.

11. A. Elaiwi, P. B. Hitchcock, K. R. Seddon, N. Srinivasan, Y.-M. Tan, T. Welton and J. A. Zora, *J. Chem. Soc., Dalton Trans.*, 1995, 3467.

12. T. Van Mourik and F. B. Van Duijneveldt, *J. Mol. Struct. (THEOCHEM)*, 1995, **341**, 63.

13. V. K. Arggarwal, I. Emme and A. Mereu, *Chem. Commun*, 2002, 1612.

14. C. K. Ingold, *Structure and Mechanism in Organic Chemistry*, 2nd Edition, Bell, London, 1969.

15. (a) E. D. Hughes and C. K. Ingold, *J. Chem. Soc.*, 1935, 244; (b) E. D. Hughes, *Trans. Faraday Soc.*, 1941, **37**, 603; (c) E. D. Hughes and C. K. Ingold, *Trans. Faraday Soc.*, 1941, **37**, 657; (d) K. A. Cooper, M. L. Dhar, E. D. Hughes, C. K. Ingold, B. J. MacNulty and L. I. Woolf, *J. Chem. Soc.*, 1948, 2043.

16. Alluni, S.; Pero, A.; Reichenbach, G. *J. Chem. Soc., Perkin Trans. 2*, **1998**, 1747-1750.

17. Hardacre, C.; Holbrey, J. D.; McMath, S. E. J.; Bowron, D. T.; Soper, A. K. *J. Chem. Phys.*, 2003, **118**, 273.

# Chapter 18

# Effect of Oxygen-Containing Functional Groups on Protein Stability in Ionic Liquid Solutions

Megan B. Turner[1,2], John D. Holbrey[1], Scott K. Spear[1], Marc L. Pusey[3], and Robin D. Rogers[1,2]

[1]Center for Green Manufacturing and [2]Department of Chemistry, The University of Alabama, Tuscaloosa, AL 35487
[3]NASA Marshall Space Flight Center, Huntsville, AL 35812

The ability of functionalized ionic liquids (ILs) to provide an environment o f i ncreased s tability f or biomolecules has been studied. Serum albumin is an inexpensive, widely available protein that contributes to the overall colloid osmotic blood pressure w ithin t he v ascular s ystem ( *1*). A lbumin i s used in the present study as a marker of biomolecular stability in the presence of various ILs in a range of concentrations. The incorporation of hydroxyl functionality into the methylimidazolium-based cation leads to increased protein stability detected by fluorescence spectroscopy and circular dichroic (CD) spectrometry.

# Introduction

Ionic liquids (ILs) are a class of solvents generally composed of large, complex organic cations associated with inorganic anions that are liquids below 100 °C. The most extensively studied ILs are systems derived from 1,3-dialkylimidazolium, tetraalkylammonium, tetraalkylphosphonium, and N-alkylpyridinium cations (2). Derivitization of the cations can include the addition of alkyl chains, introduction of branching and/or chirality, fluorination, or addition of specific active functions. Common anions that allow for the formation of low melting ILs are typically charge diffuse, and can range from simple inorganic anions such as chloride (Cl⁻), bromide (Br⁻), and iodide (I⁻), through larger pseudo-spherical polyatomic anions including hexafluorophosphate $[PF_6]^-$ and tetrafluoroborate $[BF_4]^-$ to larger, flexible fluorinated examples such as bis(trifluoromethanesulfonyl)amide $[(CF_3SO_2)N]^-$. The anionic component of the ionic liquid typically controls the reactivity of the solvent with water, coordinating ability, and hydrophobicity (2). Anions can also contain chiral components or be catalytically active, i.e. carboranes, polytungstate, or tetrachloroaluminate anions (3). Manipulation of the rheological properties of an IL system can be preformed either by; (i) functionalization and/or cation/anion substitution or (ii) through mixing of cation/anion pairs. The ability to design the solvent system to support and/or enhance reactions makes the use of ILs in chemical reactions an intriguing alternative to conventional solvents (4).

Interest in ILs can be attributed to the unique combination of qualities inherent to these materials. First, the liquids are entirely ionic in nature making them conducting materials that can be used in a number of electrochemical applications including, battery production and metal deposition (5). In fact, ILs were first designed to be used in the electrochemical field to exploit this feature. Second, in most cases ILs exhibit a wide range of liquid character while possessing negligible measurable vapor pressure making them an attractive replacement for volatile organic solvents (6). Third, ILs are composed of ions so the combinations of properties and characteristics that can be obtained differ from many physical and chemical properties to conventional solvents. The ability to 'fine-tune' the solvent properties of ILs has been exploited to create solvents ideal for a range of chemical processes. Lastly, ILs exhibit varying degrees of solvation and solubility in a range of organic solvents establishing a simple method for separations and extractions (7).

ILs have been increasingly studied for their use as solvents in biocatalytic reactions and have shown increased solubility of polar substrates resulting in increased reactivity, in some cases selectivity, and stability of the biocatalyst when compared to typical organic solvents (8). To date, reactions involving the use of biomolecules have primarily been carried out in water immiscible, or hydrophobic, ILs or systems thereof. Addition of other components, i.e. water, to the IL leads to the creation of either salt solutions or biphasic systems in which the IL acts as a co-solvent. However, the environment surrounding the biomolecule, in the presence of an IL, has been largely unstudied. How does the addition of a hydrophilic IL to an aqueous solution affect biomolecules? Do all hydrophilic ILs additives affect protein structure and conformation in the same manner? And, can an IL be synthesized to provide the benefits previously described without adversely affecting it?

Here, we explore the stability of serum albumin in a range of aqueous imidazolium IL solutions using fluorescence spectroscopy and circular dichroic (CD) spectrometry. The ILs used in this study include oxygen-containing functional groups attached to the cationic portion of the solvent (Figure 1), and have varying degrees of hydrophobicity. The aim of this work is to evaluate the role of imidazolium cation functionality on biomacromolecule stability and structure in the presence of different IL forming salts.

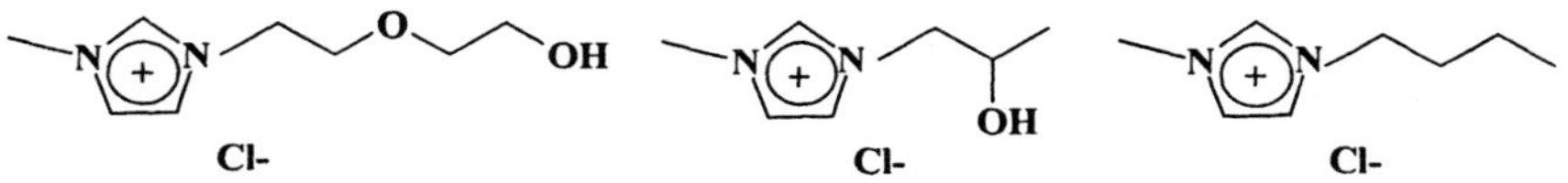

*Figure 1. Chemical structures of ILs: from left, 1-ethoxyethanol-3-methylimidazolium chloride ([EOmim]Cl), 1-(2-hydroxypropyl)-3-methylimidazolium chloride ([C₃OHmim]Cl), and 1-butyl-3-methylimidazolium chloride ([C₄mim]Cl).*

Previous studies have shown that cellulase is denatured in the presence of hydrophilic 1-butyl-3-methylimidazolium chloride ([C₄mim]Cl) even at very low concentrations (9). While it is plausible that the high Cl⁻ ion concentration contributes largely to biomolecule denaturation, it was determined that the cationic portion of the solvent also played a role. In an attempt to increase compatibility, the structure of the imidazolium cations was modified through the addition of hydroxyl functionality (4b). We postulated that the addition of oxygen would render the IL more 'water-like' based on a higher degree of intermolecular hydrogen bonding. The increase in hydrogen bonding should therefore heighten the hydrophobic effect of the solvent on the protein inducing native, or native-like, folding to a higher degree. Analytical techniques including fluorescent spectroscopy, and CD spectrometry have been used to determine if the presence of oxygen molecules, incorporated into the ILs, contribute to stabilizing protein structure.

# Experimental

Bovine serum albumin (BSA) and human serum albumin (HSA) were purchased in a lyophilized form from Sigma (Milwaukee, WI) and used as received. All other chemicals used in the study were of analytical grade a nd used without further purification.

ILs used in this study were prepared according to literature methods (*4b,10*). Purity of each IL was confirmed through $^{1}$H and $^{13}$C NMR spectroscopy.

Fluorescence data was collected with a Horiba Jobin Yvon Fluoromax-3 fluorescence spectrometer (Edison, NJ) using 1 cm path length quartz fluorescent cuvettes (Fisher Scientific; Pittsburg, PA). Preliminary excitation scans of BSA in buffer solution were taken and resulted in an observed $\lambda_{max}$ = 285 nm. Subsequent emission scans were taken over the range 300-555 nm with a slit width of 2 nm. Integration was set to 1 nm while increment time was set at 0.50 s. Each solution tested was prepared to a final volume of 2.0 mL containing 80 to 1 0 % v/v c o-solvent i n 0 .05 M c itrate buffer. 0.1 mL aliquots of BSA (2.23 mg mL$^{-1}$) were added and gently homogenized.

Samples for circular dichroism measurements were prepared containing a final concentration of 18.75 mg mL$^{-1}$ HSA in 0.05 M citrate buffer (pH 5.0), and IL i n a 1 :100 d ilution i n D I H $_2$O. T he m olar r atio o f p rotein t o IL was held constant in each sample. Measurements were taken on an Olis CD Spectrometer (Bogart, GA) using a Hinds PEM-90 photoelastic modulator (Hillsboro, OR) scanning a range from 290–190 nm. Measurements were taken at constant temperature (25 °C) under a nitrogen atmosphere using 100 μm path length quartz cuvettes.

# Results and Discussion

The fluorescence response of protein in an aqueous environment is directly proportional to the number of fluorescent tryptophan amino acid residues within the protein (*11*). Bovine serum albumin (BSA) used in this study contains three tryptophan sites within the molecule. Each of the tryptophan residues will affect the measured fluorescence intensity in each solvent and are compared to a blank consisting of BSA in an aqueous environment without the addition of IL.

The fluorescence of aqueous BSA solution was measured upon the addition of t he I Ls ([C$_4$mim]Cl, [C$_3$OHmim]Cl, and [EOmim]Cl), and with the organic co-solvent ethanol. The maximum absorbance intensity was found at 348 nm for each co-solvent except the [EOmim]Cl in which the maxima was shifted to 425 nm. The effects of changing cation components of the ILs (as salt modifiers) was investigated to determine whether certain structural themes could be observed to have significant positive or detrimental i mpact o n n ative B SA folding and fluorescence.

The addition of [C₄mim]Cl, [C₃OHmim]Cl, [EOmim]Cl or ethanol to 0.05 M citrate buffer in increasing concentrations effectively lowers the fluorescence intensity of the BSA, corresponding to reduced stability of its native folding motif (Figure 2). Addition of [C₄mim]Cl, even at 10 % wt/wt significantly lowers the intensity of the protein compared to the protein in an aqueous environment. This effect can be attributed to fluorescence quenching of the tryptophan residues as the protein becomes unfolded in the presence of the dehydrating [C₄mim]Cl. When ethanol (a well known denaturant of biomolecules at high concentrations) was added to the system in concentrations higher than 40 % wt/wt the protein precipitated from solution as aggregates formed when the conditions of the solution became unfavorable. This provides a further visual affirmation that higher fluorescent intensities are directly related to increased shielding of the tryptophan residues by the contracting of the protein.

Comparison between an IL solution versus that of a simple salt, with comparable i onic s trengths, c an g ive u s s ome i nsight i nto p ossible differences between each salts effect on a protein's structure. Figure 2b compares the simple salt, NaCl, to [C₄mim]Cl showing a vast difference in the protein's fluorescence intensity as the molarity of the solutions increase. Although the intensity of the BSA in the NaCl solution is lower than in the [C₄mim]Cl at 0.7 M, the intensity remains constant showing virtually no loss of intensity through 5.5 M. Increasing amounts of [C₄mim]Cl in the solution drastically decrease the intensity of the protein reflecting certain loss in structure.

The structure of a protein generally dictates its function and can be defined on four successive levels (12). Primary structure is defined through the unique amino acid sequence of a protein which can be obtained through nucleotide analysis. The secondary structure of a protein is described in terms of α-helices, β-sheets, turns, and unordered segments, and is dictated by interactions between the side chains of amino acids as w ell a s i nteractions w ith s olvent m olecules. The orientation of these folding motifs for the protein on a whole is considered the tertiary structure. Finally, the quaternary structure of a protein is defined as the spatial relationship between subunits.

CD spectrometry was collected to support the fluorescence data and determine the effect(s) of each IL on the secondary structure of the protein. Human serum albumin (HSA) was used as a handle in this study to avoid interference from fatty acids that are typically associated with lyophilized BSA direct from the manufacturer (13). The secondary structure of a protein can be classified as all α, all β, α+β, α/β, or unordered. Protein classified as either all α or all β consist largely, but not necessarily exclusively, of these folding motifs (13). As the name implies, α+β and α/β proteins are made up of both α-helices and β-sheets, however in α+β proteins the helices and sheets are usually confined to separate domains, whereas α/β proteins contain both structures (13).

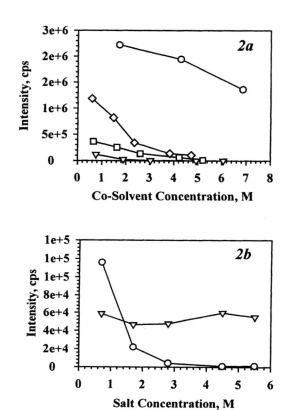

*Figure 2. Maximum fluorescence intensity vs. solvent concentration. (2a)* O: *ethanol;* ◇: *[EOmim]Cl;* □: *[C₃OHmim]Cl;* ∇: *[C₄mim]Cl. (2b)* O: *[C₄mim]Cl;* ∇: *NaCl.*

A protein, like HSA, that is classified as 'all α' produces a signature CD spectra with a strong double negative feature at 222 and 208-210 nm and a more intense positive feature at 191-193 nm (*13*). The intensities of the bands reflect the degree of helicity in the protein studies; using these features as a standard we are able to determine if additives, ILs in this case, effect the secondary structure of the protein.

From the experimental fluorescence data we saw that increasing concentrations of each co-solvent was directly related to increasing deleterious effects to protein structure represented by decreasing intensity. The CD spectra collected for each additive is able to reinforce this observation (Figure 3). The spectrum for HSA, without an additive, reflects the standard for all α proteins as described above. The addition of ILs, having varying degrees of hydrogen bond donor and acceptor ability, tend to affect the secondary structure in a predictable manner. The presence of [C₄mim]Cl, having the lowest hydrogen bond donor/acceptor ability, causes the greatest change in the HSA structure. The double negative feature is reduced to a single band at 220 nm. The stronger positive feature normally seen between 191-193 nm is completely diminished.

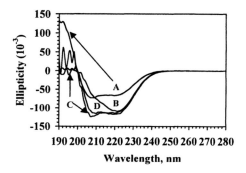

*Figure 3. Circular dichroic (CD) spectra of HSA in various IL containing solutions: A: HSA with no IL added; B: HSA with [C₄mim]Cl; C: HSA with [C₃OHmim]Cl; D: HSA with [EOmim]Cl.*

Addition of [C₃OHmim]Cl seemingly has the least effect on protein structure. This observation is somewhat unexpected owing to the fact that its potential for hydrogen bond interaction is higher than [C₄mim]Cl, but lower than [EOmim]Cl. The strong double negative feature is intact in the presence of this IL, however, the more intense positive feature is disrupted. The same trend is seen upon addition of [EOmim]Cl, but the positive feature centered at 191 nm is flat. Reduction of the positive feature *may* be attributed solely to IL-induced denaturation of the protein or may be caused by interference from the imidazolium based cation of the IL at wavelengths below ~200 nm, as the $\lambda_{max}$ of imidazolium is 211 nm. (Interference from $O_2$ absorption can be ruled out as measurements were taken under a nitrogen atmosphere with exclusion of $O_2$.)

Both fluorescence and circular dichroic studies have shown that increasing the hydrogen bonding capacity of the IL effectively increases the stability of protein in the solution. Recently, Infantes and Motherwell have averaged the number of hydrogen bond donor and acceptor contacts for some common functional groups in over forty thousand organic crystal structures that have been deposited in the Cambridge Structural Database (*14*). A comparison between this information and experimental data shows that the IL with the lowest average number of hydrogen bond interactions, [C$_4$mim]Cl, appears to have the largest adverse effect on protein stability: [C$_4$mim]Cl – 2.68 hydrogen bonding interactions < [C$_3$OHmim]Cl – 3.97 hydrogen bonding interactions < [EOmim]Cl – 4.13 hydrogen bonding interactions (*14*).

# Conclusions

Our results indicate that [C$_4$mim]Cl causes denaturation of serum albumin through either ionic strength or specific binding to the protein surface. The incorporation of oxygen molecules to the cationic portion of an IL causes an increase in Trp fluorescence intensity as compared to its non-oxygen containing counterpart. This increase in intensity may be attributed to the hydroxyl group's ability to cause less interaction between the solvent and the protein or its ability to interact with the protein in a more favourable way. The IL, containing two oxygen molecules, provides the largest opportunity for hydrogen bond interactions which is reflected through the highest measured fluorescence intensity. The trend follows as expected – the second highest intensity was measured in the [C$_3$OHmim]Cl, having one oxygen molecule, followed by the [C$_4$mim]Cl with no oxygen molecules. It is also apparent that all of the ILs tested had an adverse effect on the serum albumin at concentrations as low as 20 % wt/wt. Continued improvements to the IL, especially the anionic component, may prove to be beneficial to the overall stability of the protein at higher IL concentrations.

CD spectrometry was used to determine the effect of each IL on the secondary structure of the protein. Serum albumin, being an all α protein, was used as a handle to detect IL-induced modifications to the α-helix folding motif. CD data shows, as expected, that the largest loss in secondary structure is a result of the addition of [C$_4$mim]Cl to the protein in aqueous solution. The [C$_4$mim]Cl reduces the double negative feature of the native protein to a single peak centered around 220 nm as well as completely diminishing the positive feature at 191 nm. Both the [C$_3$OHmim]Cl and the [EOmim]Cl ILs affect the protein to a lesser extent and appear to effect it in the same manner. Both additions result in conservation of the double negative feature but diminish the positive feature. Using an imidazolium-based solvent may have contributed to the loss of the peak at 191 nm through interference. To determine if the effect is real it is necessary to test analogous imidazole-free ILs.

Results of these studies show that the cation-incorporated hydroxyl functionality may be responsible for the increased stability of serum albumin over other non-oxygenous ILs. While it is imperative to increase the data set (both ILs and proteins) these preliminary results are promising for biochemical processes in non-aqueous media. Research in this area reflects the continued importance of understanding the affects various IL features have on biomolecules to suppress unwanted reaction routes, provide higher selectivities and reactivities, and provide simple methods for the recovery and recyclability of biocatalysts. It now becomes important to relate protein stability data to protein activity data to delineate any correlations.

## Acknowledgments

This research has been supported by the U.S. Environmental Protection Agency's STAR program through grant number RD-83143201-0. (Although the research described in this article has been funded in part by EPA, it has not been subjected to the Agency's required peer and policy review and therefore does not necessarily reflect the views of the Agency and no official endorsement should be inferred.) The authors would like to thank the NASA Marshall Space Flight Center for use of their CD spectrometer facility.

## References

*1*  Carter, D. C.; Ho, J. X. *Adv. Protein Chem.* **1994**, *45*, 153-203.

*2*  (a) Welton, T. Room-Temperature Ionic Liquids. Solvents for Synthesis and Catalysis, *Chem. Rev.,* **1999**, *99*, 2071-2084. (b) *Ionic Liquids in Synthesis;* Wasserscheid, P.; Welton, T. Eds.; Wiley-VCH: Weinheim, 2002. (c) *Ionic Liquids; Industrial Applications for Green Chemistry*; Rogers, R. D.; Seddon, K. R., Eds.; ACS Symposium Series 818; American Chemical Society: Washington, DC, 2002. (d) *Green Industrial Applications of Ionic Liquids*; Rogers, R. D.; Seddon, K. R.; Volkov, S., Eds.; NATO Science Series II; Mathematics, Physics, and Chemistry–Vol. 92; Kluwer: Dordrecht, 2003. (e) *Ionic Liquids as Green Solvents: Progress and Prospects;* Rogers, R. D.; Seddon, K. R., Eds.; ACS Symposium Series 856; American Chemical Society: Washington, DC, 2003. (f) *Molten Salts XIII*; Proceedings of the International Symposium; Trulove, P. C.; De Long, H. C.; Mantz, R. A.; Stafford, G. R.; Matsunaga, M., Eds.; The Electrochemical Society, Inc.: Pennington, NJ, 2002.

*3*  (a) Abdul-Sada, A. K.; Avent, A. G.; Parkington, M. J.; Ryan, T. A.; Seddon, K. R.; Welton, T. *J. Chem. Soc., Dalton Trans.* **1993**, *22*, 3283-3286. (b) Wasserscheid, P.; Keim, W. *Angew. Chem. Int. Ed.* **2000**, *39*, 3772-. (c) Sheldon, R. *Chem. Comm.* **2001**, 2399-2407. (d) Gordon, C. M.

242

*Appl. Catal. A* **2001**, *222*, 101-117. (e) Olivier-Bourbigou, H.; Magna, L. *J. Mol. Catal. A: Chem.* **2002**, *182-183*, 419-437. (f) Dupont, J.; de Souza, R. F.; Suarez, P. A. Z. *Chem. Rev.* **2002**, *102*, 3667-3691. (g) Zhao, D. B.; Wu, M.; Kou, Y.; Min., E. *Catal. Today* **2002**, *74*, 157-189.

4   Holbrey, J. D.; Turner, M. B.; Reichert, W. M.; Rogers, R. D. *Green Chem.* **2003**, *5*, 731-736.

5   (a) Abbott, A. P.; Capper, G.; Davies, D. L.; Munro, H.; Rasheed, R. K.; Tambyrajah, V. In *Ionic Liquids as Green Solvents Progress and Prospects*; Rogers, R.D.; Seddon, K.R., Eds.; ACS Symposium Series *856*; American Chemical Society: Washington, DC, 2003; pp. 439-452. (b) De Long, H. C.; Carlin, R. T. *Proc. Electrochem. Soc.* **1994**, *94-13*, 736-743. (c) Lewandowski, A.; Swiderska, A. *Solid State Ionics* **2003**, *161* , 243-249. (d) Quinn, B. M.; Ding, Z.; Moulton, R.; Bard, A. J. *Langmuir* **2002**, *18*, 1734-1742. (e) Wilkes, J. S.; Levisky, J. A.; Wilson, R. A.; Hussey, C. L. *Inorg. Chem.* **1982**, *21*, 1263-1264. (f) Yoshizawa, M.; Narita, A.; Ohno, H. *Aust. J. Chem.* **2004**, *57*, 139-144. (g) Wilkes, J. S. *Green Chem.* **2002**, *4*, 73-80. *Note: Green Chem.* **2002**, *4(2) is entirely dedicated to ionic liquids.*

6   (a) Benton, M. G.; Brazel, C. S. Effect of Room-Temperature Ionic Liquids as Replacements for Volatile Organic Solvents in Free-Radical Polymerization. In *Ionic Liquids: Industrial Applications to Green Chemistry;* Rogers, R. D.; Seddon, K. R., Eds.; ACS Symposium Series *818*; American Chemical Society: Washington, DC, 2002; pp. 125-133. (b) Cull, S. G.; Holbrey, J. D.; Vargas-Mora, V.; Seddon, K. R.; Lye, G. J. *Biotechnol. Bioeng.* **2000**, *69*, 227-233.

7   (a) Visser, A. E.; Swatloski, R. P.; Reichert, W. M.; Willauer, H. D.; Huddleston, J. G.; Rogers, R. D. Room temperature ionic liquids as replacements for traditional organic solvents and their applications towards 'green chemistry' in separation processes. In *Green Industrial Applications of ionic Liquids;* R ogers, R . D .; S eddon, K . R .; V olkov, S ., E ds.; NATO Science Series, II: Mathematics, Physics and Chemistry *92*, Kluwer Academic Publishers: Dordrecht, 2003; pp. 137-156. (b) Visser, A. E.; Rogers, R. D. *J. Solid State Chem.* **2003**, *171*, 109-113.

8   (a) Sheldon, R. A.; Lau, R. M.; Sorgedrager, M. J.; van Ranywijk, F.; Seddon, K. R. *Green Chem.* **2002**, *4*, 147-151. (b) Kragl, U.; Eckstein, M.; Kaftzik, N. *Curr. Opinion Biotechnol.* **2002**, *13*, 565-571. (c) van Rantwijk, F.; Lau, R. M.; Sheldon, R. A. *Trends Biotechnol.* **2003**, *21*, 131-138.

9   Turner, M. B.; Spear, S. K.; Huddleston, J. G.; Holbrey, J. D.; Rogers, R. D. *Green Chem.* **2003**, *5*, 443-447.

10  Huddleston, J. G.; Visser, A. E.; Reichert, W. M.; Willauer, H. D.; Broker, G. A.; Rogers, R. D. *Green Chem.* **2001**, *3*,156-164.

11  Woodward, J.; Lee, N. E.; Carmichael, J. S.; McNair, S. L.; Wichert, J. M. *Biochem. Biophys. Acta* **1990**, *1037*, 81-85.

*12* (a) Fasman, G. D., Ed. *Circular Dichroism and the Conformational Analysis of Biomolecules*; Plenum Press: New York, NY, 1996. (b) Elysee-Collen, B.; Lencki, R. W. *Biotechnol. Prog.* **1997**, *13*, 849-856.

*13* Chen, R. F. *J. Biol. Chem.* **1967**, *242*, 173-181.

*14* Infantes, L.; Motherwell, W. D. S. *Chem. Commun.* **2004**, 1166-1167.

# Chapter 19

# Chloride Determination in Ionic Liquids

Constanza Villagrán[1], Craig E. Banks[2], Maggel Deetlefs[1],
Gordon Driver[1], William R. Pitner[1], Richard G. Compton[2],
and Christopher Hardacre[1,*]

[1]The QUILL Centre and The School of Chemistry, Queen's University
at Belfast, Belfast BT9 5AG, Northern Ireland, United Kingdom
[2]Physical and Theoretical Chemistry Laboratory, University of Oxford,
South Parks Road, Oxford, OXI 3QZ, United Kingdom

The determination of chloride impurities in water miscible
and water immiscible ionic liquids has been explored using
ion chromatography (IC) and cathodic stripping voltammetry
(CSV). This paper shows the first quantification of chloride
in [NTf₂]⁻ based ILs. The parameters investigated include
sample preparation, solvent effect, sample stability, and limit
of quantification (LOQ).

# Introduction

Ionic liquids (ILs) have been studied extensively as solvents for a wide range of chemical processes. The interest is a direct result of the diverse physical properties of these liquids and the way in which they may be systematically varied, *e.g.* the density, viscosity and water miscibility. Since they also have effectively zero vapor pressure, this makes them ideal engineering solvents for reactive chemistry allowing direct distillation of solutes from the solvent and simple solvent recycle without the production of VOC's (*1*) However, due to their low vapor pressure these solvents cannot be purified by simple distillation unlike conventional molecular liquids. Moreover, their synthesis inherently results in low levels of impurities, which influence both their physico-chemical properties (*2,3*) as well as their performance as solvents, particularly for catalytic reactions (*4-6*)

Despite the impact of halide content of ionic liquids being generally known, few analytical methods have been reported for its quantification in ionic liquids. These include Volhard and chloride-selective electrodes (*2*), a spectrophotometric method with fluorescent indicators (*7*), ion chromatography (IC) (*8,9*) and electrochemistry (*10*). These methods have concentrated on water miscible ionic liquids and on [BF$_4$]$^-$ based ionic liquids in particular.

This paper compares two general methods for the determination of chloride in both water miscible and water immiscible ionic liquids using ion chromatography(*11*) and cathodic stripping voltammetry (*12*) reported previously.

# Ion Chromatography

## Experimental

Unless otherwise stated, the chromatographic analysis were performed at room temperature using a Metrohm Model 761 Compact IC (Metrohm, Herisau, Switzerland) with suppressor module, equipped with an ICSep AN2$^{TM}$ analytical column (250 x 4.6 mm) and an ICSep AN2$^{TM}$ guard column (50 x 4.6 mm). The injection volume was 20 µL. The eluent used was a 3.2 mM Na$_2$CO$_3$ + 1.0 mM NaHCO$_3$ mixture and the suppressor regenerating solution was 0.1 M H$_2$SO$_4$. The eluent was prepared daily, filtered through a 20 µm filter and degassed by bubbling nitrogen IC samples were prepared by dissolving 0.11 g of IL in either 10 mL of deionised water or in 2 mL of acetonitrile before being diluted to 10 mL with deionised water.

**Results**

*Cation Effect*

To investigate the effect of the cation on chloride content determinations using ion chromatography, four aqueous solutions of choline chloride, [C₄dmim]Cl, [C₄Py]Cl, and [C₄mim]Cl, containing 5.1 ± 0.4 ppm of chloride were measured and compared with an aqueous solution containing 5.1 ppm NaCl. This selection of cations was chosen in order to examine whether IC is versatile over the wide range of cation types which are available. For example, choline chloride is acyclic and saturated with a hydroxyl-functionalised side chain, [C₄dmim]Cl and [C₄mim]Cl are unsaturated five-membered heteroaromatic rings, with the former deprotonated at the C(2) position and replaced with a methyl group, and [C₄Py]Cl is a six-membered unsaturated ring which is more basic than the imidazolium cations.

The ion chromatograms of these four ionic liquids and NaCl show that the retention times and peak area of chloride are essentially independent of the sample matrices with the relative standard deviations (RSD) of the retention times and peak area determined as 0.3 % and 5.3 %, respectively.

In general, for an analyte concentration below 10 ppm, a relative standard deviation of ≤10 % is an acceptable value for the concentration (*13*), and ≤2 % is an acceptable value for the retention time (*14*).

*Effect of IC sample diluent type*

The solubility of the ionic liquids in water is strongly dependent on both the length of the alkyl chains attached to the cation and the anion used. For example, whilst [C₂mim][BF₄] is totally soluble in water, [C₆mim][BF₄] has very limited solubility in water at room temperature and [C₄mim][BF₄] only has good solubility with water at room temperature; on cooling to close 0 °C the solubility decreases significantly (*15*). The equivalent hexafluorophosphate ionic liquids are significantly less soluble in water in comparison. For fluorinated anions there is also a complication that hydrolysis to fluoride may occur and therefore different solvents were tested in order to find an appropriate solvent which can prevent the formation of a high concentration of fluoride ions and solubilize the water miscible and immiscible ionic liquids.

Figure 1 shows the formation of fluoride ions with time determined by ion chromatography using a sample of 9.7 g L⁻¹ of [C₄mim][BF₄] dissolved in four different solvents, *i.e.* water, eluent, 20 vol% acetonitrile in water and 20 vol% methanol in water. Clearly the rate of hydrolysis is strongly dependent on the protic nature of the diluent used. Acetonitrile solubilizes most ionic liquids and also limits the hydrolysis of hydrolytically unstable anions and therefore the acetonitrile-water diluent was used for all subsequent analyses.

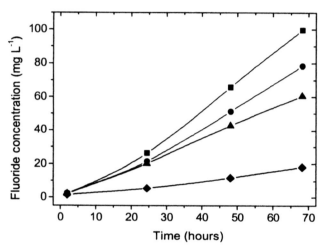

*Figure 1. Time variation of the formation of fluoride ions during*
*[C₄mim][BF₄] hydrolysis in water (■), 20 vol% methanol in water (●), 3.2 mM Na₂CO₃ +*
*1.0 mM NaHCO₃ eluent (▲) and 20 vol% acetonitrile in water (◆).*

Compared with [BF₄]⁻ based ionic liquids, ionic liquids containing [PF₆]⁻ showed little hydrolysis at room temperature when dissolved in 20 vol% acetonitrile in water. Figure 2 shows ion chromatograms for a sample of 11.2 g L⁻¹ of [C₄mim][PF₆] spiked with 0.10 ppm of chloride over 19 hours. It is clear that, over this time period, the fluoride peak does not significantly change. It is interesting to note that, although several papers have reported on the hydrolysis problems of hexafluorophosphate (*16,17*) there has been much less attention on the problems associated with tetrafluoroborate even though it is clear that significant decomposition still occurs and may even be more problematic than for hexafluorophosphate.

*Chromatographic system*

Since the chloride is an impurity in a medium containing only ions which includes a high concentration of, in general, bulky and polarizable anions with corresponding long retention times, the development of a method for the detection of trace amounts of chloride has to balance the separation efficiency of the analyte of interest with the retention time of elution of the main anion from the ionic liquid.

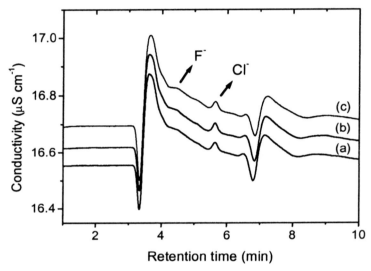

*Figure 2. Ion chromatograms of 11.2 g L⁻¹ [C₄mim][PF₆] dissolved in 20 vol%*
*acetonitrile in water following (a) 1 hr. of preparation,*
*(b) 10 hrs. of preparation and (c) 19 hrs. of preparation.*

An eluent concentration mixture of 3.2 mM $Na_2CO_3$ + 1.0 mM $NaHCO_3$ was found to be the optimum concentration to enable good separation of chloride peak from the other features observed in the traces, *e.g.* interference from the water dip, carbonate dip and fluoride. In order to have good separation efficiency at the retention time corresponding to the chloride peak, the samples were run at 1 mL min⁻¹ during the first 10 min, and thereafter the flow rate was raised up to 1.6 mL min⁻¹ to speed up the elution of the main anion from the ionic liquid. A representative chromatogram is shown in Figure 3 for [C₄mim][BF₄].

The addition of organics to the eluent can be used to change the retention time of anions. However, the addition of 10 vol% of acetone to the eluent did not improve the elution of the main anion significantly and reduced the separation of the fluoride and chloride peaks. The retention time for the different anions depends of the charge, size and polarizability of the anion and were found to follow the order $[BF_4]^- \sim [EtSO_4]^- < [OTf]^- \ll [PF_6]^- \ll [NTf_2]^-$. The hexafluorophosphate took 2 hours to begin to elute while no elution of the bis(trifluoromethanesulfonyl)imide was observed after 5 hours. Preliminary data using a gradient elution system and a hydroxide column has shown that this limitation may be overcome (Figure 4). A range of cation-anion combinations were studied and, in each case, the chromatograms were found to be independent of the cation, in agreement with the chloride samples described above.

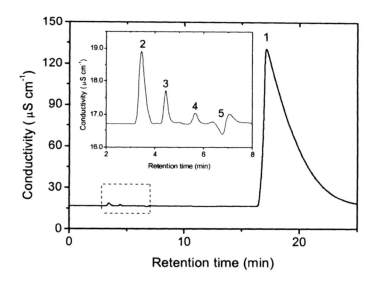

*Figure 3. Ion chromatogram of [C₄mim][BF₄] using an eluent containing 3.2 mM Na₂CO₃ + 1.0 mM NaHCO₃ and a flow rate of 1 mL min⁻¹ during the first 10 min and thereafter 1.6 mL min⁻¹. The peaks highlighted are due to (1) [BF₄] anion, (2) displacement of eluent anions in equilibrium with the column into the mobile phase due to the injection of large amount of anions, (3) fluoride, (4) chloride, and (5) the carbonate dip. The insert shows an expanded view of the trace highlighted by the dashed box.*

*Matrix effect*

Due to the complexity of the matrix and the presence of large amount of ions, the possible effect of the matrix on the detection of chloride was investigated using a chloride free ionic liquid, [C$_2$mim][EtSO$_4$]. This ionic liquid is prepared by direct alkylation of 1-methylimidazole with diethyl sulfate which eliminates the need for a metathesis step (*18*) and, therefore, any halide contamination.

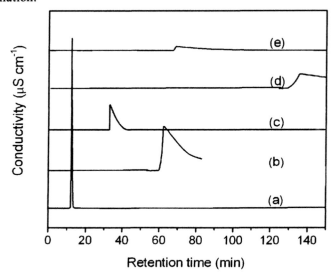

*Figure 4. Ion chromatograms of (a)/(b) [C$_4$mim][OTf], (c)/(d) [C$_4$mim][PF$_6$] and (e) [C$_4$mim][NTf$_2$]. Traces a,c and e were performed using a gradient elution system (Dionex DX120)[19] whereas traces b and d were performed using the Metrohm 761 system outlined above. It should be noted that due to the difference in suppression methods between the instruments, the intensity may only be compared for samples analysed by the same instrument. In addition, the chemical suppression used in the Metrohm 761 switches at 90 mins resulting in a dip in the trace, hence the cut off in trace b, this does not affect the peak in trace d and has been removed for clarity.*

Figure 5 compares the area-concentration profile for samples of 11.3 g L$^{-1}$ of [C$_2$mim][EtSO$_4$] ionic liquid with a known added concentration of sodium chloride and samples of water with equivalent concentrations of sodium chloride. All the recoveries were found to be between 91-105 %, *i.e.* within the acceptable limit. For analyte concentrations $\leq 0.5$ mg L$^{-1}$, the recovery percentage expected is between 60-115 % (*13*).

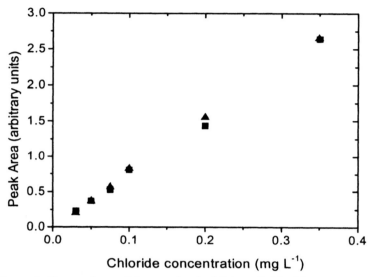

*Figure 5. Comparison of the standard addition of chloride variation in water(▲) and 11.3 g L$^{-1}$ [C$_2$mim][EtSO$_4$](■).*

*Limit of quantification (LOQ)*

As described above, due to the preparation method, ionic liquids cannot, in general, be completely halide free, and in practice LOQ values depend on the analyte and the matrix (*20*). To estimate the LOQ for the method described herein, the method of known additions was used to determine the LOQ for the different ionic liquids. In this case, the limit of quantification of the method is given by the onset of the deviation from linearity at low concentrations in the area *vs.* concentration of the chloride curve. This is an effect caused by the matrix. To illustrate the method, the area *vs.* concentration of chloride curve corresponding to 11.3 g L$^{-1}$ of [C$_4$mim][OTf] ionic liquid compared with the calibration curve at low chloride concentration is shown in Figure 6. The LOQ in this case is the lowest concentration at which the area difference between the calibration and the sample remains constant. Unlike for [C$_4$mim][OTf], no deviation from linearity was observed above the chloride impurity level found in either [C$_4$mim][BF$_4$] or [C$_4$mim][PF$_6$] ionic liquids. In these cases, the LOQ was estimated as the lowest chloride concentration measured in the ionic liquid, *i.e.* without any addition of chloride. This is obviously an upper limit to the LOQ. The LOQ values obtained for the different ionic liquids are shown in Table 2 and compared with NaCl. These values show the LOQ in the ion chromatography sample, *i.e.* the diluted ionic liquid sample (LOQ[sample]), and the value in the ionic liquid (LOQ[IL]).

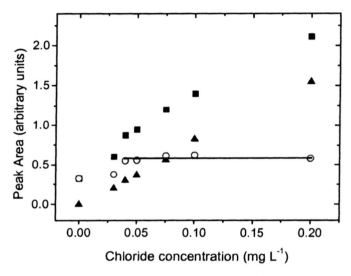

*Figure 6. Comparison of the standard addition of chloride variation in water(▲) and 11.3 g L⁻¹ [C₄mim][OTf](■), and the corresponding difference in area between the two plots(○ ).*

Clearly the LOQ [IL] is dependant on the dilution factor and lower values may be obtained if larger volumes of ionic liquid are used. However, there is a compromise between the elution time of the main anion of the ionic liquid, and hence sample turnaround, and the detection limit for a particular ionic liquid.

**Table 2.** LOQ of different ionic liquids compared with aqueous NaCl.

| Ionic liquid | LOQ[sample] | LOQ[IL] (ppm) |
|---|---|---|
| [C$_4$mim][OTf] | 0.078 | 6.9 |
| [C$_4$mim][BF$_4$] | 0.051 | 4.5 |
| [C$_4$mim][PF$_6$] | 0.075 | 7.1 |
| [C$_2$mim][EtSO$_4$] | 0.030 | 2.7 |
| NaCl | 0.030 | - |

# Cathodic stripping voltammetry

## Experimental

Electrochemical experiments were performed with a three electrode arrangement with a silver (1 mm diameter) serving as the working electrode, with a bright platinum wire used as the counter electrode. A pseudo silver wire reference electrode completed the circuit. Voltammetric measurements were carried out on a µ–autolab (Eco-Chemie) potentiostat.

## Results

Cathodic stripping using a silver electrode was investigated in order to develop a more sensitive method for chloride detection. Initially silver chloride is formed by holding the silver electrode at a positive potential in a chloride solution, and thereafter the potential is swept in an anodic direction to reveal a stripping peak corresponding to the reduction of silver chloride back to chloride anions in solution. To determine the anodic potential at which solvent breakdown occurs, a cyclic voltammogram of in [C$_4$mim][BF$_4$] containing 0.45 ppm chloride was performed. During all subsequent depositions, the potential was held slightly negative of this solvent breakdown window. The deposition time was also investigated in order to optimise the accumulation time with respect to the peak area obtained during the reduction of the silver chloride. A plot of peak area (charge) *vs.* deposition time resulted in a maxima occurring at *ca.*120 seconds. This accumulation time was used in all subsequent experiments.

Using the optimised conditions of 120 seconds at + 2.0 V (*vs.* Ag wire) followed by a potential sweep from + 2.0 to -1.7 V to strip the chloride back into the ionic liquid, [C₄mim][BF₄] was investigated with increasing concentration of chloride added. In general, multiple peaks were observed during the stripping process which transform into three peaks above 1.1 ppm of added chloride (Figure 7).

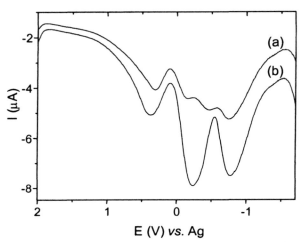

*Figure 7. Stripping of chloride in [C₄mim] [BF₄] at a 1 mm diameter silver working electrode, deposition time 120 s at +2 V (vs. Ag), square wave amplitude 25 mV, step voltage 5 mV, frequency 25 Hz.*
*(a) 0.62 ppm, and (b) 1.77 ppm of chloride added.*

In Figure 7b, the peak at +0.32 V corresponds to the stripping of silver from the electrode:

$$Ag^+ (IL) + e \rightarrow Ag$$

and the peaks at -0.21 V and -0.76 V are probably due to the oxidation of different silver chloride states, for example:

$$AgCl(S) + e \rightarrow Ag + Cl^-$$
$$AgCl_2^- (IL) + e \rightarrow Ag + 2Cl^-$$

Suturović *et al* (*21*) also reported a similar splitting of the chloride signal into two waves at chloride concentrations higher than 1 ppm in the determination of chloride by stripping chronopotentiometry in aqueous solution using a silver-film electrode. The LOQ values were determined by plotting the peak area corresponding to the chloride stripping the concentration of chloride added to [C₄mim][BF₄] (Figure 8), giving an LOQ of 0.74 ppm for chloride.

Similarly, the LOQ values for [C₄mim][NTf₂] and [C₄mim][PF₆] were determined using the protocol described above. In the case of [C₄mim][NTf₂], analogous stripping voltagrams were found compared with [C₄mim][BF₄] whereas in [C₄mim][PF₆], only one peak was observed. Strong linear correlations were obtained for the stripping signal versus added chloride for all three ionic liquids for concentrations below *c.a.* 1.5 ppm chloride (Table 3). Depending on the ionic liquid, after the addition of *c.a.* 2 ppm of chloride, the surface coverage of the electrode saturated, causing a deviation in the peak area *vs.* concentration of chloride added curve (Figure 8).

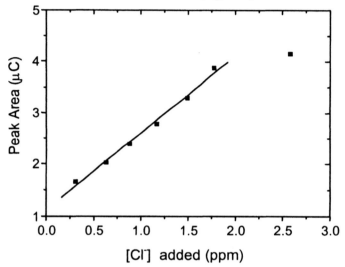

*Figure 8. Stripping peak area vs. concentration of chloride added to [C₄mim][BF₄], R² = 0.9959.*

The LOD and LOQ values found for all the ionic liquids (Table 3) were at the ppb level, with the lowest LOD (90 ppb) obtained for [C₄mim][NTf₂], *i.e.* the ionic liquid with the lowest viscosity . The chloride concentration found through cathodic stripping voltammetry in [C₄mim][BF₄], [C₄mim][NTf₂], and [C₄mim][PF₆] ionic liquids resulted in values of 0.74, 2.38, and 0.67 ppm, respectively.

**Table 3. Linear range, LOD and LOQ of cathodic stripping voltammetry for [C₄mim][BF₄], [C₄mim][NTf₂] and [C₄mim][PF₆].**

|  | Linear range / ppm of Cl⁻ added | ΔLinear range / ppm | LOD / ppm | LOQ / ppm |
|---|---|---|---|---|
| [C$_4$mim][BF$_4$] | 0.31-1.78 | 1.47 | 0.16 | 0.31 |
| [C$_4$mim][NTf$_2$] | 0.30-2.25 | 1.95 | 0.09 | 0.30 |
| [C$_4$mim][PF$_6$] | 0.25-1.59 | 1.35 | 0.16 | 0.25 |

## Comparison of Methods

IC and cathode stripping are versatile methods for the determination of chloride in ionic liquids. Overall, the cathodic stripping method is much more sensitive than IC; however, it suffers from a narrow range of concentrations over which it may be applied. IC allows a much wider range of concentrations to be analysed and also enables simultaneous determination of other impurities, for example fluoride and phosphate. Both methods can be used on water miscible and water immiscible ionic liquids. Commonly a "silver nitrate" test has been used to determine whether ionic liquids are halide free (*22*). It is performed on the water layer during the extraction of halide from the ionic liquid after the metathesis step. This test is a positive-negative test and does not allow the level of impurity present to be determined. Comparing the silver nitrate test on the washings with IC measurements on both the water and ionic liquid layers shows that a negative result can be observed below ~10 ppm chloride within the ionic liquid. It should be noted; however, that this is dependent on the ionic liquid being prepared since for water miscible ionic liquids, such as [C₄mim][BF₄], significant concentrations of the ionic liquid are washed into the water layer and this changes the ionic strength of the media and, therefore, the point at which precipitation occurs.

## Conclusions

The results of this study show that IC with suppressed conductivity detection is a reliable technique to determine chloride for a range of water miscible and immiscible ionic liquids. In terms of accuracy, ion chromatography is well suited to serve as a quality control method in the production of ionic liquids. Furthermore, cathodic stripping voltammetry at a silver disc electrode was explored and found to be optimal for the electroanalysis of sub-ppm levels of chloride.

## Acknowledgements

CV acknowledges support from QUILL and the School of Chemistry, QUB, CEB thanks the EPSRC for funding via a project studentship (GR/R14392/01), MD and GD acknowledge support from QUILL, and WRP acknowledges support from the EPSRC under grant GR/R42078. The School of Chemistry and ASEP, QUB are acknowledged for support associated with the purchase and maintenance of the IC.

## References

1 Freemantle, M. *Chem. Eng. News* **1998**, *76*, March 30; Holbrey, J. D.; Seddon, K. R. *Clean products and processes* **1999**, *1*, 223.
2 Seddon, K.R.; Stark, A.; Torres, M.J. *Pure Appl. Chem.* **2000**, *72*, 2275.
3 Trulove, P.C.; Mantz, R.A. *Ionic Liquids in Synthesis*, Wasserscheid, P.; Welton T. Eds., Wiley-VCH Verlag,Weinheim, 2003.
4 Wasserscheid, P.; Keim, W. *Angew. Chem., Int. Ed.* **2000**, *39*, 3772.
5 Owens, G.S.; Mahdi, M.A. In *Ionic Liquids - Industrial Applications to Green Chemistry*; Rogers, R.D.; Seddon, K.R., Eds; ACS Symposium Series 818; American Chemical Society: Washington, DC, pp. 321, 2002.
6 Anderson, K.; Goodrich, P.; Hardacre, C.; Rooney, D.W. *Green Chemistry* **2003**, *5*, 448.
7 Anthony, J.L.; Maginn, E.J.; Brennecke, J.F. *J. Phys. Chem. B* **2001**, *105*, 10942.
8 Billard, I.; Moutiers, G.; Labet, A.; El Azzi, A.; Gaillard, C.; Mariet, C.; Lutzenkirchen, K. *Inorg. Chem.* **2003**, *42*, 1726.
9 Anderson, J.L.; Ding, J.; Welton, T.; Amstrong, D.W. *J. Am. Chem. Soc.* **2002**, *124*, 14247.
10 Xiao, L.; Johnson K.E. *J. Electrochem. Soc.* **2003**, *150*, E307.
11 Villagran, C.; Deetlefs, M.; Pitner, W.R.; Hardacre, C.; *Anal. Chem.* **2004**, *76*, 2118.

12 Villagran, C.; Banks, C.E.; Hardacre, C.; Compton, R.G. *Anal. Chem.* **2004**, *76*, 1998.

13 Fajgelj, A.; Ambrus, A. Eds, *Principles and Practices of Method Validation,* The Royal Society of Chemistry, 2000.

14 *Validation of Metrohm Ion Chromatography Systems by using Standard Operation Procedures* (SOP), Application Bulletin 277/1 e, Metrohm.

15 Holbrey, J.D.; Seddon, K.R. *J. Chem. Soc. Dalton Trans.* **1999**, 2133.

16 Visser, A.E; Swatloski, R.P.; Reichert, W.M; Griffin, S.T.; Rogers, R.D. *Ind. Eng. Chem. Res.* **2000**, *39*, 3596.

17 Swatloski, R.P.; Holbrey, J.D.; Rogers, R.D. *Green Chemistry* **2003**, *5*, 361.

18 Holbrey, J.D.; Reichert, W.; Swatloski, R.P.; Broker, G.A.; Pitner, W.R.; Seddon, K. R.; Rogers, R.D. *Green Chemistry* **2002**, *4*, 407.

19 The analyses were performed at room temperature using an Ionpac AS16 analytical column (250 x 4 mm) and an Ionpac AG16 guard column (50 x 4 mm). The injection volume was 20 µL. The eluent used was a 10 mM KOH for the first 20 mins and thereafter 90 mM KOH at a flow rate of 1 mL min[-1]. The IC samples were prepared by dissolving 0.11 g of IL in 2 mL of acetonitrile before being diluted to 10 mL with deionised water.

20 *Standard Methods for the Examination of Water and Wastewater,* 20th Edition, APHA-AWWA-WEF, 1998.

21 Suturović, Z.J.; Marjanović, N.J.; Dokić, P.P. *Electroanalysis* **1997**, *9*, 572.

22 Hilgers, C.; Wasserscheid, P. *Ionic Liquids in Synthesis*, Wasserscheid, P.; Welton T. Eds., Wiley-VCH Verlag,Weinheim, 2003.

Chapter 20

# (1R)-4-Amino-1,2,4-triazolium Salts: New Families of Ionic Liquids

Gregory Drake[1,3], Tommy Hawkins[1], Kerri Tollison[2], Leslie Hall[2], Ashwani Vij[2], and Sarah Sobaski[2]

[1]Space and Missile Propulsion Division, Propellants Branch and [2]ERC, Incorporated, 10 East Saturn Boulevard, Air Force Research Laboratory, Edwards Air Force Base, CA 93524–7680
[3]Current address: XD22 Propulsion Research Center, NASA, Building 4205, MSFC, AL 38812

New classes of ionic liquids based upon the halide and nitrate salts of 1-alkyl substituted-4-amino-1,2,4-triazolium cations (n-alkyl = methyl -decyl, isopropyl, allyl, and methylcylcopropyl) have been synthesized, characterized by vibrational spectra, multinuclear nmr, elemental analysis, and DSC studies. Single crystal x-ray diffraction studies were carried out on 1-isopropyl-4-amino-1,2,4-triazolium bromide, 1-ethyl-4-amino-1,2,4-triazolium bromide, 1-n-propyl-4-amino-1,2,4-triazolium bromide, 1-n-hexyl-4-amino-1,2,4-triazolium bromide, and 1-n-heptyl-4-amino-1,2,4-triazolium bromide, as well as 1-isopropyl-4-amino-1,2,4-triazolium nitrate and 1-methylcyclopropyl-4-amino-1,2,4-triazolium nitrate. The details of similarities, differences, and the effects of strong hydrogen bonding in the all of the structures will be discussed.

## Introduction

The field of ionic liquids is a rapidly growing area of interest in many areas of chemistry.(2-8) Ionic liquids were originally investigated as possible replacement electrolytes in battery applications by pioneering efforts of Wilkes and Hussey.(9) With the discovery of water stable systems(10,11), recent efforts have been rapidly expanding the boundaries of applications for ionic liquids rather than on the materials themselves.     Much of the chemistry is based upon

N-N'-disubstituted imidazolium cations paired with anions such as hexafluorophosphate(*11*), tetrafluoroborate(*10*), and triflate(*6, 7*) anions. Ionic liquids have found a wide array of applications from reaction media(*4, 6-8, 12-19*), in separation sciences(*6, 7, 20-22*), and in many kinds of catalyses reactions(*6, 7, 23-28*).

At the Air Force Research Laboratory, we have been investigating low melting salts and have found that the behavior is not restricted to the well-known 1,3-di-alkyl substituted imidazolium and 1-alkyl-pyridinium cation based salts. In efforts at identifying additional classes of low melting salts, other five-membered azole rings were considered as possibly being substituted in a 1,4 conformation, similar to that found in the related 1,3-disubstituted imidazolium cation systems, might lead to similar ionic liquid properties. Examples of 1,4-disubstituted-triazolium salts have been known for quite sometime(*29-31*) and recently, Shreeve(*32-34*) has demonstrated this behavior with a large new family of ionic liquids based on fluoro-alkylated 1,4-disubstituted-1,2,4-triazolium cations. Extending the premise that asymmetric 5-membered heterocyclic cations will have poor packing in three-dimensional space, which would result in new classes of low melting materials, we have found this behavior in a similarly shaped heterocycle, 4-amino-1,2,4-triazole. The 4-amino-1,2,4-triazole ring system has a much higher nitrogen content as well as a non-basic amine group, which we felt would make for unusual physical properties. As is often true in many research arenas our predecessors have often demonstrated chemistries for materials in completely different scientific pursuits. However, no one has considered the thought of this heterocycle as an ionic liquid building block with the 4-amino-1,2,4-triazolium cationic species serving as the heterocyclic platform. There are several brief reports throughout the last several decades describing various 1-substituted-4-amino-1,2,4-triazolium salts(*35-40*), and recently former Soviet Union research groups have produced interesting examples such as 1-substituted-4-amino-1,2,4-triazolium nitrates using the powerful amination reagent, picryloxyamine, in reactions with 1-R-1,2,4-triazoles(*41, 42*) and subsequent N-amino nitration forming highly unusual 1-R-substituted-4-nitramino-1,2,4-triazole zwitterions(*42, 43*).

We have been able to synthesize and characterize a large new family of low melting salts based upon 1-substituted-4-amino-1,2,4-triazolium cation species as halide and nitrate salts through simple reactions with commercially available materials in high yields and purities. New species include 1-n-alkyl (methyl-decyl, isopropyl, 1-methylcyclopropyl-, and 2-propenyl)-1-substituted-4-amino-1,2,4-triazolium cations. The syntheses, physical properties, spectral data, as well as several x-ray diffraction crystal structures of the salts will be discussed.

## Experimental

The starting materials, 4-amino-1,2,4-triazole, methyl iodide, n-alkyl bromides (ethyl-decyl), allyl bromide, and bromomethylcyclopropane were purchased from Aldrich Chemical Company, Inc, and their purities checked by $^1$H and $^{13}$C NMR prior to use. Methanol, $CH_3OH$; Ethanol, $CH_3CH_2OH$; and 2-Propanol, $(CH_3)_2CH$-OH, (ACS reagent grade; distilled from sodium metal), and acetonitrile, $CH_3CN$ (HPLC grade; distilled from calcium hydride) were purchased from Aldrich Chemical Company, and all solvents were degassed using a liquid nitrogen freeze-thaw vacuum procedure. Diethyl ether was dried through a pre-activated alumina column prior to use. All solvents were stored inside glass vessels, which were sealed with teflon screw-cap plugs, and were equipped with #15 O-ring fittings. Infrared spectra were recorded as KBr disks (using a KBr disk as a reference background) on a Nicolet 55XC FT-IR spectrometer from 4000-400 cm$^{-1}$. Raman spectra were recorded in pyrex melting point capillaries on Bruker Model FRA 106/S Equinox 55 Raman spectrometer equipped with a 1.06 micron IR excitation laser. NMR experiments were carried out by dissolving the salts in d$_6$-dmso in 5mm nmr tubes inside a drybox, and the $^1$H and $^{13}$C spectra recorded on a Bruker Spectrospin DRX 400 MHz Ultrashield$^{TM}$ NMR. Thermal analyses were carried out in hermetically sealed, coated aluminum pans on a Thermal Analyst 200, Dupont Instruments 910 Differential Scanning Calorimeter. DSC samples were prepared and sealed inside a nitrogen-filled glove box, and once the pans were inside the DSC cell, the cell was flushed with 10 mL per minute of nitrogen gas purge during heating cycles. Elemental analyses were carried out on a Perkin Elmer Series II CHNS/O Analyzer 2400 elemental analysis instrument equipped with AD6 Autobalance and by Desert Analytics, Inc. of Tucson, AZ. Densities were measured using helium displacement techniques in a calibrated cell using a Quantachrome Ultrapycnometer 1000 instrument. The synthesis of 1-n-butyl-4-amino-1,2,4-triazolium bromide was described previously.(*38*)

1-methyl-4-amino-1,2,4-triazolium iodide (I): 4-amino-1,2,4-triazole, 5.1833g., 61.6 mmoles, was weighed out and placed in a 250 ml round-bottomed flask with a Teflon stir bar. Isopropyl alcohol, 200ml, was added and the mixture stirred for a short period of time to completely dissolve the 4-amino-1,2,4-triazole. Methyl iodide, 26.5143 g., 186 mmoles, was then added to the vigorously stirred solution. The flask was then protected from light with a black bag, and stirred for seven days at ambient temperature. At the end of this time an additional 1.50g of methyl iodide was added and the reaction mixture stirred for five additional days. The solution was pale yellow with white precipitate in the bottom of the flask. The precipitate was filtered and washed four aliquots, 50 ml each, of cold isopropyl alcohol, followed by four washings, 50 ml each, of cold diethyl ether. The white powder was then transferred to a preweighed Schlenk flask and evacuated overnight to leave 10.1840 g, 45.0 mmoles of 1-methyl-4-amino-1,2,4-triazolium iodide. Melting point 98°C, DSC onset beginning at 136°C. $^1$H NMR(d$_6$-dmso): 4.024 (singlet, area 3.067), 6.938 (singlet, area 1.661), 9.161 (singlet, area 0.932), 10.115 (singlet, area 1.000). $^{13}$C NMR (d$_6$-dmso): 39.107, 143.002, 145.109. Elemental analysis: %C: 15.94 (theory); 16.19 (found); %H: 3.12 (theory); 3.04 (found); %N: 24.79 (theory); 24.59 (found).

1-ethyl-4-amino-1,2,4-triazolium bromide (II): A 500 ml round-bottom flask equipped with an overhead stirrer was charged with 10.00 g. of 4-amino-1,2,4-triazole and 200 ml of acetonitrile. Ethyl bromide, 45 ml, 65.0 g., was added to the vigorously stirred reaction mixture. The reaction was stirred for 8 days at ambient temperature at which time, thin layer chromatography showed that all of the 4-amino-1,2,4-triazole had been consumed. The resultant solution was then rotary evaporated down leaving a colorless ionic liquid which slowly crystallized. The solid material was heated to 60°C for 5 hours under high vacuum, whereupon it melted, lost the remaining solvent, and re-solidified as highly crystalline 1-ethyl-4-amino-1,2,4-triazolium bromide in essentially quantitative yield and high purity, 22.94 g., 117 mmoles. Melting point: 63-67°C; DSC onset: 150°C. $^1$H NMR (d$_6$-dmso): 1.402, 1.420, 1.438 (triplet, area 3.000); 4.359, 4.377, 4.395, 4.413 (quartet, area 2.003); 7.084 (broad singlet, area 1.648); 9.202 (singlet, area 0.959); 10.325 (singlet, area 1.000). $^{13}$C NMR (d$_6$-dmso): 13.768, 47.335, 142.289, 145.114. Elemental analysis: %C: 24.95 (theory); 24.73 (found); %H: 4.74 (theory); 4.73 (found); %N: 29.21(theory); 29.09 (found).

1-n-propyl-4-amino-1,2,4-triazolium bromide (III): In the typical manner as cited for II above, 10.005 g., 118 mmoles, of 4-amino-1,2,4-triazole was reacted with 1-bromopropane, 58.865g., 478 mmoles in acetonitrile at 50°C, yielding highly crystalline 1-n-propyl-4-amino-1,2,4-triazolium bromide, 23.9584 g, 115 mmoles. Melting point: 63°C; DSC onset 145°C. $^1$H NMR (d$_6$-dmso): 0.806, 0.823, 0.836 (triplet, area 3.000); 1.818, 1.834 (broad multiplet, area 2.013); 4.362, 4.373 (broad multiplet, area 1.999); 7.126 (broad singlet, area 1.816); 9.244 (singlet, area 0.928); 10.440 (singlet, area 0.957).$^{13}$C NMR (d$_4$-MeOH): 11.051, 23.313, 55.538, 144.586, 146.908. Elemental analysis, %C: 29.00 (theory); 28.96 (found); %H: 5.35 (theory); 5.46 (found); %N: 27.06 (theory); 27.48 (found).

1-isopropyl-4-amino-1,2,4-triazolium bromide (IV): In the manner of II above, 10.122 g., 120 mmoles, of 4-amino-1,2,4-triazole was reacted at 50°C with 2-bromopropane 58.560 g., 476 mmoles yielding on work-up, 1-isopropyl-4-amino-1,2,4-triazolium bromide, 17.421 g., 84 mmoles. Melting point: 92°C; DSC onset 145°C. $^1$H NMR (d$_6$-dmso): 0.461, 0.472, 0.557, 0.576 (complex multiplet, relative area 4.000); 1.273, 1,283, 1.293, 1.302, 1.313, 1.320, 1.332

(complex multiplet, area 0.992); 4.261, 4.279 (doublet, area 2.032); 7.088 (broad singlet, area 1.309); 9.231 (singlet, area 0.971); 10.412 (singlet, area 1.309). $^{13}C$ NMR ($d_6$-dmso): 3.849, 10.082, 56.162, 142.150, 145.179. Elemental analysis: %C: 32.89 (theory); 32.73 (found); %H: 5.06 (theory); 5.07 (found); %N: 25.57 (theory); 25.27 (found).

1-(2-propenyl)-4-amino-1,2,4-triazolium bromide (V): In the manner of II above 4-amino-1,2,4-triazole, 10.000 g., 118 mmoles, was reacted with allyl bromide, 43.10 g., 356 mmoles resulting in 1-allyl-4-amino-1,2,4-triazolium bromide, 12.657 g., 62 mmoles. Melting point: 59-62° C; DSC onset: 130° C. $^1H$ NMR ($d_6$-dmso): 5.068, 5.288, 5.316, 5.328 5.325 (broad multiplets, area 3.033); 5.993, 5.946, 5.958, 5.974, 5.987, 6.000 (broad multiplet, area 0.890); 7.126 (broad singlet, area 1.962); 9.234 (singlet, area 0.938); 10.426 (singlet, area 1.000). $^{13}C$ NMR ($d_6$-dmso): 53.736, 120.912, 130.255, 142.572, 145.292. Elemental analysis: %C: 29.29 (theory); 29.37 (found); %H: 4.42 (theory); 4.59 (found); %N: 27.32 (theory); 27.04 (found).

1-methylcyclopropyl-4-amino-1,2,4-triazolium bromide (VI): In the manner of II above 4-amino-1,2,4-triazole 3.2154 g., 38.2 mmoles, was reacted with bromomethylcyclopropane, 10.7539 g., 79.6 mmoles, yielding 1-methylcyclopropyl-4-amino-1,2,4-triazolium bromide, 4.8167 g., 22 mmoles. Melting point: 71-73°C; DSC onset: 150° C. $^1H$ NMR ($d_6$-dmso): 0.461, 0.472, 0.557, 0.576 (complex multiplet, area 4.000); 1.27, 1,283, 1.293, 1.302, 1.313, 1.320, 1.332 (complex multiplet, area 0.992); 4.261, 4.279 (doublet, area 2.032); 7.088 (broad singlet, area 1.309); 9.231 (singlet, area 0.971); 10.412 (singlet, area 1.309). $^{13}C$ NMR ($d_6$-dmso): 3.849, 10.082, 56.162, 142.150, 145.179. Elemental analysis: %C: 32.89 (theory); 32.73 (found); %H: 5.06 (theory); 5.07 (found); %N: 25.57 (theory); 25.27 (found).

1-n-pentyl-4-amino-1,2,4-triazolium bromide (VII): In the manner of II above 4-amino-1,2,4-triazole, 2.000 g., 23.7 mmoles, was reacted with n-pentyl bromide 7.22 g., 47.8 mmoles, resulting in 1-n-pentyl-4-amino-1,2,4-triazolium bromide, 4.4737 g., 19.0 mmoles. Melting point: 54°C; DSC onset: 130°C. $^1H$ NMR ($d_6$-dmso): 0.819, 0.838, 0.855, 0.880, 0.897 (broad multiplet, area 2.321); 1.180, 1.194, 1.202, 1.215, 1.233, 1.249, 1.261, 1.278, 1.296, 1.313 (complex multiplet, area 3.438); 1.788, 1.805, 1.823, 1.842, 1.859 5.994 (complex multiplet, area 1.889); 4.343, 4.361, 4.378 (triplet, area 2.000); 7.041 (broad singlet, area 2.121); 9.217 (singlet, area 0.914); 10.357 (singlet, area 0.996). $^{13}C$ NMR ($d_6$-dmso): 13.691, 21.423, 27.453, 27.728, 51.675, 142.602, 145.196. Elemental analysis: %C: 35.76 (theory); 35.25 (found); %H: 6.43 (theory); 6.45 (found); %N: 23.83 (theory); 24.25 (found).

1-n-hexyl-4-amino-1,2,4-triazolium bromide(VIII): In a similar method to that used for II above, 2.000 g., 23.7 mmoles of 4-amino-1,2,4-triazole was reacted with n-hexyl bromide 8.080 g., 48.9 mmoles, at 60°C for 40 hours, resulting 4.924 g., 19.7 mmoles, of 1-n-hexyl-4-amino-1,2,4-triazolium bromide. Melting point: 76°C; DSC onset: 130 C. $^1$H NMR ($d_6$-dmso): 0.842, (broad singlet, area 2.377); 1.253 (broad singlet, area 5.597); 1.825 (broad singlet, area 1.757); 4.371 (broad singlet, area 2.000); 7.082 (broad singlet, area 1.966); 9.236 (singlet, area 0.859); 10.363 (singlet, area 0.927). $^{13}$C NMR ($d_6$-dmso): 13.797, 21.821, 24.985, 27.993, 30.461, 51.962, 142.587, 145.182. Elemental analysis: %C: 38.57 (theory); 38.32 (found); %H: 6.43 (theory); 6.45 (found); %N: 22.49 (theory); 22.88 (found).

1-n-heptyl-4-amino-1,2,4-triazolium bromide(IX): In the manner stated in II above, 2.000 g., 23.7 mmoles, of 4-amino-1,2,4-triazole was reacted with n-heptyl bromide 10.650 g., 59.1 mmoles, at 60°C for 40 hours, resulting 5.7460 g., 21.7 mmoles, of 1-n-heptyl-4-amino-1,2,4-triazolium bromide. Melting point: 94°C; DSC onset: 130°C. $^1$H NMR ($d_6$-dmso): 0.807, 0.824, 0.840 (triplet, area 3.000); 1.221, 1.235, 1.244 (broad multiplet, area 8.966); 1.797, 1.813, 1.830 (multiplet, area 2.146); 4.346, 4.364, 4.381 (triplet, area 2.358); 7.090 (broad singlet, area 2.122); 9.220 (singlet, area 1.059), 10.375 (singlet, area 1.094). $^{13}$C NMR ($d_6$-dmso): 13.587, 21.927, 25.270, 27.941, 28.046, 30.959, 51.672, 142.560, 145.153. Elemental analysis: %C: 41.07 (theory); 41.24 (found); %H: 7.28 (theory); 7.05 (found); %N: 21.29 (theory); 21.13 (found).

1-n-octyl-4-amino-1,2,4-triazolium bromide (X): In essentially the same manner as in II above, 2.000 g., 23.7 mmoles, of 4-amino-1,2,4-triazole was reacted with n-octyl bromide 9.790 g., 50.7 mmoles, at 60°C for 40 hours, yielding 5.4768 g., 19.7 mmoles, of 1-n-octyl-4-amino-1,2,4-triazolium bromide. Melting point: 81°C; DSC onset: 130°C. $^1$H NMR($d_6$-dmso): 0.815(broad singlet), 1.227 (broad singlet), 1.806 (broad singlet), area of all three peaks 17.211), 4.369 (broad singlet, area 2.000), 7.112 (broad singlet, area 1.993), 9.225 (singlet, area 0.880), 10.404 (singlet, area 0.868). $^{13}$C NMR ($d_6$-dmso): 13.848, 22.000, 25.319, 28.055, 28.257, 28.405, 31.102, 51.660, 142.525, 145.117. Elemental analysis: %C: 43.32 (theory); 43.57 (found); %H: 7.64 (theory); 7.56(found); %N: 20.21(theory); 20.16(found).

1-n-nonyl-4-amino-1,2,4-triazolium bromide (XI): As in II above, 2.000 g., 23.7 mmoles, of 4-amino-1,2,4-triazole was reacted with n-nonyl bromide 12.319 g., 59.1 mmoles, at 60°C for 40 hours, yielding 6.3680 g., 21.7 mmoles, of 1-n-nonyl-4-amino-1,2,4-triazolium bromide. Melting point: 81°C; DSC onset: 130°C. $^1$H NMR ($d_6$-dmso): 0.823, 0.839, 0.854 (triplet, area 3.266); 1.230 (broad, 13.571); 1.827, 1.843 (broad, area 2.165); 4.363, 4.380, 4.397 (triplet, area 2.191); 7.090 (broad singlet, area 1.873), 9.24 (singlet, area 1.069); 10.397 (singlet, area 1.111). $^{13}$C NMR (d$_6$-dmso): 13.913, 22.072, 25.363, 28.089, 28.346, 28.590, 28.758, 31.242, 51.698, 53.865, 142.580, 145.166. Elemental analysis: %C: 45.37 (theory); 45.61 (found); %H: 7.96 (theory); 8.06 (found); %N: 19.24 (theory); 19.21 (found).

1-n-decyl-4-amino-1,2,4-triazolium bromide (XIII): As in II above 2.000 g., 23.7 mmoles, of 4-amino-1,2,4-triazole was reacted with n-decyl bromide 10.770 g., 48.7 mmoles at 60°C for 40 hours, yielding 6.3227 g., 20.7 mmoles, of 1-n-decyl-4-amino-1,2,4-triazolium bromide. Melting point: 90°C; DSC onset: 135°C. $^1$H NMR (d$_6$-dmso): 0.821, 0.838, 0.853 (triplet, area 2.621), 1.226 (broad singlet, area 15.002) 1.820, 1.836 (broad doublet, area 2.039), 4.334, 4.351, 4.368 (triplet, area 2.000), 7.055 (broad singlet, area 1.887), 9.212 (singlet, area 0.984), 10.321 (singlet, area 1.027). $^{13}$C NMR (d$_6$-dmso): 13.916, 22.064, 25.348, 28.042, 28.307, 28.648, 28.785, 28.869, 31.250, 51.700, 142.592, 145.176. Elemental analysis: %C: 47.22 (theory); 47.31 (found); %H: 8.25 (theory); 8.04 (found); %N: 18.35 (theory); 18.40 (found).

1-ethyl-4-amino-1,2,4-triazolium nitrate (XIV): In a 250 mL beaker, 1-ethyl-4-amino-1,2,4-triazolium bromide, 3.9543 g., 20.5 mmoles was dissolved of methanol. In a 1 liter round bottomed flask, silver nitrate, 3.4794 g., 20.5 mmoles, was dissolved in 250 ml of methanol at 60°C in a darkened room. The methanolic solution of 1-ethyl-4-amino-1,2,4-triazolium bromide was slowly added with a disposable pipette over a 10 minute period to the vigorously stirred silver nitrate solution (which had been removed from the heating bath), and after the addition was completed the reaction mixture was stirred until it cooled to room temperature (30 minutes). It was then filtered through a plug of celite, the flask and celite plug were subsequently washed with fresh aliquots of methanol to insure complete transfer of product through filter. The resultant filtrate was rotary evaporated down to a minimum volume, transferred to a pre-weighed Schlenk flask with fresh methanol (30 ml). The flask was then immersed in a heating bath (60°C) for 16 hours, with frequent agitation to remove all of volatiles resulting in the ionic liquid product of 1-ethyl-1,2,4-triazolium nitrate, 3.5467g., 20.2 mmoles, (98%). After the mass was recorded a very small portion of the product salt was added to a concentrated silver nitrate solution, with no cloudiness or precipitation that were observable that would be indicative of bromide anion. Melting point: -50°C; DSC onset: 160°C. $^1$H NMR (d$_6$-dmso): 1.410, 1.428, 1.446 (triplet, area 3.000); 4.334, 4.352, 4.370, 4.388 (quartet, area 2.023); 7.016 (singlet, area 1.991); 9.164 (singlet, area 0.961); 10.200 (singlet, area 0.978). $^{13}$C NMR (d$_6$-dmso): 13.834, 47.465, 142.605, 145.366. Elemental analysis: %C: 27.43 (theory); 27.27 (found); %H: 5.17 (theory); 5.55 (found); %N: 39.98 (theory); 39.83 (found).

1-methyl-4-amino-1,2,4-triazolium nitrate (XV): In a similar manner to that used for XIV above, 1-methyl-4-amino-1,2,4-triazolium iodide, 1.3063g., 5.78 mmoles, was reacted with silver nitrate, 0.9825 g., 5.78 mmoles, yielding 1-methyl-4-amino-1,2,4-triazolium nitrate, 0.8925 g., 5.54 mmoles, 95%. Melting point: -54°C; decomposition onset 175°C. $^1$H NMR(d$_6$-dmso): 4.051 (singlet, area 2.974); 6.991 (singlet, area 1.719); 9.079 (singlet, area 0.954); 10.116 (singlet, 0.997). $^{13}$C NMR(d$_6$-dmso): 39.208, 143.584, 145.615. Elemental analysis: %C: 22.22 (theory); 22.06 (found); %H: 4.35 (theory); 4.67 (found); %N: 43.19 (theory); 43.17 (found).

1-n-propyl-4-amino-1,2,4-triazolium nitrate (XVI): In the manner of XIV above, 1-n-propyl-4-amino-1,2,4-triazolium bromide, 1.3736 g., 6.63 mmoles, was reacted with silver nitrate, 1.1272 g., 6.63mmoles, resulting in 1-n-propyl-4-amino-1,2,4-triazolium nitrate, 1.2488 g., 6.60 mmoles, 99%. Melting point: 33°C; decomposition onset: 175°C. $^1$H NMR (d$_6$-dmso): 0.814, 0.833, 0.851 (triplet, area 3.000); 1.790, 1.808, 1.826, 1.844, 1.862, 1.880 (sextet, area 2.031); 4.286, 4.303, 4.321 (triplet, area 2.043); 7.338 (broad, area 1.957); 9.169 (singlet, area 0.963); 10.225 (singlet, area 0.969). $^{13}$C NMR (d$_6$dmso): 10.465, 21.824, 53.437, 143.052, 145.572. Elemental analysis: %C: 31.74 (theory); 31.61 (found); %H: 5.86 (theory); 6.08 (found); %N: 37.02 (theory); 37.24 (found).

1-n-butyl-4-amino-1,2,4-triazolium nitrate (XVII): In the manner of XIV above, 1-n-butyl-4-amino-1,2,4-triazolium bromide 1.3063 g., 5.91 mmoles, was reacted with silver nitrate, 1.0041 g., 5.91 mmoles, resulting in 1-n-butyl-4-amino-1,2,4-triazolium nitrate, 1.0510 g., 5.17 mmoles, 88%. Melting point: -50°C (g.t.); decomposition onset 170°C. $^1$H NMR(d$_6$-dmso):0.856, 0.874, 0.892 (triplet, area 3.140); 1.213, 1.231, 1.249, 1.268, 1.287, 1.306 (sextet, area 2.047); 1.772, 1.790, 1.809, 1.827, 1.845 (pentet, area 2.092); 4.334, 4.352, 4.369 (triplet, area 2.101); 7.026 (singlet, area 2.061); 9.176 (singlet, area 0.959); 10.253 (singlet, area 1.000). $^{13}$C NMR(d$_6$-dmso): 13.263, 18.731, 30.139, 51.566, 142.883, 145.399. Elemental analysis: %C: 35.46 (theory); 35.75 (found); %H: 6.44 (theory); 6.98 (found); %N: 34.46 (theory); 34.53 (found)

1-isopropyl-4-amino-1,2,4-triazolium nitrate (XVIII): In the manner of XIV above, 4.0560 g., 19.6 mmoles, 1-isopropyl-4-amino-1,2,4-triazolium bromide was reacted with silver nitrate 3.3270 g., 19.6 mmoles, yielding on work-up, 1-isopropyl-4-amino-1,2,4-triazolium nitrate, 3.6160 g., 19.1 mmoles, 97%. Melting point: 66°C; DSC onset: 180°C. $^1$H NMR (d$_6$-dmso): 1.419, 1.469, 1.474, 1.485, 1.490 (multiplet, area 6.095); 4.741, 4.755, 4.768, 4.785, 4.801, 4.817, 4.834, 4.261, 4.279 (heptet, area 1.000); 6.989 (broad singlet, area 2.000); 9.170 (singlet, area 0.986), 10.244 (singlet, area 0.997). $^{13}$C NMR (d$_6$-dmso):

21.287, 55.420, 141.854, 145.301. Elemental analysis: %C: 31.74 (theory); 31.47 (found); %H: 5.86 (theory); 6.13 (found); %N: 37.02 (theory); 36.61 (found).

1-(2-propenyl)-4-amino-1,2,4-triazolium nitrate (XIX): In the manner of XIV above 1-(2-propenyl)-4-amino-1,2,4-triazolium bromide, 1.1348 g., 5.53 mmoles, was reacted with silver nitrate, 0.9976 g., 5.53 mmoles, resulting in 1-allyl-4-amino-1,2,4-triazolium nitrate, 0.9976 g., 5.33 mmoles, 96%. Melting point: -50°C (g.t.); DSC onset: 140°C. $^1$H NMR (d$_6$-dmso): 5.004, 5.018 (doublet, area 2.348); 5.321, 5.361, 5.362, 5.386 (multiplet, area 2.273); 5.982, 5.995, 6.008, 6.024, 6.037, 6.050 (multiplet, area 1.092); 7.024 (broad singlet, area 2.000); 9.206 (singlet, area 1.046); 10.217 (singlet, area 1.080). $^{13}$C NMR (d$_6$-dmso): 53.923, 120.985, 130.498, 143.035, 145.545. Elemental analysis: %C: 32.08 (theory); 31.84 (found); %H: 4.84 (theory); 5.15 (found); %N: 37.41 (theory); 37.19 (found).

1-methylcyclopropyl-4-amino-1,2,4-triazolium nitrate (XX): In the manner of XIV above 1-methylcyclopropyl-4-amino-1,2,4-triazolium bromide, 1.2467 g., 5.69 mmoles, was reacted with silver nitrate, 0.9666 g., 5.69 mmoles, yielding 1-methylcyclopropyl-4-amino-1,2,4-triazolium nitrate, 1.1309 g., 5.62 mmoles, 98%. Melting point: 56°C; DSC onset: 185° C. $^1$H NMR (CD$_3$CN): 0.411, 0.446, 0.457, 0.461, 0.469, 0.472, 0.484, 0.594, 0.613, 0.630, 0.641, 0.645, 0.650, 0.661, 0.665, 0.667 (complex multiplet, area 4.000); 1.298, 1.305, 1.309, 1.317, 1.325, 1.329, 1.336, 1.344, 1.348, 1.356, 1.368 (complex multiplet, area 1.006); 4.181, 4.200 (doublet, area 2.016), 6.637 (broad singlet, area 2.001); 8.750 (singlet, area 0.890); 10.412 (singlet, area 0.930). $^{13}$C NMR (CD$_3$CN): 4.615, 10.528, 58.173, 143.507, 146.083. Elemental analysis: %C: 35.82 (theory); 35.54 (found); %H: 5.51 (theory); 5.53 (found); %N: 34.81 (theory); 34.66 (found).

1-n-pentyl-4-amino-1,2,4-triazolium nitrate (XXI): In the manner of XIV above, 1-n-pentyl-4-amino-1,2,4-triazolium bromide, 1.2505 g., 5.31 mmoles, was reacted with silver nitrate, 0.9024 g., 5.31 mmoles, resulting in 1-n-pentyl-4-amino-1,2,4-triazolium nitrate, 0.9819 g., 4.52 mmoles, 85%. Melting point: 29°C; DSC onset: 180° C. $^1$H NMR (d$_6$-dmso): 0.812, 0.830, 0.844 (broad multiplet, area 2.894); 1.208, 1.228, 1.251, 1.269, 1.287 (complex multiplet, area 3.868); 1.807, 1.824, 1.841 (complex multiplet, area 2.048); 4.327, 4.343, 4.359 (triplet, area 2.118); 7.038 (broad singlet, area 2.097); 9.168 (singlet, area 1.022); 10.251 (singlet, area 1.000). $^{13}$C NMR (d$_6$-dmso): 13.826, 21.644, 27.655, 27.972, 51.904, 143.021, 145.532. Elemental analysis: %C: 38.70 (theory); 38.77 (found); %H: 6.95 (theory); 6.98 (found); %N: 32.23 (theory); 32.21 (found).

1-n-hexyl-4-amino-1,2,4-triazolium nitrate(XXII): In a manner similar of XIV above, 1-n-hexyl-4-amino-1,2,4-triazolium bromide, 1.0561 g., 4.23 mmoles, was reacted with silver nitrate, 0.7195 g., 4.23 mmoles, resulting of 1-n-hexyl-4-amino-1,2,4-triazolium nitrate, 0.9635 g., 4.17 mmoles, 98%. Melting point: -2-0°C; DSC onset: 170° C. $^1$H NMR (d$_6$-dmso): 0.820, 0.829, 0.836, (broad multiplet, area 3.378), 1.242 (broad singlet, area 7.011), 1.818 (broad singlet, area 2.130), 4.326, 4.334, 4.343, 4.361 (broad multiplet, area 2.034), 7.040 (broad singlet, area 2.036), 9.157 (singlet, area 1.004), 10.248 (singlet, area 1.000). $^{13}$C NMR (d$_6$-dmso): 13.878, 21.999, 25.160, 28.207, 30.652, 51.895, 142.952, 145.450. Elemental analysis: %C: 41.51 (theory); 41.28 (found); %H: 7.40 (theory); 7.93 (found); %N: 30.28 (theory); 30.14 (found).

1-n-heptyl-4-amino-1,2,4-triazolium nitrate(XXIII): In the manner stated in XIV above, 1-n-heptyl-4-amino-1,2,4-triazolium bromide, 1.0031 g., 3.81 mmoles, was reacted with silver nitrate, 0.6470 g., 3.81 mmoles, resulting of 1-n-heptyl-4-amino-1,2,4-triazolium nitrate, 0.9115 g., 3.72 mmoles, 97%. Melting point: 35°C; DSC onset: 165°C. $^1$H NMR (d$_6$-dmso): 0.823, 0.825, 0.840, 0.844, 0.857 (multiplet, area 3.303); 1.236, 1.247, 1.258, 1.356, 1.378, 1.396 (multiplet, area 9.072); 4.320, 4.338, 4.355 (triplet, area 2.261); 7.012 (broad singlet, area 2.000), 9.175 (singlet, area 1.065), 10.230 (singlet, area 1.099). $^{13}$C NMR (d$_6$-dmso): 13.918, 21.973, 25.327, 27.965, 28.034, 31.012, 51.784, 142.730, 145.298. Elemental analysis: %C: 44.07 (theory); 43.85 (found); %H: 7.80 (theory); 8.08 (found); %N: 28.55 (theory); 28.70 (found).

1-n-octyl-4-amino-1,2,4-triazolium nitrate (XXIV): In essentially the same manner as in XIV above, 1.0047 g., 3.62 mmoles, of 1-n-octyl-4-amino-1,2,4-triazolium bromide was reacted with silver nitrate, 0.6514 g., 3.83 mmoles, yielding 1-n-octyl-4-amino-1,2,4-triazolium nitrate, 0.9347 g., 3.60 mmoles, 99%. Melting point: 34°C; DSC onset: 165°C. $^1$H NMR (d$_6$-dmso): 0.769, 0.826, 0.843, 0.855 (multiplet, area 3.049); 1.235 (broad singlet, 10.221); 1.823, 1.838 (multiplet, area 2.028); 4.312, 4.329, 4.346 (triplet, area 2.121); 7.000 (broad singlet, area 2.073), 1.806 (broad singlet); 9.184 (singlet, area 0.961), 10.218 (singlet, area 1.000). $^{13}$C NMR (d$_6$-dmso): 13.988, 22.142, 25.442, 28.133, 28.365, 28.544, 31.230, 51.810, 142.845, 145.378. Elemental analysis: %C: 46.31(theory); 46.88 (found); %H: 8.16 (theory); 8.23 (found); %N: 27.00 (theory); 26.98 (found).

1-n-nonyl-4-amino-1,2,4-triazolium nitrate (XXV): As in XIV above, 1-n-nonyl-4-amino-1,2,4-triazolium bromide, 1.0039 g., 3.44 mmoles, was reacted with silver nitrate, 0.5855 g., 3.44 mmoles, , yielding 1-n-nonyl-4-amino-1,2,4-triazolium nitrate, 0.9339 g., 3.41 mmoles, 99%. Melting point: 53°C; DSC onset: 175°C. $^1$H NMR (d$_6$-dmso): 0.834, 0.841, 0.851, 0.859, 0.867 (multiplet, area 3.119); 1.241(broad, area 12.725); 1.829, 1.845 (broad, area, 2.074); 4.300,

4.307, 4.318, 4.335 (multiplet, area 2.058); 6.944 (broad singlet, area 2.000); 9.179 (singlet, area 0.959), 10.172 (singlet, area 0.998). $^{13}$C NMR (d$_6$-dmso): 13.950, 22.129, 25.419, 28.116, 28.385, 28.642, 28.813, 31.302, 51.789, 142.802, 145.322. Elemental analysis: %C: 48.33 (theory); 48.27 (found); %H: 8.48 (theory); 9.03 (found); %N: 25.61(theory); 25.44 (found).

1-n-decyl-4-amino-1,2,4-triazolium nitrate (XXVI): As in XIV above, 1-n-decyl-4-amino-1,2,4-triazolium bromide, 1.0085 g., 3.30 mmoles, was reacted with silver nitrate, 0.5606 g., 3.30 mmoles, resulting in 1-n-decyl-4-amino-1,2,4-triazolium nitrate, 0.8451 g., 2.94 mmoles, 84%. Melting point: 51°C; DSC onset: 185°C. $^1$H NMR (d$_6$-dmso): 0.828, 0.844, 0.859 (triplet, area 4.021), 1.235 (broad singlet, area 15.760) 1.826, 1.841 (broad doublet, area 2.057), 4.310, 4.327, 4.344 (triplet, area 2.042); 6.993 (broad singlet, area 2.000), 9.178 (singlet, area 0.986), 10.209 (singlet, area 0.993). $^{13}$C NMR (d$_6$-dmso): 13.948, 22.127, 25.409, 28.091, 28.365, 28.713, 28.849, 28.993, 31.317, 51.781, 142.774, 145.303. Elemental analysis: %C: 50.15 (theory); 50.26 (found); %H: 8.76 (theory); 9.40 (found); %N: 24.37 (theory); 24.48 (found).

## General Synthesis and Physical Properties

Previously, a communication reported the synthesis of one substituted-1,2,4-triazoles in a one pot scheme via the alkylation of 4-amino-1,2,4-triazole forming 1-alkyl-4-amino-1,2,4-triazolium salts, with subsequent diazotization of the amino group, forming desired 1-substituted-1,2,4-triazoles(*38*). However, the intermediates were not the focus as the real effort was directed towards the pharmaceutically active 1-substituted-1,2,4-triazoles.(*44*) Earlier efforts reported the alkylation of substituted 4-amino-1,2,4-triazoles, however, the efforts focused on 3, (5)-alkyl- and aryl- substituted 4-amino-1,2,4-triazoles(*35-40*). Very recently a report was made on the synthesis of 1-substituted-4-amino-1,2,4-triazolium nitrates involving the amination of 1-substituted-1,2,4-triazoles with picryloxyamine followed by subsequent treatment with excess nitric acid forming new nitrate salts(*41*). However this route is complicated with the use of the highly reactive picryloxyamine(*46*) and is not convenient for widespread use. It was realized that the intermediate 1-R-4-amino-1,2,4-triazolium halide salts have essentially the same overall shape as the well-known 1,3-dialkyl-imidazolium cation-based ionic liquids and might possess similar poor three-dimensional packing characteristics resulting in a new class of ionic liquids. Expanding and improving the reaction schemes previously reported(*38*), a large family of new 1-substituted-4-amino-1,2,4-triazolium salts have been synthesized and fully characterized. The synthesis of all of the materials was accomplished by the reaction of excess n-haloalkane (halo = Br, I) with 4-

amino-1,2,4-triazole (> 2:1) typically in acetonitrile. The product salts precipitated from the reaction mixture upon cooling. (Reaction 1)

Use of less than a 2:1 mole ratio of alkyl halide to 4-amino-1,2,4-triazole led to extensive reaction times, incomplete conversion to alkylated product, as well as contamination of the ionic liquid product with 4-amino-1,2,4-triazole. Most of the materials precipitated as oils initially but could be induced to solidify by complete removal of the solvent, followed by recrystallization from hot acetonitrile or isopropyl alcohol. Densities were measured for all of the halides salts and are listed in Table 1. The materials follow the expected trend, as the alkyl chains length increases, the density drops in a gradual manner.

Subsequent metathesis of the 1-alkyl-substituted-4-amino-1,2,4-triazolium salts with silver nitrate in hot methanol, filtration of the unwanted silver halide salt, and subsequent removal of all the methanol from the ionic liquid product, resulted in high yields of high purity products. (Reaction 2). The ionic liquids could be recrystallized from hot ethyl acetate/methanol solutions which ridded the product of any silver salt contamination. In all cases the product salts were subsequently purified until no precipitate was observed from mixing a small aliquot of product salt with concentrated aqueous silver nitrate solution.

All of the ionic liquids were stable at room temperature and showed no signs of decomposition even after a year of storage at ambient temperatures. Melting points of the alkyl halide salts followed the expected increasing trend as the molecular weight increased. There were some anomalies most notable being the very high melting point of the 1-isopropyl-4-amino-1,2,4-triazolium bromide that is significantly higher than the corresponding n-propyl, n-butyl, or ethyl derivatives as can be seen in Table 1. Several of the nitrate salts had melt points at or below ambient temperatures, defining them as members of room temperature ionic liquids (RTILs) (*3, 5*). While the melting point generally increased as the straight n-alkyl chain progressed from ethyl to n-decyl, one striking feature was that the salts with even numbered n-alkyl chains had lower melting points than those with odd numbered chains (see Table 2). There is no straight-forward answer for this trend as it is not observed in the corresponding halide salts or in similar imidazolium based salt systems. Nevertheless, all of the 1-substituted-1,2,4-triazolium halide and nitrate salts had melting points below 100°C defining them in the well known class of ionic liquids(*4, 8*).

All of the ionic liquids were soluble in water, methanol, ethanol, dimethylsulfoxide, dimethylformamide, hot acetonitrile and isopropanol, and were sparingly soluble in hot ethyl acetate, while insoluble in diethyl ether, tetrahydrofuran, and similarly low dielectric constant solvents. All of the materials were slightly hygroscopic, which complicated the collection of spectra. Another property unique to this family of compounds is the weakly acidic nature of all of the 1-R-4-amino-1,2,4-triazolium cations in aqueous solutions. Approximately 1M aqueous solutions of all the salts gave pHs in the range 3.5-4.5, strongly supporting the equilibrium of the 1-R-4-amino-1,2,4-triazolium cation with the 1-R-4-imino-1,2,4-triazoliuim zwitterion, giving $K_a$'s for these alkylated heterocyclic cations ranging roughly from $3.0 \times 10^{-5}$ to $3.0 \times 10^{-4}$ (Reaction 3).

This is not surprising as the parent heterocycle 4-amino-1,2,4-triazole is weakly acidic(*47-50*), with aqueous solutions having a pH of around 5. Previous investigations with other substituted 4-amino-1,2,4-triazolium salts have found a wide array of reaction products stemming from this type of zwitterionic species including ring opening reactions in the presence of various bases(*51, 52*).

Ionic liquids have significant claims for their high temperature stability(*2-6.*) However, this stability depends not only on the cation but the nature of the corresponding anion. The size, charge delocalization and basicity of the parent heterocycle and anion weigh heavily in the overall stability of resultant salts. In the materials studied here the parent heterocycle, 4-amino-1,2,4-triazole, is a significantly poorer base by several orders of magnitude than 1-alkyl-imidazoles. Hence, it would be logical to expect the thermal stability of

Table 1. Physical properties of 1-R-4-amino-1,2,4-triazolium bromides.

| 1-R-4AT salts | m.p. (°C) | decomp.(°C) | ρ (g/cm³) |
|---|---|---|---|
| 1-methyl (I-) | 92 | 135 | 1.98 |
| 1-ethyl | 67 | 130 | 1.69 |
| 1-n-propyl | 63 | 120 | 1.56 |
| 1-isopropyl | 92 | 110 | 1.60 |
| 1-allyl | 62 | 130 | 1.59 |
| 1-butyl | 48 | 130 | 1.46 |
| 1-methylcyclopropyl | 73 | 150 | 1.58 |
| 1-n-pentyl | 54 | 130 | 1.37 |
| 1-n-hexyl | 76 | 120 | 1.34 |
| 1-n-heptyl | 94 | 120 | 1.30 |
| 1-n-octyl | 80 | 135 | 1.27 |
| 1-n-nonyl | 81 | 140 | 1.26 |
| 1-n-decyl | 90 | 135 | 1.23 |

Table 2. Physical properties of 1-R-4-amino-1,2,4-triazolium nitrates.

| Salt | melting point(°C) | dec. onset(°C) |
|---|---|---|
| 1-methyl | -54(g) | 175 |
| 1-ethyl | -55(g) | 160 |
| 1-n-propyl | 33 | 175 |
| 1-isopropyl | 66 | 180 |
| 1-(2-propenyl) | -50(g) | 140 |
| 1-n-butyl | -50(g) | 170 |
| 1-methylcyclopropyl | 56 | 185 |
| 1-n-pentyl | 29 | 185 |
| 1-n-hexyl | 0 | 170 |
| 1-n-heptyl | 35 | 165 |
| 1-n-octyl | 34 | 165 |
| 1-n-nonyl | 53 | 175 |
| 1-n-decyl | 51 | 185 |

1-alkyl-4-amino-1,2,4-triazolium halide salts to be less than that of the corresponding imidazolium analogues, and this is what is observed in differential scanning calorimetry studies. In Table 1, the melting point and decomposition onset, i.e. where the trace begins to have a positive slope and not the maximum of the exotherm, are listed for the halides, and the decomposition onsets are rather low. There are two different, plausible initial decomposition routes for the 1-alkylated-4-amino-1,2,4-triazolium halide salts; (A) Simple proton transfer from heterocyclic N-amino group to the bromide anion, forming a zwitterion and HBr; and (B) Re-alkylation of the bromide anion forming the starting materials (Reaction 4). However, the actual pathway of decomposition is unknown until further studies are carried out.

In the case of the nitrate salts, the melting points were typically lower than the halide salts, and the corresponding DSC decomposition onsets were considerably higher (Table 2). One plausible explanation for this difference is the charge delocalization over the nitrate anion versus the spherical bromide anion, as it is well documented that charge delocalized anions have significantly wider liquid ranges in imidazolium and triazolium based ionic liquid systems(2-6, 32-34, 53, 54).

## Spectral Properties of Low Melting Salts

Several forms of spectroscopy were used to characterize the new salts including vibrational and multinuclear nmr spectra. $^1$H and $^{13}$C nmr revealed shifts in both the alkyl chains attached to the heterocyclic ring, as well as subsequent shifts in the heterocylic carbon and hydrogen signals. The 4-amino-1,2,4-triazole ring contains two C-H linkages in the ring as well as a pendant $NH_2$ group, which upon alkylation of the ring in the N(1) position, breaks the mirror symmetry of the ring. This is immediately evident in the proton spectra, with the disappearance of 4-amino-1,2,4-triazole's C-H singlet (+8.5 ppm), and the emergence of two, equal downfield C-H singlets (+9.2 ppm and +10.3 ppm). The $NH_2$ broad singlet has also been shifted to an average value of +7.1 ppm, a 0.7 ppm downfield shift from the neutral 4-amino-1,2,4-triazole. Also there was a 1 ppm downfield shift (average value +4.3 ppm) of the alkyl pendant group protons attached to the first carbon bonded to N(1) of the triazole ring, as compared to those of the alkyl halides. This was accompanied with subsequent smaller downfield shifts in the adjacent alkyl chain hydrogen environments. These shifts are most likely attributed to the + 1 formal charge placed upon the N(1) of 4-amino-1,2,4-triazole ring, as well as the alkyl chain being bonded to the electron withdrawing 4-amino-1,2,4-triazole ring. Such shifts are not unexpected or unreasonable and have been noted in simple protonated triazole systems(49, 54, 55),, 1-R-4-amino-(3),(5)-substituted-1,2,4-triazolium salts(35-39) 1-R-4-nitramino-1,2,4-triazole zwitterions(42-44), as well as 1-4-difluoroalkylsubstituted-1,2,4-triazolium ionic liquids(32-34)

Similar effects were observed in the $^{13}C$ spectra; the loss of symmetry of the ring upon alkylation resulted in two singlets - one slightly upfield from the starting material and one shifted downfield from the 4-amino-1,2,4-triazole singlet. The shift can be explained by the carbon atom directly attached to the N(1) atom which has undergone alkylation, forcing more p character upon the C=N bond, resulting in an upfield shift, while the carbon atom on the far side of the ring experiences a loss of electron density in the ring but no real change in its direct bonding environment. The $^{13}C$ signals of the pendant alkyl group were shifted downfield with carbon attached directly to N(1) of the 4-amino-1,2,4-triazole ring having the largest shift, usually in the +50 to +55 ppm downfield which is typical of carbon nitrogen single bond environments, mirroring those shifts observed in the $^1H$ spectra and agree well with those reported in similar systems(31-34, 38, 41-43, 50, 57, 58).

The vibrational spectra are quite complex of all the new ionic liquids especially as alkyl chain lengths increase. N-H stretches are observable in both the infra-red and Raman ranging from 3400-3200 $cm^{-1}$ as rather sharp bands and are not unusual(55). The C-H bands of both the heterocyclic and alkyl chains are apparent but making absolute assignments of these bands is nearly impossible due to extensive hydrogen bonding. From the structure of parent heterocycle with an exocyclic N-amino group as well as information found in all the crystal structures to be discussed later, there was a tremendous amount of cation anion interactions between N-H and C-H protons of the heterocycle as well as with some of the C-H protons of the pendant alkyl groups. The vibrational spectra of all of the new salts are quite intense, complex, and broad in the region of 3350-2600 $cm^{-1}$ which is strong evidence of $NH_2$---$X^-$ hydrogen bonding interactions and has been noted in many other amine systems(37-39, 59-68). Bands typical of nitrate salts, 1375 $cm^{-1}$ in the infrared and 1043 $cm^{-1}$ in the Raman are easily observed in all the spectra and match well to those observed in other nitrate salt systems(*69-71*).

## Single Crystal X-ray Diffraction Studies of New Salts

Several of the new salts were studied by single crystal x-ray diffraction studies at ambient and low temperatures and essentially gave the expected structures(*72-76*). All of the materials were substituted at the N(1) of the 1,2,4-triazole ring with the alkyl group. Each of the new salts will be discussed on an individual basis. However there were no major trends that were noted among all of the crystal structures. In all the halide structures, the N-amino protons saddle the plane of the triazolium ring (above and below), sometimes with the lone pair facing towards the side of the ring N-alkylation (ethyl, isopropyl, n-propyl) and away in the n-hexyl and n-heptyl bromide structures. It would have been thought that the lone pair of the N-amino group in the plane of the ring between atoms 3 and 4 of the 1,2,4-triazole ring would have a longer bond, but this was not the case. These structural features influence the corresponding bond lengths, as the ring C-N ring bond which faces the lone pair of N-amino group has a substantially longer bond than the ring C-N bond on the far side of the ring. Secondly, in all the structures there are significant interactions between the anions and the N-amino protons as well as the C-H protons of the heterocylic ring. Previously an x-ray diffraction study of the structure of 4-amino-1,2,4-triazole was reported(*76*), however we decided to reinvestigate the structure of the neutral heterocycle for a comparative study with the cationic structures.

4-amino-1,2,4-triazole crystallizes in a monoclinic cell with Cc symmetry and major details of the structural solution are shown in Table 3. The N-amino group in the 4 position of the ring has the $NH_2$ hydrogen atoms straddling the plane of the ring atoms in one direction. There are two asymmetric heterocyclic rings in the unit cell (Figure 1). The bond distances within the ring structure follow the trends that would be expected for the neutral 4-amino-1,2,4-triazole where the two formal double bonds of the triazole ring are between N(1) and C(5) and N(2) and C(3). The other bonds within the 1,2,4-triazole ring have distances that are slightly longer than these, however all bonds are shorter than typical C-N and N-N single bond distances(*77*), supporting the idea of ring delocalization of pi bond electrons.

Hydrogen bonding is very pronounced in the structure with several significant contacts, which are within Van der Waal distances(*78*) (Figure 2). The most significant are the N-amino hydrogen contacts to nitrogen atoms either N(1) or N(2) of the neighboring 4-amino-1,2,4-triazole ring. However, despite these contacts there are no significant variances in any of the bond distances in either asymmetric ring in the unit cell, within experimental error.

1-ethyl-4-amino-1,2,4-triazolium bromide (II) crystallized as a monoclinic crystal system with $P2_{(1)}/n$ space group symmetry, and the crystal

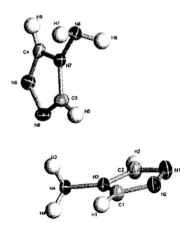

Figure 1. X-ray crystallography structure of 4-amino-1,2,4-triazole. The bond distances (Å) are: N(1)-N(2)=1.379(5); N(1)-C(2)=1.307(5); N(2)-C(1)=1.300(4); N(3)-N(4)=1.418(4); N(3)-C(1)=1.360(5); N(3)-C(2)=1.338(4); N(4)-H(3)=0.98(5); N(4)-H(4)=0.94(4); N(5)-C(4)=1.297(6); N(5)-N(6)=1.396(4); N(6)-C(3)=1.312(6); N(7)-C(4)=1.356(4); N(7)-N(8)=1.411(5); N(7)-C(3)=1.350(4); N(8)-H(7)=0.92(5); N(8)-H(8)=0.95(3); C(1)-H(1)=0.94(4); C(2)-H(2)=0.83(6); C(3)-H(5)=0.94(3); C(4)-H(6)=0.89(4).
*(See page 1 of color insert.)*

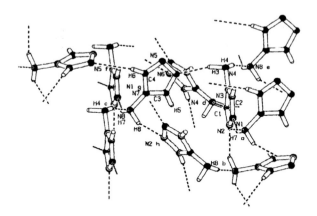

Figure 2. Structure showing significant hydrogen bonds of 4-amino-1,2,4-triazole. The interatomic distance (Å) and symmetry code are: N(8)..H(3)=2.05(5), x,y,z; n(8) e..H(4)=2.10(4), x, y, -1+z; N(1)..H(7) a=2.19, -½+x, ½+y,z; N(1) g..H(8)=2.70(5), x, 1-y, ½+z; N(2)..H(8) b=2.03(4), x, 1-y, ½+z; N(4) d..H(1)=2.51(4), -1/2+x, 3/2-y, -1/2+z; N(5) f..H(6)=2.56(4), x, 2-y, ½+z; N(6) f..H(6)=2.69(4), x, 2-y, ½+z.
*(See page 1 of color insert.)*

structure is shown in Figure 3 with details of the x-ray study summarized in Table 3. Several of the bond distances in the heterocyclic ring of the 1-ethyl-4-amino-1,2,4-triazolium cation have been affected due to the alkylation of the triazole ring. With alkylation of the ring there is a formal +1 charge placed upon the site of alkylation, the N(1) atom of the 1,2,4-triazole ring, and it would be expected that the bonds to N(1) of the 1,2,4-triazole ring would be affected. In the alkylated heterocyclic ring, the C(2)-N(1) bond distance of 1.312(2) Å and the C(1)-N(2) bond distance of 1.310(2) Å are essentially the same and reflect the expected sp$^2$ C=N bond distances. The N-amino N(1)–N(2) distance of 1.369(2) Å is slightly shorter than that observed in the neutral 4-amino-1,2,4-triazole structure and is slightly shorter than a typical N-N single bond distance (1.40 Å)(76, 79-86). The heterocyclic C-H bond distance of C(1)–H(1) is 0.99(3) Å is slightly longer than the C(2)-H(2) distance of 0.94(3) Å. The C(2) atom is bonded directly to the quarternarized N(1) of the ring which could explain the shortness of the the C(2)-H(2) bond through inductive effects. However, these C-H bond distance differences are also significantly affected by strong interactions with the bromide anions. The other major bonds of the ring C(1)-N(3) and C(2)-N(3) at 1.356(3) Å and 1.337(2) Å , respectively, vary somewhat from that observed in the neutral material. Surprisingly, the C(2)-N(3) bond is shorter despite its proximity to the N-amino lone pair which saddles the ring on the same side of the ring, while the C(1)-N(3) bond is longer despite being in vicinity of the N-amino protons. This suggests that quarternization of the neighboring nitrogen in the triazole ring plays a more important role inductively shortening the C-N bond nearest it. The N(3)-N(4) bond distance, 1.419(2) Å, is very similar to that in the neutral material. Otherwise the C-H and N-H bond distances of the heterocyclic ring are significantly different N(4)-H(3) = 0.92(2) Å while N(4)-H(4) = 0.83(2) Å is shorter which can probably be attributed to hydrogen bonding to the bromide anion. The bond distances in the alkyl chain do not vary tremendously and the corresponding angles between all of the atoms of the heterocyclic cation are within the expected values and will not be discussed.

Hydrogen bonding in ionic liquids has been a significant point of interest amongst several researchers with most of the discussion based on interactions of various substituted imidazolium chloride salts (87-91.) Another report has discussed significant interactions between 1-ethyl-3-methyl imidazolium cation and its iodide counterion(92). The only other example of a structurally characterized ionic liquid containing bromine atoms is that of the 1-ethyl-3-methylimidazolium tri-u-bromo-bis[tribromoruthenate(III)])(93) where the authors saw no significant cation-anion interactions. In Figure 4, all of the significant interactions between the cations and anion are shown by dotted lines. Also evident is the effect on the corresponding bond length of the bonded hydrogen atom. Several of the hydrogen bromide interactions are much less than the sum of the Van der Waal radii (3.3 Å)(78), The shortest contact of 2.53(2) Å

Table 3. Crystal Data and Details of the Structure Determination.

| Chemical Name | 4-Amino-1,2,4-Triazole | 1-Ethyl-4ATBr |
|---|---|---|
| Formula | $C_2 H_4 N_4$ | $C_4 H_9 N_4Br$ |
| Formula Weight | 84.09 | 193.05 |
| Crystal System | Monoclinic | Monoclinic |
| Space group | Cc     (No. 9) | $P2_1/n$    (No. 14) |
| a [Å] | 11.94(2) | 5.117(2) |
| b [Å] | 10.79(1) | 18.439(5) |
| c [Å] | 8.28(1) | 7.846(2) |
| α [deg] | 90 | 90 |
| β [deg] | 133.23(2) | 98.371(5) |
| γ [deg] | 90 | 90 |
| V [Å³] | 777(2) | 732.4(3) |
| Z | 8 | 4 |
| D(calc) [g/cm³] | 1.437 | 1.751 |
| μ(MoKa) [ /mm ] | 0.107 | 5.533 |
| F(000) | 352 | 384 |
| Temperature (K) | 100 | RT |
| Theta Min-Max [Deg] | 3.0,  25.4 | 2.8,  28.3 |
| Dataset | -12: 14 ;  -9: 12 ; -9: 9 | -6:  6 ; -24: 23 ; -10:  7 |
| Tot., Uniq. Data, R(int) | 1753,  1201, 0.030 | 4423,  1688, 0.027 |
| Observed data [I > 2.0 σ(I)] | 1155 | 1557 |
| Nref, Npar | 1201, 142 | 1688, 119 |
| R, wR2, S | 0.0412, 0.1128, 1.05 | 0.0261, 0.0722, 1.03 |
| where P= (Fo^2^+2Fc^2^)/3 | w = 1/[\s^2^(Fo^2^) +(0.0874P)^2^] | w = 1/[\s^2^(Fo^2^)+(0.0528P)^2^] |
| Max. and Av. Shift/Error | 0.00, 0.00 | 0.00, 0.00 |
| Min. and Max. Resd. Dens. [e/Å³] | -0.29, 0.25 | -0.60, 0.73 |

Table 3 (cont.). Crystal data and details of collection.

| Chemical Name | 1-n-Propyl-4ATBr | 1-Isopropyl-4ATBr |
|---|---|---|
| Formula | $C_5 H_{11} N_4Br$ | $C_5 H_{11} N_4Br$ |
| Formula Weight | 207.08 | 207.08 |
| Crystal System | Triclinic | Triclinic |
| Space group | P-1 (No. 2) | P-1 (No. 2) |
| a [Å] | 5.1237(7) | 6.4138(9) |
| b [Å] | 10.943(1) | 7.633(1) |
| c [Å] | 15.681(2) | 8.709(1) |
| α [deg] | 105.777(2) | 94.446(2) |
| β [deg] | 92.264(2) | 92.170(2) |
| γ [deg] | 99.091(2) | 97.807(2) |
| V [Å³] | 832.3(2) | 420.6(1) |
| Z | 4 | 2 |
| D(calc) [g/cm³] | 1.653 | 1.635 |
| μ(MoKa) [ /mm ] | 4.875 | 4.822 |
| F(000) | 416 | 208 |
| Temperature (K) | 100 | RT |
| Theta Min-Max [Deg] | 1.4, 25.4 | 2.7, 26.4 |
| Dataset | -6: 5 ; -13: 13 ; -18: 16 | -8: 5 ; -9: 9 ; -10: 10 |
| Tot., Uniq. Data, R(int) | 4468, 2989, 0.015 | 2349, 1676, 0.022 |
| Observed data [I > 2.0 σ(I)] | 2746 | 1600 |
| Nref, Npar | 2989, 269 | 1676, 136 |
| R, wR2, S | 0.0269, 0.0774, 1.09 | 0.0272, 0.0701, 1.06 |
| where P= (Fo^2^+2Fc^2^)/3 | w = 1/[\s^2^(Fo^2^) +(0.0556P)^2^+0.08 65P] | w = 1/[\s^2^(Fo^2^) +(0.0406P)^2^+0.10 45P] |
| Max. and Av. Shift/Error | 0.00, 0.00 | 0.00, 0.00 |
| Min. and Max. Resd. Dens. [e/Å³] | -0.71, 0.88 | -0.52, 0.47 |

*Continued on next page.*

Table 3 (cont.). Crystal data and details of collection.

| Chemical Name | 1-n-Hexyl-4ATBr | 1-n-Heptyl-4ATBr |
|---|---|---|
| Formula | $C_8 H_{17} N_4 Br$ | $C_9 H_{19} N_4 Br$ |
| Formula Weight | 249.16 | 263.18 |
| Crystal System | Monoclinic | Monoclinic |
| Space group | $P2_1/c$ (No. 14) | $P2_1/c$ (No. 14) |
| a [Å] | 19.131(5) | 20.914(4) |
| b [Å] | 6.314(2) | 6.422(1) |
| c [Å] | 9.611(2) | 9.505(2) |
| α [deg] | 90 | 90 |
| β [deg] | 98.567(4) | 95.602(3) |
| γ [deg] | 90 | 90 |
| V [Å$^3$] | 1148.0(5) | 1270.4(4) |
| Z | 4 | 4 |
| D(calc) [g/cm$^3$] | 1.442 | 1.376 |
| μ(MoKa) [ /mm ] | 3.547 | 3.210 |
| F(000) | 512 | 544 |
| Temperature (K) | RT | 100 |
| Theta Min-Max [Deg] | 2.2, 25.4 | 2.0, 25.4 |
| Dataset | -23: 22 ; -7: 7 ; -11: 9 | -23: 25 ; -4: 7 ; -11: 10 |
| Tot., Uniq. Data, R(int) | 5751, 2117, 0.034 | 6313, 2326, 0.032 |
| Observed data [I > 2.0 σ(I)] | 1824 | 2086 |
| Nref, Npar | 2117, 127 | 2326, 144 |
| R, wR2, S | 0.0410, 0.1091, 1.04 | 0.0429, 0.1258, 1.06 |
| where P= (Fo^2^+2Fc^2^)/3 | w = 1/[\s^2^(Fo^2^) +(0.0797P)^2^] | w = 1/[\s^2^(Fo^2^)+ (0.0993P)^2^ |
| Max. and Av. Shift/Error | 0.00, 0.00 | 0.00, 0.00 |
| Min. and Max. Resd. Dens. [e/Å$^3$] | -0.64, 7.92 | 0.86, 1.54 |

Table 3 (cont.). Crystal data and details of collection.

| Chemical Name | 1-Isopropyl-4AT NO$_3$ | 1-Methylcyclopropyl NO$_3$ |
|---|---|---|
| Formula | C$_5$ H$_{11}$ N$_5$ O$_3$ | C$_6$ H$_{11}$ N$_5$ O$_3$ |
| Formula Weight | 189.19 | 201.20 |
| Crystal System | Triclinic | Monoclinic |
| Space group | P-1     (No. 2) | P2$_1$/n     (No. 14) |
| a [Å] | 7.018(3) | 5.402(3) |
| b [Å] | 7.176(3) | 8.526(4) |
| c [Å] | 9.100(4) | 20.255(9) |
| α [deg] | 106.426(6) | 90 |
| β [deg] | 91.865(7) | 97.317(8) |
| γ [deg] | 93.224(7) | 90 |
| V [Å$^3$] | 438.3(3) | 925.3(8) |
| Z | 2 | 4 |
| D(calc) [g/cm**3] | 1.434 | 1.444 |
| μ(MoKα) [ /mm ] | 0.119 | 0.117 |
| F(000) | 200 | 424 |
| Temperature (K) | 100 | 100 |
| Theta Min-Max [Deg] | 2.3,  25.4 | 2.0,  25.5 |
| Dataset | -8: 4 ; -7: 8 ; -10: 9 | -6: 6 ; -10: 7 ; -24: 23 |
| Tot., Uniq. Data, R(int) | 1641,  1343,  0.015 | 4718,  1695, 0.029 |
| Observed data [I > 2.0 σ(I)] | 1234 | 1530 |
| Nref, Npar | 1343, 129 | 1695, 171 |
| R, wR2, S | 0.0562, 0.1787, 1.15 | 0.0331, 0.0904, 1.05 |
| where P= (Fo^2^+2Fc^2^)/3 | w = 1/[\s^2^(Fo^2^)+ (0.1139P)^2^+0.2471P] | w = 1/[\s^2^(Fo^2^)+ (0.0513P)^2^+0.2126P] |
| Max. and Av. Shift/Error | 0.00, 0.00 | 0.00, 0.00 |
| Min. and Max. Resd. Dens. [e/Å^3] | -0.30, 0.29 | -0.24, 0.20 |

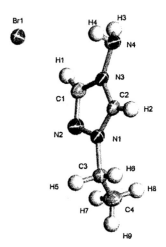

Figure 3. X-ray crystallography structure of 1-ethyl-4-amino-1,2,4-triazolium bromide (II). The bond distances (Å) are: N(1)-N(2)=1.369(2); N(1)-C(2)=1.312(2); N(1)-C(3)=1.480(2); N(2)-C(1)=1.310(2); N(3)-N(4)=1.419(2); N(3)-C(1)=1.356(3); N(3)-C(2)=1.337(2); N(4)-H(3)=0.92(2); N(4)-H(4)=0.83(2); C(3)-C(4)=1.502(3); C(1)-H(1)=0.99(3); C(2)-H(2)=0.94(3); C(3)-H(5)=0.98(3); C(3)-H(6)=0.84(2); C(4)-H(7)=0.89(2); C(4)-H(8)=0.98(3); C(4)-H(9)=0.95(3). *(See page 2 of color insert.)*

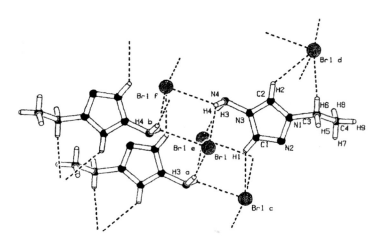

Figure 4. Structure showing significant cation-anion contacts of 1-ethyl-4-amino-1,2,4-triazolium bromide (II). The interatomic distance (Å) and symmetry code are: Br(1)..H(3) a=2.53(2), -x, -y, 1-z; Br(1)..H(4)= 2.89(2), x, y, z; Br(1) f..H(4)= 2.84(2), x-1, -y, 2-z; Br(1) c..H(1)= 2.77(3), -1+x, y, z; Br(1) d..H(2)= 2.89(3), x, y, 1+z; Br(1) d..H(6)= 3.00(2), x, y, 1+z. *(See page 2 of color insert.)*

between the anion and cation, Br(1) – H(3), of the N-amino group (with a Br-H-N angle of 164.1(2)° ) accords to the longer N-amino N-H bond distance 0.92(2) Å and is the strongest interaction of the structure. All of the cation anion interactions are listed in Figure 4. As expected, the strong bromide hydrogen interactions lead to lengthening of the associated N-H or C-H single bond distances.

1-n-propyl-4-amino-1,2,4-triazolium bromide (III) crystallized in a triclinic cell with P-1 symmetry with two asymmetric 1-n-propyl-4-amino-1,2,4-triazolium cations as well as two corresponding bromide anions. A view of the crystal structure is in Figure 5, with the details of the x-ray study in Table 3. Bond distances in the cation have not been significantly affected by the alkylation of the ring. In both cations, the nitrogen- nitrogen bond distance in the two heterocyclic rings (N(1)-N(2) = 1.372(3) Å and N(5)-N(6) = 1.374(3) Å) are nearly identical and are essentially unchanged to those found in the parent heterocyle. The carbon nitrogen bond distances in the 1,2,4 triazolium rings corresponding to C(2)-N(1) and C(1)-N(1) are fairly short and typical of carbon nitrogen double bonds(77, 94-101). The C-N bond distances corresponding to C(1) and C(2) bonded to N(4) of the parent 1,2,4-triazole ring are all significantly shorter than typical C-N single bonds yet are longer than typical C=N bond lengths and are typical to those observed in triazole systems(79-86, 94-101). The C-N bonds which are closer to the alkylated N(1) position of the 1,2,4-triazole ring are slightly shorter ( C(2)-N(3) = 1.329(3) Å and C(7)-N(7) = 1.328(3) Å) than the corresponding C-N bond distances on the far side of the ring [ C(1)-N(3) = 1.364(4) Å and C(6)-N(7) = 1.365(3) Å] and could be interpreted as a result of the alkylation and quarternization of the N(1) position of the 1,2,4-triazole ring, similar to what was observed in the 1-ethyl-4-amino-1,2,4-triazolium bromide above. The N-NH$_2$ bond lengths in both triazolium cations are essentially the same (N(3)-N(4) = 1.414(3) Å and N(7)-N-(8) = 1.418(3) Å) and vary little from the starting heterocycle.

There is a large amount of hydrogen bonding present in the unit cell between both asymmetric bromide anions and the corresponding cations that is within Van der Waal distances (3.03 A)(77) and is depicted in Figure 6. Symmetry related cations have been placed in the view to illustrate all of the asymmetric hydrogen bromide contacts that are shorter than the sum of the Van der Wall radii. The most significant contacts are between the N-amino hydrogen atoms and the bromide anions in the structure. There is also a significant effect on the N-amino hydrogen bond distances with the N(4)-H(4) bond distance of 0.79(3) Å, being significantly shorter than the N(4)-H(3) bond distance (0.87(3) Å )in the first 1-n-propyl-4-amino-1,2,4-triazolium cation. However in cation 2, both the N(8)-H(14) and N(8)-H(15) bond distances are 0.84(3) Å despite hydrogen bonding. There is no obvious reason for this bond variance present in one ring despite the fact that all the N-amino hydrogen atoms are involved in significant interactions with the bromide anions.

284

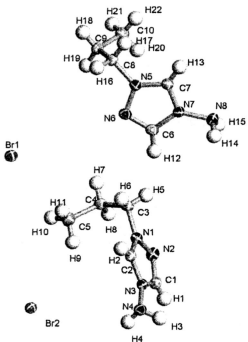

Figure 5. X-ray crystallography structure of 1-propyl-4-amino-1,2,4-triazolium bromide (III). The bond distances (Å) are: N(1)-N(2)=1.373(3); N(1)-C(2)=1.307(3); N(1)-C(3)=1.470(3); N(2)-C(1)=1.307(4); N(3)-N(4)=1.414(3); N(3)-C(1)=1.364(4); N(3)-C(2)=1.329(3); N(4)-H(4)=0.79(3); N(4)-H(3)=0.87(3); N(5)-C(7)=1.313(3); N(5)-C(8)=1.466(3); N(5)-N(6)=1.374(3); N(6)-C(6)= 1.300(3); N(7)-C(6)=1.365(3); N(7)-C(7)=1.328(3); N(7)-N(8)=1.418(3); N(8)-H(15)=0.87(3); N(8)-H(14)=0.87(3); C(3)-C(4)=1.514(3); C(4)-C(5)=1.524(4); C(1)-H(1)=0.87(3); C(2)-H(2)=0.92(3); C(3)-H(5)=0.92(4); C(3)-H(6)=0.93(3); C(4)-H(7)=0.92(3); C(4)-H(8)=0.98(3); C(5)-H(10)=1.00(3); C(5)-H(9)=1.02(3); C(5)-H(11)=0.89(4); C(8)-C(9)=1.515(4); C(9)-C(10)=1.514(4); C(6)-H(12)=0.96(3); C(7)-H(13)=0.92(3); C(8)-H(16)=0.94(3); C(8)-H(17)=0.98(3); C(9)-H(18)=0.92(3); C(9)-H(19)=0.96(3); C(10)-H(20)=0.99(3); C(10)-H(21)=0.98(3); C(10)-H(22)=0.96(3).

*(See page 3 of color insert.)*

1-isopropyl-4-amino-1,2,4-triazolium bromide(IV) crystallized in a triclinic cell with P-1 symmetry. There is one asymmetric cation and anion in the unit cell (Figure 7). The bond lengths and angles are typical as in the previous examples. The N(1)-C(2) and N(2)-C(1) bond lengths are nearly the same at 1.298(3) Å and 1.297(3) Å, which show significant C=N double bond character. The other two carbon nitrogen bonds (C(1)-N(3), distance 1.355(3) Å, and C(2)-N(3), distance 1.327(3) Å) follow the same trend as observed in the other salt structures with the C-N bond that is closer to the site of N-alkylation being the shorter bond. Once again the C-N bond lengths are between that of a typical single bond and that of a double bond, indicating charge delocalization. The N(3)-N(4) bond length is 1.400(3) Å and is essentially identical to all those in the other cation structures of this study. The bonds in the isopropyl group are all within typical C-C, C-N, and C-H bond distances as well(77).

There is a large amount of hydrogen bond contacts with the sum of the Van der Wall radii (3.03 Å)(77) in the crystal lattice and they are depicted in Figure 8, revealing a complex three dimensional structure. As in the previous structures there is a difference between the N-amino hydrogen bond distances but it is not as prominent (N(4)-H(3) = 0.86(3) Å, and N(4)-H(4) = 0.92(4) Å) since both hydrogen atoms are involved in significant interactions with the bromide counterion.

1-n-hexyl-4-amino-1,2,4-triazolium bromide (VIII) crystallized as thin plates in a monoclinic cell with $P2_1/c$ symmetry. There was one asymmetric cation and anion in the unit cell (Figure 9). The bond distances are mostly similar to the above mentioned structures but there are some significant differences. The pendant n-alkyl chain has assumed the common low energy "zig-zag" chain radiating away from the triazole ring. Once again the N-amino group is in a saddle position with the primary plane of the triazole ring, however in this structure, the N-amino group hydrogen atoms are facing the side where alkylation of the heterocyclic ring nitrogen has taken place, instead of facing away as in the proceeding structures. The bond distances follow the trend observed in the previous cation structures, where the C-N bond distance closest to the site of N(1) alkylation (C(2)-N(3) = 1.334(4) Å) is shorter than that of the C-N bond distance on the opposite side of the ring (C(1)-N(3) = 1.350(4) Å), but the difference between the bond lengths is not as pronounced as in the cations with shorter alkyl side chains.

Hydrogen bonding contacts are most prominent between the N-amino hydrogen atoms and the bromide anion as well as in a few interactions between the hydrogen atoms attached to the carbon atoms of the 1,2,4-triazole ring shown in Figure 10. The N-amino hydrogen bond distances are asymmetric with an N(4)-H(4C) distance of 0.93(3) Å and an N(4)-H(4D) distance of 0.81(3) Å, following the general trend noted in all the heterocyclic salts in this paper. With the much longer alkyl side chain, the structures are exhibiting hydrophobic/hydrophilic packing tendencies. There is an orthogonal

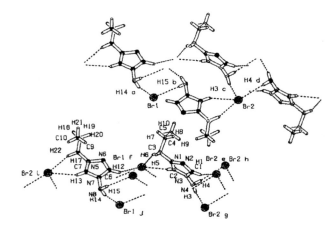

Figure 6. Structure showing significant cation-anion contacts of 1-propyl-4-amino-1,2,4-triazolium bromide (III). The interatomic distance (Å) and symmetry code are: Br(2) g..H(3)= 2.57(3), x, -1+y, z; Br(2) h..H(4)= 2.71(3), 1-x, 1-y, -z; Br(1)..H(14) a= 2.54(3), x, -1+y, z; Br(1)..H(15) b= 2.71(3), 2-x, 1-y, 1-z; Br(2) e..H(1)= 2.91(3), 2-x, 1-y, -z; Br(1) f..H(2)= 2.64(3), 1-x, 1-y, 1-z; Br(1) f..H(6)= 2.91(3), 1-x, 1-y, 1-z; Br(1) f..H(12)= 2.99(3), 1-x, 1-y, 1-z; Br(2) i..H(13)= 2.66(3), 2-x, 1-y, 1-z; Br(2) i..H(17)= 2.95(3), 2-x, 1-y, 1-z.
*(See page 3 of color insert.)*

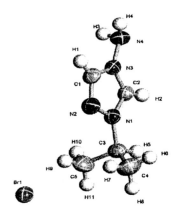

Figure 7. X-ray crystallography structure of 1-isopropyl-4-amino-1,2,4-triazolium bromide (IV). The bond distances (Å) are: N(1)-N(2)=1.361(3); N(1)-C(2) =1.298(3); N(1)-C(3)=1.477(3); N(2)-C(1)=1.297(3); N(3)-N(4)=1.400(3); N(3)-C(1)=1.355(3); N(3)-C(2)=1.327(3); N(4)-H(4)=0.92(4); N(4)-H(3)=0.86(3); C(3)-C(5)=1.504(5); C(3)-C(4)=1.500(4); C(1)-H(1)=0.92(3); C(2)-H(2)=0.90(3); C(3)-H(5)=1.00(3); C(4)-H(6)=0.96(4); C(4)-H(7)=0.96(5); C(4)-H(8)=0.95(5); C(5)-H(9)=0.96(5); C(5)-H(10)=0.96(5); C(5)-H(11)=0.92(5). *(See page 4 of color insert.)*

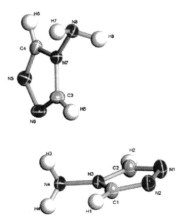

Figure 20.1. X-ray crystallography structure of 4-amino-1,2,4-triazole.

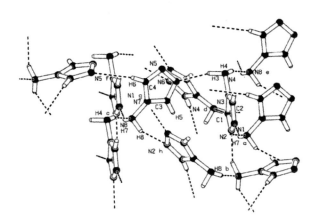

Figure 20.2. Structure showing significant hydrogen bonds of 4-amino-1,2,4-triazole.

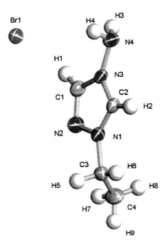

*Figure 20.3. X-ray crystallography structure of 1-ethyl-4-amino-1,2,4-triazolium bromide (II).*

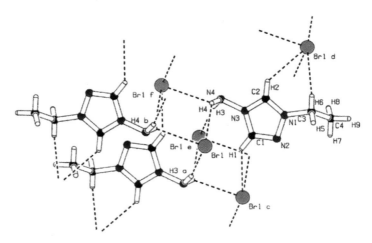

*Figure 20.4. Structure showing significant cation-anion contacts of 1-ethyl-4-amino-1,2,4-triazolium bromide (II).*

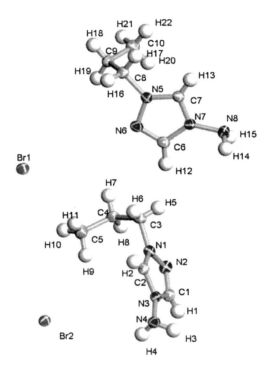

*Figure 20.5 X-ray crystallography structure of 1-propyl-4-amino-1,2,4-triazolium bromide (III).*

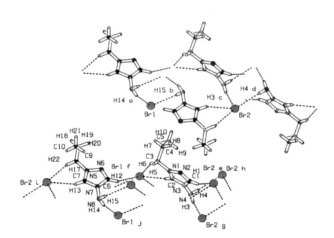

*Figure 20.6. Structure showing significant cation-anion contacts of 1-propyl-4-amino-1,2,4-triazolium bromide (III).*

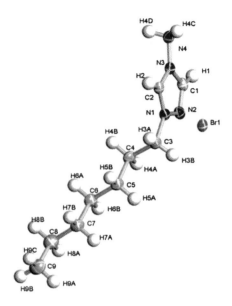

*Figure 20.11. X-ray crystallography structure of 1-heptyl-4-amino-1,2,4-triazolium bromide (IX).*

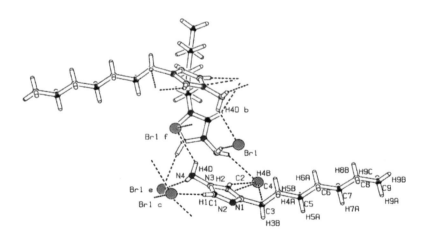

*Figure 20.12. Structure showing significant cation-anion contacts of 1-hexyl-4-amino-1,2,4-triazolium bromide (IX).*

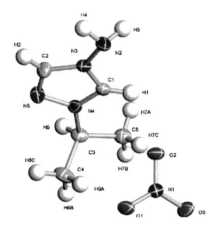

*Figure 20.13. X-ray crystallography structure of 1-isopropyl-4-amino-1,2,4-triazolium nitrate (XVIII).*

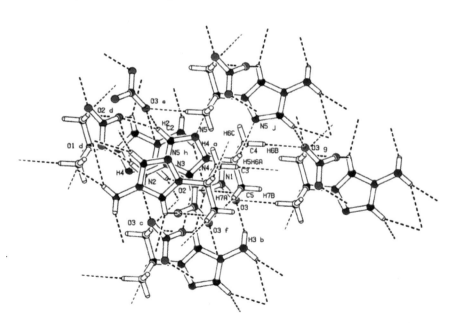

*Figure 20.14. Structure showing significant cation-anion contacts of 1-isopropyl-4-amino-1,2,4-triazolium nitrate (XVIII).*

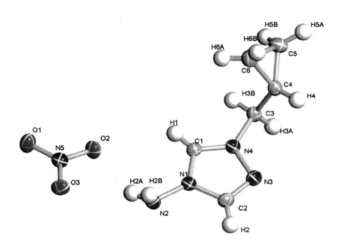

**Figure 20.15.** *X-ray crystallography structure of 1-methylcyclopropyl-4-amino-1,2,4-triazolium nitrate (XX).*

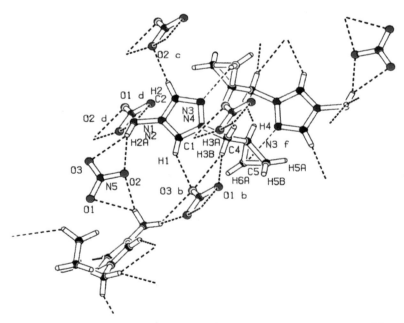

**Figure 20.16.** *Structure showing significant cation-anion contacts of 1-methylcyclopropyl-4-amino-1,2,4-triazolium nitrate (XX).*

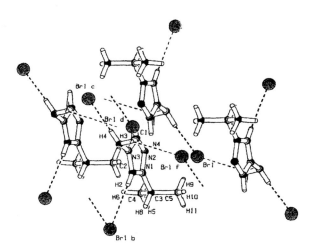

Figure 8. Structure showing significant cation-anion contacts of 1-isopropyl-4-amino-1,2,4-triazolium bromide (IV). The interatomic distance (Å) and symmetry code are: Br(1) c..H(4)= 2.50(4), -x, 2-y, 1-z; Br(1) d..H(3)= 2.98(3), x, 1+y, 1+z; Br(1) f..H(3)= 2.96(3), 1-x, 2-y, 1-z; Br(1) b..H(2)= 2.79(3), x, y, 1+z; Br(1) b..H(5)= 2.95(3), x, y, 1+z. *(See page 4 of color insert.)*

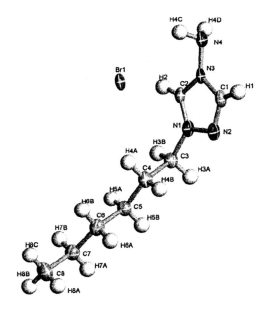

Figure 9. X-ray crystallography structure of 1-hexyl-4-amino-1,2,4-triazolium bromide (VIII). The bond distances (Å) are: N(1)-N(2)=1.373(3); N(1)-C(2)=1.308(4); N(1)-C(3)=1.456(4); N(2)-C(1)=1.293(4); N(3)-N(4)=1.394(4); N(3)-C(1)=1.350(4); N(3)-C(2)=1.334(4); N(4)-H(4D)=0.81(4); N(4)-H(4C)=0.93(3); C(3)-C(4)=1.514(4); C(4)-C(5)=1.529(4); C(5)-C(6)=1.526(5); C(6)-C(7)=1.504(5); C(7) –C(8)=1.531(5); C(1)-H(1)= 0.9304; C(2)-H(2)=0.9298; C(3)-H(3A)=0.9705; C(3)-H(3B)=0.9702; C(4)-H(4A)=0.9698; C(4)-H(4B)= 0.9701; C(5)-H(5A)=0.9697; C(5)-H(5B)=0.9704; C(6)-H(6A)= 0.9708; C(6)-H(6B)=0.9692; C(7)-H(7A)=0.9702; C(7)-H(7B)=0.9697; C(8)-H(8A)=0.9598; C(8)-H(8B)=0.9599; C(8)-H(8C)=0.9590.

*(See page 5 of color insert.)*

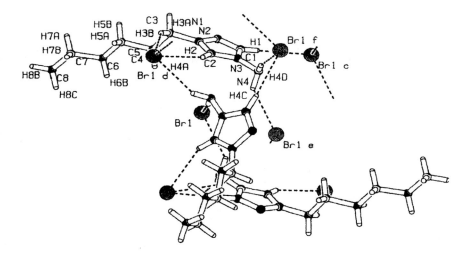

Figure 10. Structure showing significant cation-anion contacts of 1-hexyl-4-amino-1,2,4-triazolium bromide (VIII). The interatomic distance (Å) and symmetry code are: Br(1) e..H(4C)= 2.73(3), 1-x, 1-y, -z; Br(1) f..H(4D)= 2.66(4), 1-x, -½+y, ½-z; Br(1) c..H(1)= 2.7300, x, -1+y, z; Br(1) d..H(2)= 2.7600, x, 3/2-y, ½+z; Br(1) d..H(3B)= 2.8500, x, 3/2-y, ½+z. *(See page 5 of color insert.)*

arrangement of the cations with the n-hexyl side chain that radiates away from the triazolium ring to start forming a "head to tail" arrangement of the cations with the accompanying bromide anions. Thus, pockets are formed between the N-amino groups of the 4-amino-1,2,4-triazole cationic rings. In the packing diagram shown in Figure 10, this is much more apparent. This compares as well as contrasts with one dimensional layered behavior noted in long n-alkyl substituted imidazolium salt systems which display liquid crystalline behavior(102-104).

1-n-heptyl-4-amino-1,2,4-triazolium bromide (IX) crystallized in a monoclinic cell with $P2_1/c$ symmetry and is shown in Figure 11. The structure is very similar to the n-hexyl-4-amino-1,2,4-triazolium bromide salt reported above with one asymmetric cation and anion per unit cell. Once again the n-heptyl alkyl chain radiates away from the triazolium ring in the "zig-zag" structure. The N-amino group saddles the ring, and as in the n-hexyl salt, the N-amino-group hydrogen atoms are pointed towards the alkylated N(1) position of the triazole ring. The bond distances in the ring follow the trend of all the triazolium cations and are very similar in bond distances. The bonds between carbon nitrogen atoms of the heterocyclic ring (N(1)-C(2) = 1.314(4) Å and N(2)-C(1) = 1.303(4) Å) are essentially the same and their distances denote double bond character as would be expected. The other carbon nitrogen bonds (C(1)-N(3) = 1.357(4) Å and C(2)-N(3) = 1.333(3) Å), follow the trend where the shorter of the bonds, C(2)-N(3), is nearer to the point of N-alkylaton of the ring, closest to the quarternized nitrogen, N(1). However, it appears that with the N-amino hydrogen atoms facing this bond, there is not as great a difference between the carbon-nitrogen bond distances,( C(1)-N(3) and C(2)-N(3)) as those observed in the smaller alkyl side chained cations reported above. This mirrors the behavior observed in 1-n-hexyl-4-amino-1,2,4-triazolium bromide (VIII). Otherwise the bond distances are not out of the ordinary and match well to those observed in other heterocyclic systems(79-86, 94-101) and will not be discussed further.

As in the n-hexyl substituted cation, hydrogen bonding appears to strongly influence the overall packing of 1-n-heptyl-4-amino-1,2,4-triazolium bromide. The strongest interactions are between the N-amino hydrogen atoms and the bromide which leads to the formation of a two-dimensional structure of n-heptyl chains lining up, end to end, with the triazolium rings forming pockets for the bromide anions (illustrated in Figure 12). The packing structure is essentially identical to that illustrated for the 1-n-hexyl-4-amino-1,2,4-triazolium bromide in Figure 10 above. The bond distances for the N-amino hydrogen bonds are significantly affected with the N(4)-H(4C) distance of 0.90(4) Å, while the N(4)-H(4O) distance of 0.80(5) Å is considerably shorter. Previously, Seddon et al.(102-104) have described liquid crystal behavior with longer chained n-alkyl methylimidazolium salts exhibiting smectic properties. The longer n-alkyl-4-amino-1,2,4-triazolium based salts appear to have the same

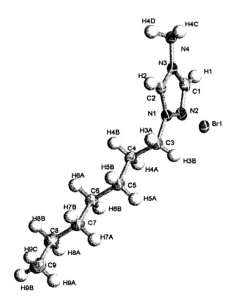

Figure 11. X-ray crystallography structure of 1-heptyl-4-amino-1,2,4-triazolium bromide (IX). The bond distances (Å) are: N(1)-N(2)=1.377(3); N(1)-C(2)=1.314(4); N(1)-C(3)=1.464(4) N(2)-C(1)=1.303(4); N(3)-N(4)=1.401(3); N(3)-C(1)=1.357(4); N(3)-C(2)=1.333(3); N(4)-H(4D)=0.90(4); N(4)-H(4C)=0.80(5); C(3)-C(4)=1.516(5); C(4)-C(5)=1.529(4); C(5)-C(6)=1.526(4); C(6)-C(7)=1.524(4); C(7)-C(8)=1.513(4); C(8)-C(9)=1.531(5); C(1)-H(1)=0.9498; C(2)-H(2)=0.9498; C(3)-H(3A)=0.97(4); C(3)-H(3B)=0.91(3); C(4)-H(4A)=0.9905; C(4)-H(4B)=0.9907; C(5)-H(5A)=0.9896; C(5)-H(5B)=0.9903; C(6)-H(6A)=0.9899; C(6)-H(6B)=0.9893; C(7)-H(7A)=0.9906; C(7)-H(7B)=0.9903; C(8)-H(8A)=0.9897; C(8)-H(8B)=0.9904; C(9)-H(9A)=0.9802; C(9)-H(9B)=0.9793; C(9)-H(9C)=0.9795.

*(See page 6 of color insert.)*

kind of physical properties, but this will have to be validated with further work that is outside the scope of this study.

Two crystal structures were obtained for the corresponding family of nitrate salts which generally were much more difficult to crystallize than their halide analogues. 1-isopropyl-4-amino-1,2,4-triazolium nitrate (XVIII) crystallized in a triclinic cell of P-1 symmetry with a view of the structure in Figure 13. There is one asymmetric cation and one nitrate anion in the structure. One important difference in the structure of the nitrate salts versus the halide salts is the relative position of the N-amino group in relation to the triazole ring. In the 1-isopropyl-4-amino-1,2,4-triazolium nitrate structure, the cation has a twisted N-amino group which doesn't saddle the triazole ring in a symmetric fashion as observed in the halide structures. The C(1)- N(4) bond distance, 1.316(3) Å, and C(2)-N(5) bond distance, 1.305(4) Å are very similar to that observed in all the structures within this study as well as several others elsewhere, revealing significant double bond character (*79-86, 94-101*). The C(1)-N(3) bond distance (1.338(3) Å) is nearest to the site of N-alkylation, is shorter than the other C-N bond (C(2)-N(3) = 1.354(3) Å) and agrees well with the trend observed here as well as in other systems where the C-N bond closest to the site of alkylation is shorter than the corresponding bond on the far side of the ring. In direct contrast, the N-amino hydrogen bond distances are essentially the same (N(2)-H(3) = 0.90(3) Å and N(2)-H(4) = 0.92(4) Å) which is somewhat surprising despite the significant amount of hydrogen bonding. All of the carbon nitrogen bond distances are within typical bond distances in comparison with other similar triazole based materials, 3,4,5-triamino-1,2,4-triazolium bromide(*83*) as well as 1,2,4-triazolium perchlorate(*56*) and several 1-N-coordinated 4-amino-1,2,4-triazole based salts(*84-86*). The nitrate anion is planar and its angles and bond distances are typical for those observed in nitrate salts(*105-108*) and they will not be discussed.

Hydrogen bonding is very pronounced in 1-isopropyl-4-amino-1,2,4-triazolium nitrate with extensive hydrogen atom contacts between the oxygen atoms of the nitrate anion and many of the hydrogen atoms present in the cation (Figure 14). However there appears to be no major perturbation in any of the bond distances despite these rather significant contacts. Extensive hydrogen bonding probably explains the "twist" of the N-amino group out of the saddle position of the triazole ring since several of the N-amino proton contacts are very short (N(2)-H(3)---O(3) = 2.11(3) Å, N(2)-H(4)---O(1) = 2.08(4) Å) and are much shorter than that of the sum of Van der Waal radii 2.72 Å (*78*). Additionally there are many contacts between the nitrate anion oxygen atoms and both the heterocyclic carbon hydrogen atoms and several of the isopropyl carbon hydrogen atoms. However, many of these carbon hydrogen bond distances have been fixed in the structure refinement process and this must be considered in the arguments.

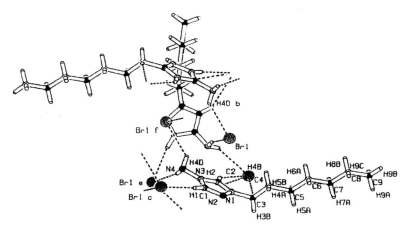

Figure 12. Structure showing significant cation-anion contacts of 1-heptyl-4-amino-1,2,4-triazolium bromide (IX). The interatomic distance (Å) and symmetry code are: Br(1) e..H(4C)= 2.66(4), 1-x, -1/2+y, 3/2-z; Br(1) f..H(4D)= 2.77(4), 1-x, 1-y, 2-z; Br(1) c..H(1)= 2.7100, x, -1+y, z; Br(1) d..H(2)= 2.7900, x, 3/2-y, -1/2+z; Br(1)..H(3A)= 2.85(3), x, 3/2-y, ½+z. *(See page 6 of color insert.)*

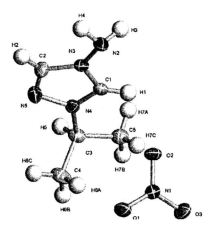

Figure 13. X-ray crystallography structure of 1-isopropyl-4-amino-1,2,4-triazolium nitrate (XVIII). The bond distances (Å) are: O(1)-N(1)=1.241(3); O(2)-N(1)=1.255(3); O(3)-N(1)=1.254(3); N(2)-N(3)=1.415(3); N(3)-C(1)=1.338(3); N(3)-C(2)=1.354(3); N(4) –N(5)=1.363(3); N(4)-C(1)=1.316(3); N(4)-C(3)=1.484(3); N(5)-C(2)=1.305(4); N(2)-H(3)=0.90(3); N(2)-H(4)=0.92(4); C(3)-C(5)=1.509(3); C(3)-C(4)=1.517(4); C(1)-H(1)=0.9294; C(2)-H(2)=0.9299; C(3)-H(5A)=0.9804; C(4)-H(6C)=0.9600; C(4)-H(6A)=0.9598; C(4)-H(6B)=0.9602; C(5)-H(7A)=0.9597; C(5)-H(7B)=0.9600; C(5)-H(7C)=0.9599. *(See page 7 of color insert.)*

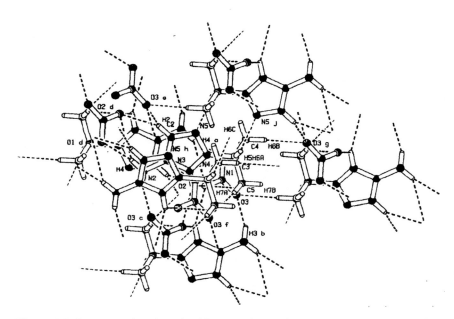

Figure 14. Structure showing significant cation-anion contacts of 1-isopropyl-4-amino-1,2,4-triazolium nitrate (XVIII). The interatomic distance (Å) and symmetry code are: O(3) c..H(3)= 2.11(3), 1-x,1-y,-z; O(1) d..H(4)= 2.08(4), 2-x,1-y,-z; O(2) d..H(4)= 2.62(4), 2-x,1-y,-z; O(2)..H(1)= 2.4700, x, y, z; O(2) c..H(1)= 2.3200, 1-x,1-y,-z; O(3) e..H(2)= 2.3900, 1+x,1+y,z; O(2) d..H(2)= 2.4300, 2-x,1-y,-z; N(5) j..H(5)= 2.6600, 2-x,2-y,1-z; O(3) g..H(6B)= 2.6300, 1-x,1-y,1-z; O(3) f..H(7A)= 2.6900, x,1+y,z; O(2)..H(7C)= 2.6900, x, y, z.

*(See page 7 of color insert.)*

1-methylcyclopropyl-4-amino-1,2,4-triazolium    nitrate    (XX) crystallized in a monoclinic cell with P2$_1$/n symmetry and is shown in Figure 15. There is one asymmetric cation and an accompanying nitrate anion in the unit cell. As in the 1-isopropyl-4-amino-1,2,4-triazolium nitrate structure, the N-amino group is "twisted" out of the saddle position with the plane of the 1,2,4-triazole ring. This, again, is most likely due to the extensive hydrogen bonding within the crystal structure. The distances in the heterocyclic ring are all within normal bonding distances there is not a lot of  variance between  the C-N bonding environments in the heterocyclic ring as noticed in all of the other structures. The carbon nitrogen bond distance closest to the site of N-alkylation, C(1)-N(4) = 1.3104(17) Å,  while the corresponding C-N bond distance on the far side of the ring, C(2)-N(3) = 1.3027(18) Å, are essentially the same which follows the common trend seen in 1,2,4-triazole systems(79-86, 94-101). As well the other C-N bond distances in the ring follow the trend where the C-N closest to the site of alkylation (C(1)-N(1) =1.3317(17) Å) is the shorter than the other C-N bond (C(2)-N(1) = 1.3543(17) Å). As was observed in 1-isopropyl-4-amino-1,2,4-triazolium nitrate (XVIII), the N-amino hydrogen bond distances are essentially the same (N(2)-H(2A) = 0.894(17) Å and N(2)-H(2B) = 0.884(17) Å) despite significant hydrogen bonding to the nitrate anion. As in the preceding nitrate structure, the nitrate anion is planar and its bond distances and angles are typical to those that have been observed in other nitrate salts(105-108) and they will not be discussed further.

Hydrogen bonding is rather significant in 1-methylcyclopropyl-4-amino-1,2,4-triazolium nitrate with many contacts between the oxygen atoms of the nitrate anion and hydrogen atoms of the N-amino group, the carbon hydrogen atoms of the heterocyclic ring as well as some contacts with the hydrogen atoms of the pendant methylcyclopropyl ring (Figure 16). Many of these are well within Van der Waal bonding distances(78) and are listed in Figure 16 caption. Nevertheless, despite this large amount of hydrogen bonding, all of the bond distances in the structure are not drastically affected from normal distances observed in other structures, and it seems to be a structure which illustrates large coulombic interactions between the cation and anion.

## Conclusions

A large new family of ionic liquids based upon 1-substituted-4-amino-1,2,4-triazolium cations with halides and nitrate anions have been synthesized and well characterized. These new ionic liquids display typical ionic liquid behaviors akin to their dialkyl-imidazolium relatives, such as low melting points and long liquidous ranges, however differ with their weak acidity, lessened thermal stability and high solubility in polar solvents. X-ray diffraction studies of several of the new salts revealed the expected structures with 1-substituted-4-

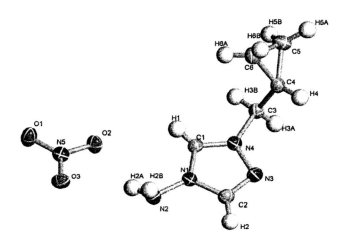

Figure 15. X-ray crystallography structure of 1-methylcyclopropyl-4-amino-1,2,4-triazolium nitrate (XX). The bond distances (Å) are: O(1)-N(5)=1.234(2);O(2)-N(5)=1.266(2); O(3)-N(5)=1.249(2); N(1)-C(2)=1.354(2); N(1)-N(2)=1.401(2); N(1)-C(1)=1.332(2); N(3)-N(4)=1.366(2); N(3)-C(2)=1.303(2); N(4)-C(1)=1.310(2); N(4)-C(3)=1.466(2); N(2)-H(2A)=0.89(2); N(2)-H(2B)=0.88(2); C(3)-C(4)=1.497(2); C(4)-C(6)=1.500(2); C(4)-C(5)=1.500(2); C(5)-C(6)=1.496(2); C(1)-H(1)=0.87(2); C(2)-H(2)=0.94(2); C(3)-H(3A)=0.96(2);C(3)-H(3B)=0.97(2); C(4)-H(4) = 0.95(2); C(5)-H(5A)=0.98(2);C(5)-H(5B)=0.95(2); C(6)-H(6A)=0.95(2); C(6)-H(6B)=0.97(2). *(See page 8 of color insert.)*

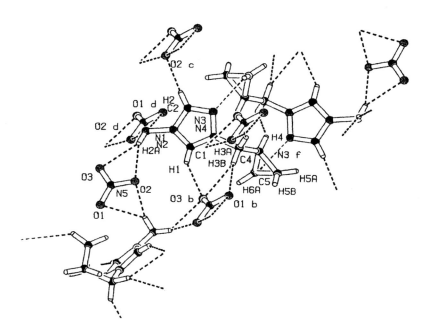

Figure 16. Structure showing significant cation-anion contacts of 1-methylcyclopropyl-4-amino-1,2,4-triazolium nitrate (XX). The interatomic distance (Å) and symmetry code are: O(2)..H(2A)=2.18(2), x, y, z; O(3)..H(2A)=2.39(2), x, y, z; O(1) d..H(2B)=2.39(2), 1/2-x, -1/2+y, 3/2-z; O(2) d..H(2B)=2.12(2), 1/2-x, -1/2+y, 3/2-z; O(3) b..H(1)=2.42(2), 1/2-x, 1/2+y, 3/2-z; O(2) c..H(2)=2.20(2), 3/2-x, -1/2+y, 3/2-z; O(3) e..H(3A)=2.44(2), 3/2-x, -1/2+y, 3/2-z; O(1) b..H(3B)=2.53(2), 1/2-x, 1/2+y, 3/2-z; O(3)b..H(3B) =2.45(2), 1/2-x, 1/2+y, 3/2-z; N(3) f..H(6B)=2.73(2), 2-x, 2-y, 2-z.

*(See page 8 of color insert.)*

amino-1,2,4-triazolium cations with bromide and nitrate anions. Extensive hydrogen bonding was very prevalent between the N-amino group protons and the corresponding anions, however bond distances in the cation were not dramatically affected in any significant manner. As well the longer alkyl chained 1-R-4-amino-1,2,4-triazolium halide salts had unusual packing properties which warrants further investigations.

## Acknowledgements

The authors would like to thank Dr. Michael Berman (AFOSR) and Mr. Michael Huggins (AFRL/PRS), Mr. Ronald Channel and Mr. Wayne Kalliomaa (AFRL/PRSP) for funding and encouragement of this research effort. As well the authors are grateful to Bret Wight for carrying out some of the elemental analyses. One author (GWD) is grateful to stimulating discussions and support from Jeff Bottaro and Mark Petrie (SRI International, Inc.) as well as Jeffrey Sheehy (NASA Marshall Space Flight Center).

## References

1. Present Address: Propulsion Research Center TD40, NASA Marshall Space Flight Center, AL, 35812.
2. Holbrey, J.D.; Seddon, K.R. *Clean Products and Processes.* **1999,** 1, 223-236.
3. Welton, T. *Chem. Rev.* **1999,** 99, 2071-2083.
4. Wasserscheid, P.; Keim, W. *Angew, Chem. Int. Engl.* **2000,** 39, 3772-3789.
5. Seddon, K. R. *Kinetics and Catalysis* **1996,** 37, 693-696.
6. Wassershceid, P.; Welton, T. Eds. *Ionic Liquids in Synthesis* Wiley-VCH **2003,** Weinheim, Fed. Rep. Germ. 7-348.
7. Rogers, R.; Seddon, K. Eds. *Ionic Liquids Industrial Applications for Green Chemistry* ACS Symposium Series 818, American Chemical Society, Wash., D.C., **2002,** 2-458.
8. Hagiwara, R.; Ito, Y. *J. Fluor. Chem.* **2000,** 105, 221-227.
9. Wilkes, J. *Green Chemistry.* **2002,** *4,* 3-10.
10. Wilkes, J.; Zaworotko, M. *J. Chem. Soc. Chem. Commun.* **1992,** 965-96.
11. Fuller, J.; Carlin, R.; De Long, H.; Haworth, D. *J. Chem. Soc. Chem. Commun.* **1994,** 299-300.
12. Sherif L. U.S. Patent 5,731,101, 1998.
13. Huddleston, J.; Willauer, H.; Swatloski, R.; Visser, A.; Rogers, R. *Chem. Commun.* **1998,** 1765-1766.
14. Green, L.; Homeon, I.; Singer, R. *Tetrahedron Letters.* **2000,** *41,* 1343-1346.

15. Abdul-Sada, A. U.S. Patent 5,994,602 1999.
16. Chauvin, Y.; Mussmann, L.; Olivier, H. *Angew. Chem. Int. Ed. Engl.* **1995,** *34,* 2698-2700.
17. Hussey, C. *Pure & Appl. Chem.* **1988,** *60,* 1763-1772.
18. Sherif, F. U.S. Patent 5,731,101 1998.
19. Dyson, P.; Grossel, M.; Srinivasan, N.; Vine, T.; Welton, T.; Williams, D.; White, A.; Zigras, T. *J. Chem. Soc. Dalton Trans.* **1997,** 3465-3469.
20. Koecher, J. U.S. Patent 5,128,267, 1992.
21. Munson K. U.S. Patent 6,339,182, 2002.
22. Koch, E. U.S. Patent 5,827,602, 1998.
23. Teles, S. U.S. Patent 5,508,422, 1996.
24. Schneider, T. U.S. Patent 5,840,894, 1998.
25. Ambler, P. U.S. Patent 5,304,615, 1994.
26. Gordon, C. *Applied Catalysis A. General* **2001,** *222,* 101-117.
27. Suarez, P.; Dullius, J.; Einloft, S.; DeSouza, R.; Dupont, J. *Polyhedron* **1996,** *15,* 1217-1219.
28. Davis, J.; Forrester, K. *Tetrahedron Letters* **1999,** *40,* 1621-1622.
29. Curphey, T. J.; Prasad, K. S. *J. Org. Chem.* **1972,** *37,* 2259-2265.
30. Vestergaard, B,; Bjerrum, N.J.; Petrushina, I.; Hjuler, H.; Berg, R..; Begtrup, M. *J. Electrochem. Soc.* **1993,** *140,* 3108-3113.
31. Surpateanu, G.; Lablanche-Combier, A.; Grandglaudon, P.; Couture, A.; Mouchel, B. *Revue Roumaine de Chemie.* **1993,** *38,* 671-681.
32. Mirzaei, Y.; Twamely, B.; Shreeve, J. *J. Org. Chem.* **2002,** *67,* 9340-9345.
33. Singh, R.; Manandhar, S.; Shreeve, J. *Tetrahedron Letters,* **2002,** *43,* 9497-9499.
34. Mirzaei, Y.; Shreeve, J. *Synthesis,* **2003,** 24-26.
35. Becker, G.; Böttcher, H.; Röthling, T.; Timpe, H. *Wissenschaftl. Zeitscher.* **1966,** *8,* 22-25.
36. Becker, H.; Timpe, H. *J. Prakt. Chem.* **1969,** *311,* 9-14.
37. Becker, H.; Nagel, D.; Timper, H. *J. Prakt. Chem.* **1973,** *315,* 1131-1138.
38. Astleford, B.; Goe, G.; Keay, J.; Scriven, E. *J. Org. Chem.* **1989,** *54,* 731-732.
39. Gromova, S.; Barmin, M.; Shylapochnikov, V.; Kolesnikova, O.; Mel'nikov, V. *Russian Journal of Organic Chemistry* **1996,** *32,* 1190-1193.
40. Gromova, S.; Barmin, M.; Mel'nikov, A.; Korolev, E.; Mel'nikov, V. *Russian Journal of Applied Chemistry* **1995,** *72,* 1409-1411.
41. Shitov, O.; Vyazkov, V.; Tartakovskii, V. *Izv. Akad. SSSR, Ser. Khim.* **1989,** *11,* 2654-2656.
42. Shitov, O.; Korolev, V.; Tartakovsky, V. *Russ. Chem. Bull.* **2002,** *51,* 499-502.
43. Shitov, O.; Korolev, V.; Bogdanvo, V.; Tartakovsky, V. *Russ. Chem. Bull.* **2003,** *52,* 695-699.

44. Kofman, T.; Kartseva, G.; Shcherbinin, M. *Russ. J. Org. Chem.* **2002**, *38*, 1343-1350.
45. Balasubramanian, M.; Keay, J.; Scriven, E. *Heterocycles* **1994**, *37*,1951-1975.
46. Tamura, Y.; Minamikawa, J.; Ikeda, M. *Synthesis.* **1977**, 1-17.
47. Milicent, R.; Redeuilh, C. *J. Heterocyclic Chem.* **1980**, *17*, 1691-1696.
48. Barszcz, B.; Gabryszewski, M.; Kulig, J.; Lenarcik, B. *J. Chem. Soc. Dalton Trans.* **1986**, 2025-2028.
49. De la Concepción, M.; Foces-Foces, M.; Cano Hernandez, F.; Claramunt, R.; Sanz, D.; Catalán, J.; Fabero, F.; Fruchier, A.; Elguero, J. *J. Chem. Soc. Perkin Trans 2.* **1990**, 237-244.
50. Claramunt, R.; Sanz, D.; Catalain, J.; Fabero, F.; Garcia, N.; Foces-Foces, C.; Llamas-Saiz, A.; Elguero, J. *J. Chem. Soc. Perkin Trans 2.* **1993**, 1687-1699.
51. Temple, C.; Montgomery, J. *The Chemistry of Heterocyclic Compounds; Triazoles 1,2,4* J. Wiley & Sons, NY, NY, 1981, 599-658.
52. Becker, V.; Nagel, D.; Timpe, H. J. *J. Prakt. Chem.* **1973**, *315*, 97-105.
53. Bonhôte, P.; Dias, A.; Papageorgiou, N.; Kalyanasundaram, K.; Grätzel, M. *Inorg. Chem.* **1996**, *35*, 1168-1178.
54. Larsen, A..; Holbrey, J.; Tham, F.; Reed, C. *JACS* **2000**, A-I.
55. Barlin, G.; Batterham, T. *J. Chem. Soc.(B).* **1967**, 516-518.
56. Drake, G.; Hawkins, T.; Brand, A.; Hall, L.; McKay, M.; Vij, A.; Ismail, I. *Prop. Expl. Pyrotech.* **2003**, *28*, 174-180.
57. Weigert, F.; Roberts, J. *JACS.* **1968**, *90*, 3543-3549.
58. Wofford, D.; Forkey, D.; Russell, J. *J. Org. Chem.* **1982**, *47*, 5132-5137.
59. Lieber, E.; Levering, D.; Patterson, L. *Anal. Chem..* **1951**, *23*, 1594-1604.
60. Biles, F.; Endredi, H.; Keresztury, G. *J. Mol. Struct.* **2000**, *530*, 183-200.
61. Edsall, J.; Scheinberg, H.. *J. Chem. Phys.* **1940**, *8*, 520-525.
62. Bellanato, J. *Spectrochimica Acta.* **1960**, *16*, 1344-1357.
63. Chenon, B.; Sandorfy, C. *Can. J. Chem.* **1958**, *36*, 1181-1206.
64. Durig, J.; Bush, S.; Mercer, E. *J. Chem. Phys.* **1966**, *44*, 4238-4247.
65. Edsall, J. *J. Chem. Phys.* **1937**, *5*, 225-237.
66. Catalano, E.; Sanborn, R.; Frazer, J. *J. Chem. Phys.* **1963**, *38*, 2265-2272.
67. Steiner, T. *Agnew. Chem. Int. Ed.* **2002**, *41*, 48-76.
68. Decius, J.; Pearson, D. *JACS.* **1953**, 75, 2436-2349.
69. Nakagawa, I.; Walter, J. *J. Chem. Phys.* **1969**, *51*, 1389-1397.
70. Rousseau, D.; Miller, R. *J. Chem. Phys.* **1968**, *48*, 3409-3413.
71. Walrafen, G. E.; Irish, D.E. *J. Phys. Chem.* **1964**, 911-913.
72. X-ray Crystallography. The single crystal X-ray diffraction data were collected on a Bruker 3-circle platform diffractometer equipped with a SMART CCD (charge coupled device) detector with the $\chi$-axis fixed at 54.74° and using $MoK_\alpha$ radiation ($\alpha = 0.71073$ Å) from a fine-focus tube. This diffractometer was equipped with KryoFlex apparatus for low temperature data collection using

controlled liquid nitrogen boil off. The goniometer head, equipped with a nylon Cryoloop with a magnetic base, was then used to mount the crystals using PFPE (perfluoropolyether) oil. Cell constants were determined from 90 ten-second frames. A complete hemisphere of data was scanned on omega (0.3°) with a run time of ten-second per frame at a detector resolution of 512 x 512 pixels using the SMART software.[1] A total of 1271 frames were collected in three sets and final sets of 50 frames, identical to the first 50 frames, were also collected to determine any crystal decay. The frames were then processed on a PC running on Windows NT software by using the SAINT software[2] to give the hkl file corrected for Lp/decay. The absorption correction was performed using the SADABS[3] program. The structures were solved by the direct method using the SHELX-90[4] program and refined by the least squares method on $F^2$, SHELXL-97[5] incorporated in SHELXTL Suite 5.10 for Windows NT.[6] All non-hydrogen atoms were refined anisotropically. For the anisotropic displacement parameters, the U(eq) is defined as one third of the trace of the orthogonalized $U_{ij}$ tensor. The hydrogen atoms were located either from difference electron density maps or generated at calculated positions . 1. *SMART V* 4.045 Software for the CCD Detector System, Bruker AXS, Madison, WI, 1999. 2. *SAINT V 4.035 Software for the CCD Detector System* Bruker AXS, Madison, WI , 1999. 3. *SADABS, Program for absorption correction for area detectors* Version 2.01, Bruker AXS, Madison, WI, 2000. 4. Sheldrick, G. M. *SHELXS-90, Program for the Solution of Crystal Structure* University of Göttingen, Germany, 1990. 5. Sheldrick, G. M. *SHELXL-97, Program for the Refinement of Crystal Structure* University of Göttingen, Germany, 1997. 6. *SHELXTL 5.10 for Windows NT, Program library for Structure Solution and Molecular Graphics* Bruker AXS, Madison, WI, 2000.

73. *SMART for WNT/2000, version 5.625* Bruker AXS, Inc., Madison, WI, 2001.

74. *SAINT PLUS, version 6.22*, Bruker AXS, Inc., Madison, WI, 2001.

75. *SHELXTL, version 6.10* Bruker AXS, Inc., Madison, WI, 2000.

76. Starova, G.; Frank-Kameneckaya, O.; Levner, M. *Vestn. Leningr. U., Fiz. Khim.* **1991**, 103.

77. *Conquest V. 1.5, Cambridge Structural Database System Version 5.24 (July Update, 2003)* CCDC, Cambridge, UK, 2003.

78. Bondi, A. *J. Phys. Chem.* **1964**, *68*, 441-451.

79. Drabent, K.; Ciunik, L. *Chem. Commun.* **2001**, 1254-1255.

80. Isaacs, N.; Kennard, C. *J. Chem. Soc. (B)* **1971**, 1270-1273.

81. Seccombe, R.; Kennard, C. *J. Chem. Soc. Perkin Trans. II* **1973**, 9-11.

82. Seccombe, R.; Tillack, J.; Kennard, C. *J.C.S. Perkin II.* **1973**, 6-8.

83. Seccombe, R.; Kennard, C. *J.C.S. Perkin II.* **1972,** 1-3.

84. Podberezskaya, N.; Pervukhina, N.; Doronin, V. *Zhurnal Strukturnoi Khimii.* **1991**, *32*, 34-39.

302

85. Liu, J.; Fu, D.; Zhuang, J.; Duan, C.; You, X. *J. Chem. Soc, Dalton Trans.* **1999**, 2337-2342.
86. Drabent, K.; Ciunik, Z. *Chem. Commun.* **2001**, 1254-1255.
87. Wilkes, J.; Levisky, J.; Wilson, R.; Hussey, C. *Inorg. Chem.* **1982**, *21*, 1263-1264.
88. Tait, S.; Osteryoung, R. *Inorg. Chem.* **1984**, *23*, 4252-4360.
89. Fannin, A.; King, L.; Levisky, J.; Wilkes, J. *J. Phys. Chem.* **1984**, *88*, 2609-2614.
90. Wilkes, J.; Hussey, C.; Sanders, J. *Polyhedron.* **1986**, *5*, 1567-1571.
91. Dymek, C.; Stewart, J. *Inorg. Chem.* **1989**, 28, 1472-1476.
92. Abdul-Sada, A.; Greenway, A.; Hitchcock, P.; Mohammed, T.; Seddon, K.; Zora, J. *J. Chem. Soc. Chem. Commun.* **1986**, 1753-1754.
93. Appleby, D.; Hitchcock, P. ; Seddon, K.; Turp, J.; Zora, J.; Hussey, C.; Sanders, J.; Ryan, T. *J. Chem. Soc. Dalton Trans.* **1990**, 1879-1887.
94. Gorter, S.; Engelfriet, D. Acta *Crystallogr., Sect. B. Struct. Crystallogr. Cryst. Chem.* **1981**, 37, 1214-1216.
95. Haasnoot, J.; De Keyzer, G.C.M.; Verschoor, G.C. *Acta Crystallogr., Sect. C:Cryst. Struct. Commun.* **1983**, 39, 1207-1210.
96. Driessen, W.; De Graaff, R.; Vos, J. *Acta Crystallogr., Sect. C:Cryst. Struct. Commun.* **1983**, 39, 1635-1639.
97. Lynch, D.; Smith, G.; Byriel, K.; Kennard, C. *Acta Crystallogr., Sect. C:Cryst. Struct. Commun.* **1994**, 50, 1291-1297.
98. Engelfriet, D.; Groeneveld, W. ; Nap, G. *Z. Naturforsch., Teil A* **1980**, *35*, 1387-1392.
99. Lynch, D.; Latif, T.; Smith, G.; Byriel, K.; Kennard, C.; Parsons, S. *Aust. J. Chem.* **1998**, *51*, 403-411.
100. Lynch, D.; Dougall, T.; Smith, G.; Byriel, K.; Kennard, C. *J. Chem. Cryst.* **1999**, 29, 67-72.
101. Goldstein, P.; Ladell, J.; Abowitz, G. *Crystallogr., Sect. B. Struct. Crystallogr. Cryst. Chem.* **1969**, 25, 135-146.
102. Bowlas, C.; Bruce, D.; Seddon, K. *Chem. Commun.* **1996**, 1625-1626.
103. Gordon, C. ; Holbrey, J. ; Kennedy, A.; Seddon, K. *J. Mater. Chem.* **1998**, *8*, 2627-2636.
104. Holbrey, J.; Seddon, K. *J. Chem. Soc., Dalton Trans.* **1999**, 2133-2139.
105. Oyumi, Y.; Brill, T.; Rheingold, A.; Lowe-Ma, C. *J. Phys. Chem.* **1985**, *89*, 2309-2314.
106. Bracuti, A. *J. Crystallogr. Spectrosc. Res.* **1993**, *23*, 669-673.
107. Mylrajan, M.; Srinivasan, T.; Sreenivasamurthy, G. *J. Crystallogr. Spectrosc. Res.* **1985**, *15*, 493-502.
108. Doxsee, K.; Francis, P.; Weakley, T. *Tetrahedron* **2000**, 56, 6683-6688.

Chapter 21

# Liquid-Crystalline Phase of Phosphonium Salts with Three Long *n*-Alkyl Chains as Ordered Ionic Fluids

David J. Abdallah[1,4], Hui C. Wauters[1], Dylan C. Kwait[1],
C. L. Khetrapal[2], G. A. Nagana Gowda[2], Allan Robertson[3],
and Richard G. Weiss[1,*]

[1]Department of Chemistry, Georgetown University,
Washington, DC 20057–1227
[2]Sanjay Gandhi Post Graduate Institute of Medical Sciences,
Lucknow-226 014, India
[3]Cytec Industries Inc., Niagara Falls, Ontario, L2E 6T4, Canada
[4]Current address: Clariant Corporation, 70 Meister Avenue,
Somerville, NJ 08876

We describe the syntheses and properties of liquid-crystalline phosphonium salts, especially those with 3 long *n*-alkyl chains, a shorter substituent, and a variety of anions.

Many ionic liquid crystals are known. Most may be divided into two classes: (1) the organic part is attached to a negatively charged head group, especially as in metal alkanoates;[1] (2) the organic parts are attached to a positively charged head group, especially quaternary salts comprised of a Group VA (Group 15) cationic head group (N.B., N or P). Of the latter variety, those with *one*[2] or *two*[3] long *n*-alkyl chains[a] have been studied extensively, and many form either smectic phases when neat or other assemblies, such as micelles,

---

(a) In this work, the arbitrary definition of a 'long chain' is ≥ 10 carbon atoms.

© 2005 American Chemical Society　　　　　　　　　　**303**

when mixed with water.[4] We have found interesting 'biradial' and 'tetraradial' packing arrangements for *crystalline phases* of several Group VA salts containing *four* equivalent *n*-alkyl chains with 10 to 18 carbon atoms.[5,6] All of the salts crystallize as stacked monolayers with an 'ionic plane' consisting of an array of anions and positively-charged N or P atoms in the middle of each layer, but none exhibits a liquid-crystalline phase.

Recently, we have found that many phosphonium salts (**nPmA**) with *three* long equivalent *n*-alkyl chains (containing **m** = 10, 14, or 18 carbon atoms each), one shorter chain with **n** = 0-5 carbon atoms or a benzyl group (**Bz**), and monovalent anions (**A**) of various types and sizes do form a liquid-crystalline phase. The greater thermal stability and generally wider mesophase temperature ranges of the phosphonium salts are two important reasons why we have focused our attention on them rather than the more accessible and more easily synthesized ammonium salts. Some ammonium salts have larger liquid-crystalline ranges than their corresponding phosphonium salts. An example is the **BzN18Br** ($T_{K-SmA2}$ 78.3 °C; smectic range = 12.5 °C) and **BzP18Br** ($T_{K-SmA2}$ 70.8 °C; smectic range = 8 °C) pair. When heated to 140°C, **BzN18Br** exhibits a broad exothermic transition centered around 130 °C, corresponding to what we believe is a Hofmann-type elimination reaction.[9] Furthermore, differential scanning calorimetry thermograms indicate that large molecular reorganizations accompany the crystal-to-liquid crystal transitions of these ($\Delta H$ = 145.5 KJ/mol for **BzN18Br** and 117.7 KJ/mol for **BzP18Br** during the first heatings) and the other salts discussed here.

The additional stability and lack of highly conjugated groups make the **nPmA** salts excellent candidates to be ordered ionic fluids for performing thermal and photochemical reactions of solute molecules.

## Synthetic Aspects

The phosphonium salts are easily prepared by $S_N2$ reactions from the corresponding phosphines and alkyl halides (chlorides and bromides) in the absence of oxygen. The phosphines react rapidly with molecular oxygen, yielding phosphoranes and other oxidized species. To minimize exposure to molecular oxygen, reagent transfers were performed in a dry box under a nitrogen atmosphere and the solvents employed were saturated with molecular nitrogen prior to their use. Once the phosphorus atoms are quaternized, the salts can be handled in air without problem. However, especially those **1PmA** with shorter **m** chains are hygroscopic and must be handled in a dry atmosphere to ensure that hydrates are not formed.

Quantitative anion exchange was usually achieved by converting the chloride or bromide salt to its ethyl xanthate and then adding a stong acid containing the desired anion type (Scheme 1).[7] Details of the procedures for purification and characterization of the salts are included within the references cited.[5,6, 8,9,10,11]

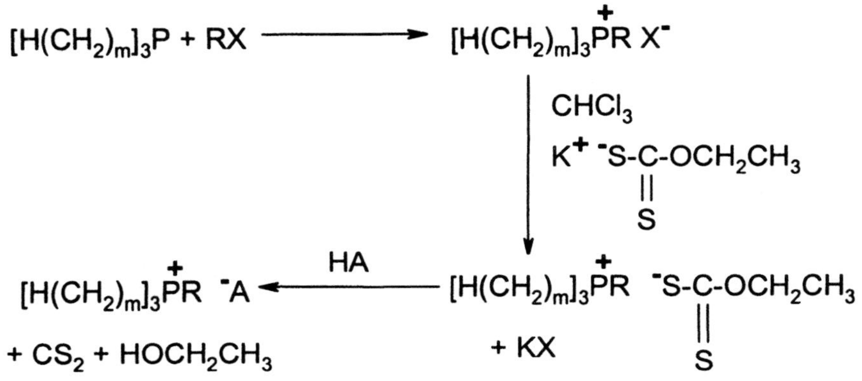

$$[H(CH_2)_m]_3P + RX \longrightarrow [H(CH_2)_m]_3\overset{+}{P}R\ X^-$$

$$\downarrow \begin{array}{l} CHCl_3 \\ K^+\ {}^-S\text{-}C\text{-}OCH_2CH_3 \\ \quad\quad \underset{S}{\parallel} \end{array}$$

$$[H(CH_2)_m]_3\overset{+}{P}R\ {}^-A \xleftarrow{\ HA\ } [H(CH_2)_m]_3\overset{+}{P}R\ {}^-S\text{-}C\text{-}OCH_2CH_3$$

$$+ CS_2 + HOCH_2CH_3 \qquad\qquad + KX \qquad\qquad \underset{S}{\parallel}$$

*Scheme 1. General synthetic route to **nPmA** salts.*

## Structure and packing within crystalline phases

From single-crystal X-ray crystallographic measurements, all of the phosphonium and ammonium salts with four equivalent long $n$-chains that we have investigated adopt packing arrangements with two chains projected along one side and the other two chains projected along the opposite side of a rough plane defined by a mosaic of cationic head group atoms and anions.[5] When the $n$-alkyl chains consist of < 12 carbon atoms, the packing arrangement is 'tetraradial (i.e., with the four chains of each cation projected roughly in a tetrahedral arrangement); for salts with > 12 carbon atoms per chain, the arrangement is 'biradial' (i.e., pairs of chains lie next to each other). We have found several variants of the 'biradial' arrangement (Figures 1 and 2), but in all of them, one chain of each pair adopts some gauche bends in order to redirect it along the axis of its partner.

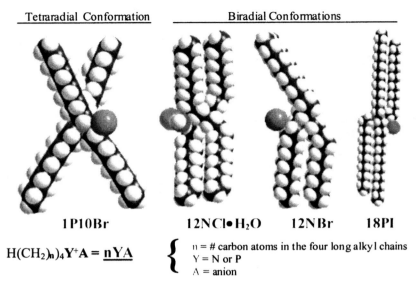

Tetraradial Conformation          Biradial Conformations

1P10Br          12NCl•H₂O          12NBr          18PI

$$H(CH_2)_n)_4 Y^+ A = \underline{nYA}$$

$\Big\{$
n = # carbon atoms in the four long alkyl chains
Y = N or P
A = anion

*Figure 1. A tetraradial and three biradial forms of nYA salts.*

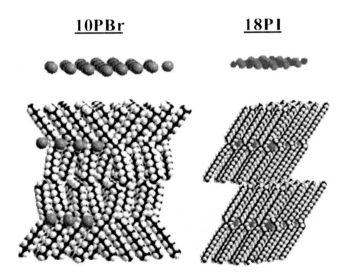

**10PBr**          **18PI**

*Figure 2. Packing arrangements of crystalline 10PBr and 18PI showing the distance between ionic layers and the distribution of ionic centers within.*

Fewer examples of the crystal packing by salts with three long *n*-alkyl chains and one shorter chain are known. In our hands, they are more difficult to crystallize into single crystals suitable for diffraction experiments than salts with four equivalent long chains. A part of this difficulty may be related to the greater complexity of their packing arrangements. The three **nPmA** and **nNmA** of this type we have examined thus far[9,12] adopt a bent "h" shape in which two chains are projected, in parallel, along one side of the rough plane defined by the ionic centers and the other chain is interdigitated with a chain from another molecule. For example, molecules of **BzN18Br** (Figure 3) pack in alternating interdigitated and non-interdigitated regions of alkyl chains separated by roughly defined ionic planes. Spacings between ionic planes alternate between 19.8 Å (an interdigitated segment) and 30.4 Å (two non-interdigitated segments) and their sum constitutes the length of the *c*-axis in a unit cell.

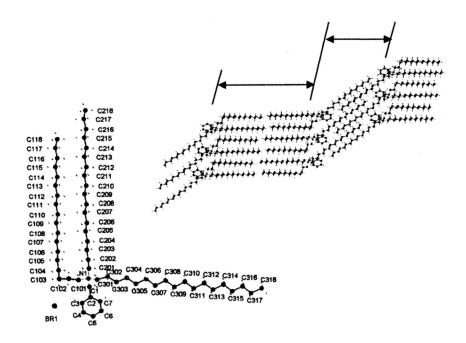

*Figure 3. Crystal packing showing two layer distances and structure of*
***BzN18Br.***

Powder X-ray diffraction patterns of most of the salts at room temperature consisted of peaks that could be indexed as 00*l* at low angles and other relatively sharp peaks at higher angles, frequently superimposed on a broad peak. The latter feature indicates that the alkyl chains of the salts are somewhat disordered in the solid phases. It may also explain why so few of the **1PnA** are easily crystallized. In fact, the solids are not mechanically strong; they are easily distorted by applying pressure and differentiation of some of them from liquid crystals is very difficult by polarizing optical microscopy alone. The examples of **1P10Cl** (that passes directly from the solid to the isotropic phase) and its hydrate, **1P10Cl•H₂O** (that is smectic at room temperature), are shown in Figure 4.

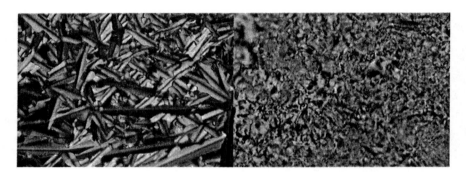

*Figure 4. Left: Batonnets forming in a fan-like texture in **1P10Cl** at 40 °C upon cooling (solid phase). Right: Spherulite texture of **1P10Cl•H₂O** at room temperature (smectic phase).*

### Structure and packing within liquid-crystalline phases

Based primarily on information from X-ray diffractometry (N.B., one sharp low-angle peak corresponding to a distance greater than one but less than two times the extended length of a single molecule, and a broad high-angle peak) and polarizing optical microscopy (N.B., optical patterns like those reported for other smectic A phases and the appearance of oily streak patterns when the samples are sheared), all of the liquid-crystalline phases of the **nPmA** salts with 'small' **n** groups are the smectic $A_2$ (SmA$_2$) type.[13]   A general structural description of this phase is shown in Figure 5.

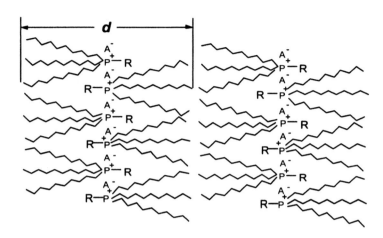

*Figure 5.   Cartoon representation of the smectic A$_2$ packing phases of **nPmA** salts. R represents the short **n** group.*

Because the sum of the cross-sectional areas of the three long chains is much larger than that of an N or P centered head group, assemblies somewhat like those of the symmetrical *tetra-n*-alkyl salts, in which the ionic parts are arranged in planes that bisect pairs of chains on each molecule, were anticipated.   Additionally, we considered incorrectly that nematic phases, in which the molecules retain only orientational order and are *not* arranged in layered assemblies,   were present based on the ease with which these liquid crystals can be aligned in magnetic fields;[14] smectic phases are usually difficult to orient in this way.[15] However, despite their ease of alignment, all of the liquid-crystalline **nPmA** and the corresponding **nNmA** investigated are smectic A$_2$ phases (i.e., mesophase *bi*layered assemblies).[5]

A typical thermogram of **1P14ClO$_4$** is displayed in Figure 6. Thermograms from the first heating of the solvent-crystallized morph of most salts contain an additional low $\Delta$H solid-solid[b] transition that is not present in

---

(b) Abreviations for phase transitions are: K–K $\Rightarrow$ solid-solid; K-SmA$_2$ $\Rightarrow$ solid-Smectic A$_2$; K-I $\Rightarrow$ solid–isotropic; SmA$_2$-I $\Rightarrow$ Smectic A$_2$–isotropic.

the second heating scan. The reproducibility of the second and subsequent heating thermograms and all cooling thermograms provide compelling evidence for the aforementioned thermal stability of the salts. They also demonstrate that the morphs obtained by solvent recrystallization and melt cooling are not always the same.

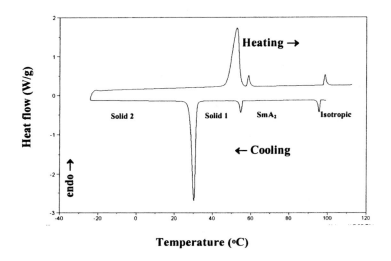

*Figure 6. DSC thermogram of 1P14ClO₄. The sample was cooled from the isotropic melt and then heated at a rate of 5 °C/min.*

Although enthalpies of the K→SmA₂ transitions of the enantiotropic (*e*) liquid-crystalline salts (**0P18I, 1P18A (A = Br, I, NO₃, BF₄, ClO₄, PF₆), 2P18A (A = Br, I), and BzP18Br**) vary over a wide range, most are near 80 KJ/mol, and those with lower values have an additional K→K transition near in temperature. The enthalpies of all of the SmA₂→I transitions are much lower, <5 KJ/mol, as expected.

Figure 7 provides insights into the relationship between structures and liquid-crystalline properties (especially transition temperatures) of the *tri-n-*octadecyl salts, **nP18Br, nP18I** and **nN18I**.[14,16] The **nP18I** consistently have lower T_K→SmA2 and broader mesophase ranges than the corresponding **nN18I**. When **nP18A** anions are bromide, the mesophases either exhibit much narrower ranges or are not present. Trends in the entropies of reversible

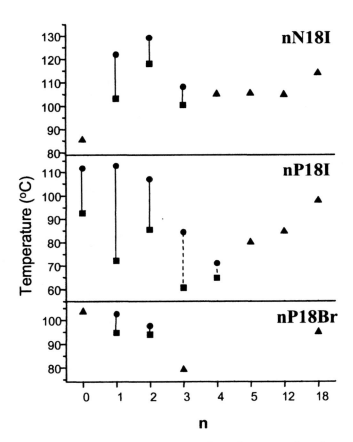

*Figure 7. Phase transition temperatures from the onset of peaks in DSC thermograms of **nN18I** and **nP18A**. Transition temperatures are for melting (▲, $T_m$), solid-$SmA_2$ (■, $T_{K-SmA2}$), and $SmA_2$-isotropic (♦, $T_{SmA2-I}$) transitions. Solid lines indicate temperature ranges of enantiotropic phases from first heating and dotted lines indicate temperature ranges of monotropic phases from first cooling.*

312

mesophase transitions, $\Delta S_{SmA2\rightarrow I}$, indicate that the size of **n** has a large influence on the order within an $SmA_2$ phase.

Because liquid-crystalline ranges are largest when the fourth substituent is methyl (**n** =1), we have focused our efforts primarily on salts with **1Pm** cations. The **1P18A** with large anions (*i.e.*, nitrate, perchlorate, tetrafluoroborate, and hexafluorophosphate) have similar mesophase ranges and $T_{SmA2-I}$ values that differ from those of the bromide salt (Figure 8). Furthermore, our initial studies with **1PmBF₄** salts indicated that the onset of liquid crystallinity approaches room temperature as the length of **m** decreases from 18 to 10. They also suggested that some **1P10A** might be liquid crystalline at even subambient temperatures.

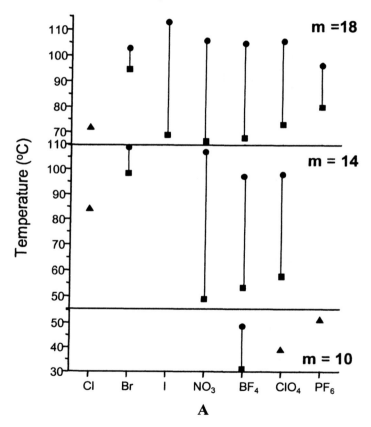

*Figure 8. Phase transition temperatures from the onset of peaks in DSC thermograms of 1PmA. Transition temperatures are for melting (▲, $T_m$), solid-SmA₂ (■, $T_{K-SmA2}$), and SmA₂-isotropic (♦, $T_{SmA2-I}$) transitions. Solid lines indicate temperature ranges of enantiotropic phases from second heating.*

Those indications were borne out in subsequent experiments with five other **1P10A** salts in which **A** is chloride, chloride (with a molecule of water), bromide, bromide (with a molecule of water), or nitrate.[8] Of these, four are completely saturated molecules, two (the hydrates) are liquid crystalline over a more than 90 degree range that commences below room temperature and ends above 90 °C, and one (the nitrate) is in an enantiotropic liquid-crystalline phase at ca. 60-83 °C. The two anhydrous halide salts pass directly from their (soft) solid to isotropic phases when heated, and no monotropic mesophases were detected.

Somewhat surprisingly, X-ray diffraction patterns at different temperatures cannot be used to distinguish between the liquid-crystalline and isotropic phases because significant aggregation and ordering remain within the latter. The presence or absence of sharp peaks in the high angle regions of the X-ray diffractograms do, however, differentiate the solid and liquid-crystalline phases. Additionally, the low angle peaks of the X-ray diffractograms indicate that the distances between the ionic layers within the smectic phases decrease as temperature is raised but the interlayer distances are constant within the solid phases (Figure 9).

As noted for the **1P18A**, the heats of the $SmA_2 \rightarrow I$ transitions (<3 KJ/mol) are much lower than those of the corresponding $K \rightarrow I$ transitions (>8 KJ/mol), and the onset of the isotropic phases can be located unabmiguously from polarizing optical micrographs.

## Preliminary Applications

The hydrates are the first two completely saturated, ordered, room-temperature ionic liquid crystals that we are aware of. Their lack of strongly absorbing chromophores in the near UV or visible regions and facile alignments within magnetic fields should make them interesting media for spectroscopic or photochemical studies of solutes and as switchable conductors of electrical current. Initial efforts have been directed mostly toward spectroscopic studies. Thus, the locations of two solvatochromic dyes, Nile Red and 1,1-dicyano-2-[6-(dimethylamino)naphthylen-2-yl]propene (**DDNP**), within the ionic assemblies have been approximated from UV/vis absorption (N.B., the $E_{NR}$ polarity scale for Nile Red[17]) and fluorescence spectra (N.B., emission maxima from **DDNP**[18]). The $E_{NR}$ values indicate that molecules of Nile Red in the **1P10A** salts are in sites whose polarity is somewhat lower than those afforded by the more conventional (isotropic) 1-alkyl-3-methylimidazolium ionic liquids ([Cm-mim][A] in Figure 10, where m is the length of the n-alkyl chain at carbon-1 and A is the counterion),[19] and the emission spectra from **DDNP** are consistent with this conclusion. Both sets of data suggest that the probe molecules reside preferentially in regions somewhat

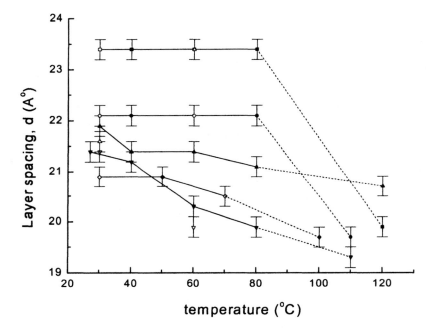

*Figure 9. Dependence of the layer spacings on temperature in 1P10A salts.*
*(□) 1P10Cl; (○) 1P10Br; (△) 1P10Cl H₂O; (▽) 1P10Br H₂O; (◇)1P10NO₃.*
*Solid symbols correspond to data obtained on heating to the isotropic phase*
*and open ones to cooling from it. Solid lines connect points within the*
*anisotropic (solid or liquid-crystalline) phases. Dashed lines connect the*
*highest temperature points within the anisotropic phases and the*
*corresponding isotropic phases (points at highest temperature).*

removed from the ionic layers of the head groups and their anions in the ordered phases of the **1P10A**. However, the electronic nature of the dyes must be an important factor in determining their regions of incorporation, and there is no reason to believe that probes of different structure and polarity cannot reside within the ionic layers.[20]

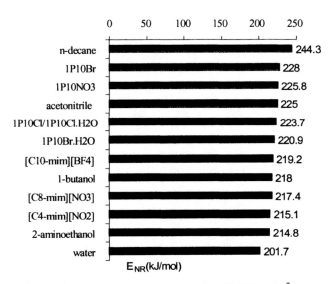

*Figure 10. $E_{NR}$ values at room temperature for **1P10A** salts[8] and other media.[17,19]*

Figure 11 shows the temperature dependence of the AC conductivity of **1P10A** salts at 20 Hz (i.e., near the DC limit).[21] The conductivity of **1P10Cl·H₂O** increases regularly with increasing temperature, while that of **1P10Cl** reaches a maximum ($2.9 \times 10^{-6}$ S/m) at 70 °C and then decreases. The conductivity of **1P10Br** is reasonably constant throughout its soft solid temperature range, and increases in the isotropic melt where ions become more mobile. The phase transition between the soft solid and smectic phase of **1P10NO₃** near 60 °C is discernible in the conductivity measurements. The conductivity rises between 30 and 50 °C, falls at 70 °C, and then rises thereafter. Although detailed interpretations of these data must await further analyses, it is clear that ionic mobility within the phases, whether liquid-crystalline or soft solids, generally increases with increasing temperature. Although the measured conductivities, $\sim(1\text{-}5) \times 10^{-6}$ S m$^{-1}$, are $10^5$-$10^6$ times smaller than the highest values reported for stable, room-temperature, isotropic

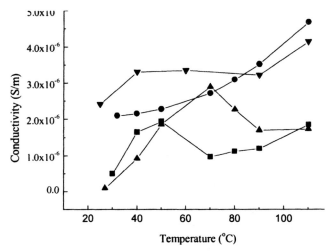

*Figure 11. Temperature dependence of the AC conductivity of 1P10A salts at 20 Hz. (■) 1P10NO3; (●) 1P10Cl•H2O; (▲) 1P10Cl; (▼) 1P10Br.*

ionic liquids,[22] they may be sufficiently large for specific applications. More importantly, the conductivities may become exceedingly large if the liquid-crystalline phases are aligned in the conductance cells so that the ionic planes are perpendicular to the cell plates. In such a case, switching between perpendicular and parallel alignment should result in an 'on-off' switch.

Liquid-crystalline ionic salts that easily aligned in magnetic fields may be useful matrices in which to determine the conformations of solutes by NMR measurements.[23] To that end, we have begun investigations of the structures of simple molecules in the smectic phases of the **1P10A** salts. [24] Figures 12 and 13 are spectra of deuterated and protiated acetonitrile within the liquid-crystalline phase of **1P10NO₃**. Figure 12 demonstrates that the acetonitrile molecules are weakly oriented (order parameter ~0.01) as required to reduce the splitting frequencies from dipolar couplings and make their analyses easily tractable.[25] Figure 13 is a typical 2D spectrum from which the structural parameters for the molecule can be extracted. For instance, at 1 wt% of acetonitrile, its HCH angle is calculated to be 110.7°. Efforts to reduce the order parameters (and line widths) further by judicious choice of the ionic liquid crystal structure, so that more sophisticated structural information like that being obtained from NMR spectra in aqueous 'bicells',[26] are ongoing.

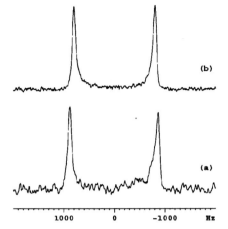

*Figure 12.* $^2H$ *NMR spectra of (a) 1.5 wt% and (b) 10 wt% CD$_3$CN in* **1P10NO$_3$** *at room temperature showing quadrupole split doublets due to orientation.*

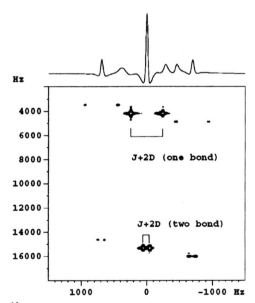

*Figure 13.* $^1H$ - $^{13}C$ *HMBC spectrum of 3 wt% CH$_3$CN in* **1P10NO$_3$** *showing one and two bond $^1H$-$^{13}C$ dipolar splittings.*

318

## Conclusions and the Future

Many **nPmA** salts are now well characterized. They are thermally more stable than the corresponding ammonium salts, are stable in dry air (some are deliquescent), and lack chromophores and unsaturated groups that may interfere with measurements of many solute properties and reactions. Their thermal properties, especially the presence of smectic liquid-crystalline phases at convenient temperatures in several of the salts, and simple structures are attractive features for many applications.

Some important questions about the salts remain to be answered. Two of them involve recyclability:[1]

"Are the salts 'green' solvents?"

"If so, how green are they?"

If the **nPmA** are recyclable and conditions can be found for their easy separation from reactants and products without extraction, they may be extremely useful for conducting selective reactions on the industrial scale. We anticipate that others and we will exploit these salts in the near future.

**Acknowledgments.** We thank several collaborators whose names appear in the references for their participation in this research. We are grateful to the National Science Foundation and the Petroleum Research Fund administered by the American Chemical Society for their financial support.

## References

[1] Mirnaya, T. A.; Volkov, S. V. in *Green Industrial Applications of Ionic Liquids*; Rogers, R. D., *et al.*, Eds; Kluwer: Dordrecht, 2003, p 439.

[2] (a) Gault, J. D.; Gallardo, H. A.; Muller, H. J. *Mol. Cryst. Liq. Cryst.* **1985**, *130*, 163. (b) Busico, V.; Cernicchiaro, P.; Corradini, P.; Vacatello, M. *J. Phys. Chem.* **1983**, *87*, 1631. (c) Busico, V.; Corradini, P.; Vacatello, M. *J. Phys. Chem.* **1982**, *86*, 1033. (d) Margomenou-Leonidopoulou, G.; Malliaris, A.; Paleos, C. M. *Thermochimica Acta* **1985**, *85*, 147.

[3] (a) Kanazawa, A.; Tsutsumi, O.; Ikeda, T.; Nagase, Y. *J. Am. Chem. Soc.* **1997**, *119*, 7670. (b) Alami, E.; Levy, H.; Zana, R.; Weber, P.; Skoulios, A. *Liq. Cryst.* **1993**, *13*, 201. (c) Kanazawa, A.; Ikeda, T. in *Liquid Crystal Materials and Devices*; Bunning, T. J.; Chen, S. H.; Chien, L.-C.; Kajiyama, T.; Koide, N.; Lien, S.-C. A., Eds.; Materials Research Society: Warrendale, PA; 1999, p 201.

[4] (a) Margomenou-Leonidopoulou, G. *J. Thermal Anal.* **1994**, *42*, 1041. (b) Margomenou-Leonidopoulou, G. *ICTAC News* **1993**, *26*, 24. (c) Tschierske, C. *J. Mater. Chem.* **1998**, *8*, 1485. (d) Tschierske, C. *Prog. Polym. Sci.* **1996**, *21*, 775.

[5] Abdallah, D. J.; Bachman, R. E.; Perlstein, J.; Weiss, R. G. *J. Phys. Chem. B* **1999**, *103*, 9269.

[6] Abdallah, D. J.; Kwait, D. C.; Bachman, R. E.; Weiss, R. G. *Cryst. Growth Design* **2001**, *1*, 267.

[7] Sepulveda, L.; Cabrera, W.; Gamboa, C.; Meyer, M. *J. Colloid Interface Sci.* **1987**, *117*, 460.

[8] Chen, H.; Kwait, D. C.; Gönen, Z. S.; Weslowski, B. T.; Abdallah, D. J.; Weiss, R. G. *Chem. Mater.* **2002**, *14*, 4063.

[9] Abdallah, D. J.; Lu, L.; Cocker, T. M.; Bachman, R. E.; Weiss, R. G. *Liq. Cryst.* **2000**, *27*, 831.

[10] Abdallah, D. J.; Weiss, R. G. *Chem. Mater.* **2000**, *12*, 406.

[11] Abdallah, D. J.; Robertson, A.; Hsu, H.-F.; Weiss, R. G. *J. Am. Chem. Soc.* **2000**, *122*, 3053.

[12] Lee, G.-M.; Weiss, R. G. unpublished results.

[13] Demus, D.; Goodby, J.; Gray, G. W.; Spiess, H.-W.; Vill, V. *Handbook of Liquid Crystals, High Molecular Weight Liquid Crystals*, vol 3, Wiley-VCH: New York, 1998, p 312.

[14] Lu, L.; Sharma, N.; Nagana Gowda, G. A.; Khetrapal, C. L.; Weiss, R. G. *Liq. Cryst.* **1997**, *22*, 23.

[15] Keller, H.; Hatz, R. *Handbook of Liquid Crystals* Verlag Chemie:Weinheim, 1980, pp 155ff.

[16] Lu, L. *Ph. D. Thesis*, Georgetown University, Washington, DC, 1997.

[17] Reichardt, C. *Solvents and Solvent Effects in Organic Chemistry*, 2nd ed.; VCH: Weinheim, 1988, p 359.

[18] Jacobson, A.; Petric, A.; Hogenkamp, D.; Sinur, A.; Barrio, J. R. *J. Am. Chem. Soc.* **1996**, 118, 5572.

[19] Carmichael, A. J.; Seddon, K. R. *J. Phys. Org. Chem.* **2000**, *13*, 591.

[20] Baker, S. N.; Baker, G. A.; Kane, M. A.; Bright, F. V. *J. Phys. Chem. B* **2001**, *105*, 9663.

[21] Atkins, P. W. *Physical Chemistry*, 5th ed.; W. H. Freeman: New York, 1994, pp 757, 834.

[22] (a) Bonhôte, P.; Dias, A.-P.; Papageorgiou, N.; Kalyanasundaram, K.; Grätzel, M. *Inorg. Chem.* **1996**, *35*, 1168. (b) Matsumoto, H.; Yanagida, M.; Tanimoto, K.; Nomura, M.; Kitagawa, Y.; Miyazaki, Y. *Chem. Lett.* **2000**, 922. (c) MacFarlane, D. R.; Meakin, P.; Amini, M.; Forsyth, M. *J. Phys.: Condens. Matter* **2001**, *13*, 8257.

[23] Lu, L.; Nagana Gowda, G. A.; Suryaprakash, N.; Khetrapal, C. L.: Weiss, R. G. *Liq. Cryst.* **1998**, *25*, 295.

[24] Nagana Gowda, G. A.; Chen, H.; Khetrapal, C. L.; Weiss, R. G. *Chem. Mater.* submitted.

[25] Diehl, P.; Khetrapal, C. L. in *NMR Basic Principles and Progress*; Vol 1; Diehl, P.; Fluck, E.; Kosfeld, R., Eds.; Springer-Verlag: Heidelberg, 1969, p 1.

[26] (a) Tjandra, N.; Bax, A. *Science* **1997**, *278*, 1111. (b) Bax, A. *Protein Science* **2003**, *12*, 1.

# Reactions

Chapter 22

# Hydrogenation Reactions in Ionic Liquids: Finding Solutions for Tomorrow's World

Paul J. Dyson[1,*], Tilmann Geldbach[1], Federico Moro[1], Christoph Taeschler[2], and Dongbin Zhao[1]

[1]Institut de Chimie Moléculaire et Biologique, Ecole Polytechnique Fédérale de Lausanne, EPFL-BCH, CH–1015 Lausanne, Switzerland
[2]Lonza Ltd., 3930 Visp, Switzerland

Ionic liquids are excellent solvents for conducting biphasic hydrogenation reactions offering many advantages over both related homogeneous processes and biphasic reactions in other alternative solvents. The relevant physicochemical properties of ionic liquids that lend themselves to hydrogenation catalysis are summarised and key hydrogenation reactions/protocols conducted using ionic liquids are discussed. Two new heterogeneous catalyst systems for the hydrogenation of benzene are described. One employs a well known rhodium nanoparticle catalysts, and compared to its use in molecular solvents, reaction rates, ease of use and recyclability, are all improved. The other comprises a new simple platinum-chloroaluminate catalyst system. The preparation of the chloroaluminate ionic liquid on a moderate scale is described together with the chloride salt precursor and the tetrafluoroborate salt. Potentially these ionic liquids (and others) could be prepared by Lonza on a 100 ton scale.

Ionic liquids are without doubt excellent solvents for conducting multiphasic catalysed reactions, including hydrogenation reactions (1, 2, 3, 4). They compete well with other alternative solvents such as water, fluorous phases and supercritical fluids and can even be combined with these alternative solvents in order to conduct reactions without necessitating the use of volatile organic solvents (5). A remarkable amount has been achieved in the field of ionic liquids, which, although may be traced back to 1914, only really took off in the 1990's (6). In a very short time a considerable amount of basic and applied data has been published and the first industrial scale ionic liquid based process has gone on-line (7). One of the chief concerns regarding the large scale implementation of ionic liquids is their cost. However, some ionic liquids can now be produced on a multi-tonne scale for a few dollars per litre (see below). Combined with the fact that special ligands are not essential for solubilising catalysts in ionic liquids and the fact that ionic liquids are not lost to the environment through evaporation, they could prove to be less expensive than the other 'green' alternatives.

## Properties of Ionic Liquids Relevant to Hydrogenation Catalysis

The solubility of hydrogen gas in a range of different ionic liquids has been determined, initially using an electronic flow mass controller for [bmim][BF$_4$] and [bmim][PF$_6$] (8) and subsequently for a larger range of ionic liquids using high pressure $^1$H NMR spectroscopy (9). The data from these latter determinations are displayed in Figure 1.

The solubility of hydrogen in most of the ionic liquids studied is similar to that in water, although in [P(C$_6$H$_{13}$)$_3$(C$_{14}$H$_{29}$)][PF$_3$(C$_2$F$_5$)$_3$] it is considerably higher. However, compared to molecular solvents the solubility of hydrogen in ionic liquids tends to be low. It has been noted that hydrogen solubility in ionic liquids, like water, increases with increasing temperature, unlike molecular solvents where the solubility of gases decreases with increasing temperature. Despite the low solubility of H$_2$ in ionic liquids the rate of mass transfer of hydrogen into ionic liquids appears to be rapid, and consequently, hydrogen concentration is unlikely to be rate limiting in catalyzed reactions such as hydrogenation (9) and hydroformylation (10).

Another feature of ionic liquids that makes them ideally suited to hydrogenation reactions is that polar compounds are more soluble in ionic liquids than non-polar ones, and in general, unsaturated organic substrates are more polar (and hence soluble) than their hydrogenated counterparts. As such, many substrates are partially (or completely) soluble in the ionic liquid-catalyst phase, whereas the products are far less soluble, giving rise to *pseudo-*

homogeneous reactions conditions with all the benefits of a highly efficient biphasic separation. Since the polarities, or at least the solubility properties, of ionic liquids can be tuned, then ionic liquids can be optimized for specific substrates and products (see below). It is worth noting that the low solubility of many unsaturated organic substrates in water is restrictive and also in fluorous phases it is non-polar substrates that are most soluble and polar substrates which are least soluble, making ionic liquids the immobilization solvents of choice for biphasic hydrogenation reactions.

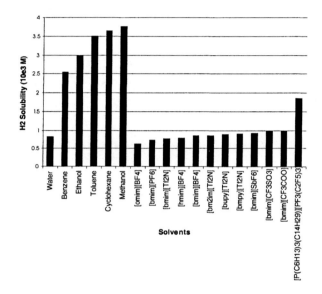

*Figure 1. Solubility of $H_2$ in ionic liquids together with some molecular solvents for comparison purposes at 0.101 MPa (1 atm).*

Other benefits of using ionic liquids that have been noted in hydrogenation reactions, but are not necessarily restricted to hydrogenation reactions, and are equally applicable to many other catalyzed reactions, include increased turnover frequencies and numbers and good catalyst recycling and reuse.

## Hydrogenation Reactions in Ionic Liquids

Many different substrates have been hydrogenated using catalysts immobilzed in ionic liquids including alkenes, alkynes, arenes and polymers as well as prochiral substrates (*11*). The first successful examples of hydrogenation reactions using ionic liquids to immobilse the catalyst were only demonstrated as recently as 1995 and 1996 by Chauvin (*12*) and de Souza, Dupont and co-workers (*13*), respectively. In the former study pentene was hydrogenated using [Rh(nbd)(PPh$_3$)$_2$]$^+$ (nbd = norbornadiene), immobilized in [bmim][BF$_4$], [bmim][PF$_6$] or [bmim][SbF$_6$], and activity was higher than when carried out in propanone. It was also shown that ionic liquids could have a marked influence on the selectivity of reactions (see Figure 1). For example, the hydrogenation of cyclohexadiene to cyclohexene was achieved with 98 % selectivity and 96 % conversion in [bmim][SbF$_6$]. The selectivity was ascribed to the fact that the cyclohexadiene substrate was five times more soluble in [bmim][SbF$_6$] than cyclohexene. In the second study [bmim]-AlCl$_3$ (mole fraction 0.45), [bmim][BF$_4$] and [bmim][PF$_6$] were used to dissolve Wilkinson's catalyst, RhCl(PPh$_3$)$_3$, and cyclohexene was reduced to cyclohexane under 10 atmospheres of H$_2$ at 25°C. It was noted that good biphasic separation of the product was possible. In addition, [Rh(cod)$_2$]$^+$ (cod = 1,5-cyclooocatadiene) was also used as a catalyst precursor in [bmim][BF$_4$], which in keeping with the related study by Chauvin, was found to be more active than the corresponding reaction in molecular solvents. These nascent papers set the scene for the subsequent research on hydrogenation catalysis in ionic liquids.

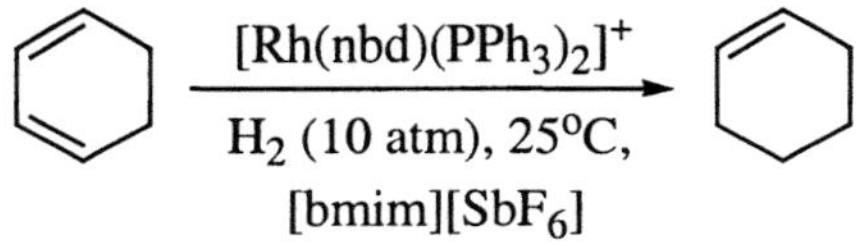

$$\text{[Rh(nbd)(PPh}_3)_2]^+$$
$$\xrightarrow{\hspace{2cm}}$$
$$\text{H}_2 \text{ (10 atm), 25°C,}$$
$$\text{[bmim][SbF}_6]$$

*Figure 2. One of the first examples of a hydrogenation reaction carried out in a room temperature ionic liquid.*

The hydrogenation of cyclohexene (and cyclohexadiene) in ionic liquids has subsequently been described in a number of other papers. [Rh(PPh$_3$)$_2$(nbd)][PF$_6$] was used as the catalyst in [bmim][PF$_6$], but significantly, the ionic liquid and catalyst could be immobilized on a silica gel support material (Figure 3), which reduced the amount of ionic liquid required and also significantly increased the activity of the catalyst (*14*). At the time, the preparation of ionic liquids on a large scale had not been achieved so reducing the amount required for any potential industrial process was important. The resulting fixed-bed technology can also be advantageous for reactions involving volatile substrates and other examples are also known (*15*). Extremely efficient catalyst immobilization could also be obtained in a nitrile-functionalized ionic liquid (*16*).

Iridium nanoparticles have also been used to catalyze the hydrogenation of cyclohexene in [bmim][PF$_6$] (*17*). The nanoparticles, prepared in the ionic liquid by reduction of [IrCl(cod)]$_2$ using hydrogen, contained on average 300 iridium atoms although nanoparticles ranging from 60 to 800 atoms were present. Palladium nanoparticles protected by a phenanthroline sheath also prepared in [bmim][PF$_6$] have been shown to effect the hydrogenation of cyclohexene to cyclohexane under mild conditions (*18*).

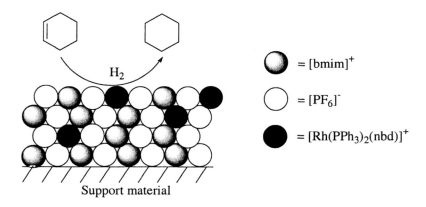

*Figure 3. Representation of a supported ionic liquid-catalyst phase used to hydrogenate alkenes.*

It is worth pointing out that in all these processes no organic co-solvents are required, only the ionic liquid used to immobilize the catalyst and the neat substrate are present. In a way, however, solid-supported catalysts could arguably be as effective and environmentally benign as no organic solvents would be required for the hydrogenation of a substrate such as cyclohexene. The real advantage of ionic liquids becomes apparent when the related cyclohexadiene substrate is used, since it can be hydrogenated to cyclohexene with extremely high selectively as can a number of other diene substrates (*19, 20, 21*). The role of the ionic liquid appears to be critical in this process as described above. Benzene and other arenes – notoriously difficult to hydrogenate – may also be hydrogenated using catalysts immobilized in ionic liquids such as [bmim][BF$_4$] and [bmim][PF$_6$] (*22, 23*). In addition, the chloroaluminates, often considered less useful today because of their sensitivity to air and moisture, are highly effective solvents for reduction of polyaromatic compounds (*24*).

A wide range of other substrates have been hydrogenated using catalysts immobilized in ionic liquids including prochiral substrates (*25*), substrates with various functional groups, and regioselective hydrogenations have also been accomplished (*26*). Other substrates studies include polymers (*27*) and CO$_2$ (*28*), the latter being an extremely important substrate in the context of green chemistry. Ionic liquids have also been combined with $CO_2$ (*29*) and water (*30*) to facilitate extraction of hydrogenated products while avoiding the use of organic solvents.

From these studies, it would appear that the most useful catalysts are either ionic homogeneous complexes or nanoparticles. However, catalyst identification in ionic liquids is not trivial (*31*) and in many cases the compound added as the catalyst may turn out not to be the active catalyst species. Overall, hydrogenation reactions in ionic liquids compare favorably to homogeneous reactions. Catalyst activity and retention in the ionic liquid tends to be high giving rise to excellent turnover frequencies and numbers, and hence excellent recycling and reuse.

## Hydrogenation of benzene in ionic liquids

As mentioned above, a number of ionic liquid systems used to hydrogenate benzene and other arenes have already been reported. Arene hydrogenation is an area of considerable interest (*32*, *33*, *34*) and biphasic strategies using soluble nanoparticles catalysts have been highlighted as an area with considerable potential (*35*). The problem with molecular catalysts is that many seem to be unstable and are rapidly converted to nanoparticle catalysts under the reducing hydrogen environment (*35*, *36*). As such, carefully designed nanoparticles catalysts should be, and invariable are, more efficient. One highly effective nanoparticle catalyst obtained from $RhCl_3$ and Aliquat® 336 has been used in water to hydrogenate various aromatic substrates including benzene, arenes (*37*) and polyaromatic compounds (*38*). Originally this catalyst was presumed to be composed of the solvated ion pair $[(C_8H_{17})_3NMe]^+[RhCl_4]^-$ (*39*), but a detailed mechanistic study demonstrated that the active catalyst was in fact colloidal (*40*), (other arene hydrogenation catalysts reputed to be homogeneous have since been shown to be heterogeneous). Using the literature protocol, except replacing the usual molecular solvent with 1-methyl-3-alkylimidazolium tetrafluoroborate ionic liquids, it was possible to generate surfactant stabilized rhodium nanoparticles which were evaluated in the hydrogenation of benzene. In a typical experiment, $RhCl_3.3H_2O$ ($4.92 \times 10^{-2}$ mmol), Aliquat® 336 ($5.39 \times 10^{-2}$ mmol), the ionic liquid (1 ml), trioctylamine (100 µl), benzene (89 µl) and water (50 µl) was stirred under hydrogen at atmospheric pressure at 30°C. The progress of the reaction was monitored by removal of a sample (2 µl) every 10 minutes which was analyzed by GC. A typical conversion-time profile is shown in Figure 4 for the reaction conducted in [omim][BF$_4$]. It is noteworthy that there is an incubation period, corresponding to the time taken to generate the active nanoparticles catalyst, which is comparable to that when the reaction is carried out under homogeneous conditions in thf. The rate of reaction is approximately the same in the three ionic liquids examined, viz. [bmim][BF$_4$], [hmim][BF$_4$] and [omim][BF$_4$] with the final conversion after 190 minutes being between 83 and 87%. In contrast, the analogous reaction in thf afforded only 63% cyclohexane after the same time period. The increased reaction rate in the ionic liquids could be due to a number of reasons, and since the solubility of both substrates, benzene and hydrogen, is lower in the ionic liquids than in thf, perhaps the most likely reason is that the nanoparticles generated in the ionic liquid are smaller than those formed in thf, although, as yet, we do not have any evidence to confirm this hypothesis.

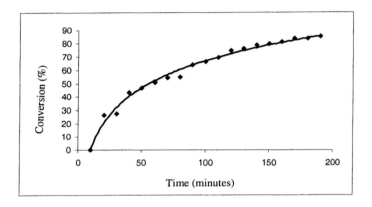

*Figure 4. Graph showing the conversion-time profile of benzene to cyclohexane in [omim][BF$_4$] employing the RhCl$_3$–Aliquat® 336 nanoparticle catalyst.*

It should be noted that under the conditions employed all the rate studies were conducted under homogeneous conditions. However, it is possible to scale-up the process to give a more typical biphasic system, and under such conditions, removal of the cyclohexane product is facile and the ionic liquid-catalyst mixture can be recycled and reused.

The acidic chloroaluminate ionic liquid [emim]Cl-AlCl$_3$ (x = 0.67), combined with electropositive metals such as aluminium, zinc and lithium together with a proton source, have been used to hydrogenate aromatic compounds, and partial reduction of polyaromatic compounds was even achieved (24). The partial reduction of benzene to cyclohexene is an extremely important reaction as cyclohexene is a precursor to cyclohexanol (41). The most effective catalyst for the conversion of benzene to cyclohexene which is used by the Asahi Chemical Co. employs a ruthenium metal catalyst, ZrO$_2$, and ZnSO$_4$ and operates in water (42, 43). The process affords cyclohexene with 60%

selectivity after 90% conversion of benzene, and presently it produces 50,000 tons of cyclohexene per year. Use of water is essential and while various roles for the water are proposed, one important function is to facilitate elimination of the cyclohexene from the catalyst surface. It is not unreasonable to postulate that ionic liquids might also lend themselves to such processes since partial hydrogenation frequently takes place in ionic liquids (see above).

We have found that heterogeneous catalysts can be generated *in situ* in the chloroaluminate ionic liquid [N-octyl-3-picolinium]Cl-AlCl$_3$ (see below for preparation) which are active benzene hydrogenation catalysts. The ionic liquid, transition metal chloride and benzene substrate are mixed together and heated at 50°C under 50 atmospheres of hydrogen. The system comprises of a single phase and as the benzene is converted to cyclohexane (thus far partial hydrogenation products have not been observed) a second phase forms. Using a multicell reactor it is possible to rapidly screen a large series of transition metal salts as visual inspection shows if reaction takes place, and the conversion can be rapidly quantified by measuring the volume of the product phase. The most effective system discovered so far employs K$_2$PtCl$_4$. In a typical experiment, [N-octyl-3-picolinium]Cl-AlCl$_3$ (1.5 ml), K$_2$PtCl$_4$ (21 mmol) and benzene (3 ml) were pressurised with hydrogen (50 atm.) and heated at 50°C for 3 hours. Under these conditions a conversion of 99.3% is obtained. While a kinetic analysis of the system gas yet to be conducted a mercury poisoning experiment indicates that the active catalyst is heterogeneous.

## Large scale preparation of the [N-octyl-3-picolinium] ionic liquids

The N-octylpicolinium-based ionic liquids were prepared according to the route shown in Figure 5. In the first step, 3-picoline was reacted with 1-chlorooctane to afford [N-octyl-3-picolinium]Cl in high yield. Subsequent reaction of the [N-octyl-3-picolinium]Cl with either AlCl$_3$ or HBF$_4$ affords [N-octyl-3-picolinium]Cl-AlCl$_3$ and [N-octyl-3-picolinium][BF$_4$], respectively.

In a typical experiment, 3-picoline (1.68 kg) and 1-chlorooctane were placed in a 5 l reaction vessel affording a two phase reaction mixture which was heated to 130°C within 2.5 hours. After about 2 to 4 hours at 130°C, the two phases unify and a pale brown solution forms. The solution slowly solidifies on standing at room temperature for several hours. The melting point of the product is between 40 and 45°C and the content of N-octyl-3-picolinium chloride exceeds 96 %. In order to remove the brownish colour of the product, the N-octyl-3-picolinium chloride was dissolved in water (13 l) and treated with activated charcoal (320 g) and stirred at 50°C for 30 minutes and then filtered. For the preparation of the chloroaluminate liquid N-octyl-3-picolinium chloride

(1 kg) was melted at 70°C and aluminum chloride (1.1 kg) added at such a rate so that the reaction temperature was maintained between 70 and 80°C. The resulting solution consisted of > 95 % of the desired N-octyl-3-picolinium aluminum chloride ionic liquid. The tetrafluoroborate salt is prepared from an aqueous solution of N-octyl-3-picolinium chloride (2.0 kg in 5.5 kg water), by treatment with tetrafluoroboronic acid 50 % (1.37 kg) at 22°C, resulting in a pale yellow emulsion. During the addition of tetrafluoroboronic acid, the temperature of the reaction mixture rises to about 25°C. After stirring for 30 minutes the phases separate and the aqueous solution was extracted with dichloromethane (3.4 l). The unified organic solutions were extracted several times with 0.5 l portions of water, until the pH value of the washings was above 6. The organic phase was concentrated on a rotary evaporator at 50 to 60°C and the resulting product had a content of the desired N-octyl-3-picolinium tetrafluoroborate of over 98 % and the chloride content of the product was approximately 10 ppm. What is clear from these preparations is that the custom synthesis facilities at Lonza could produce these ionic liquids as well as a number of others on a much larger scale with similar purities. Should there be a demand for 50 to 100 tons of these ionic liquid then the cost would be around $ 15 per liter, considerably lower than current prices.

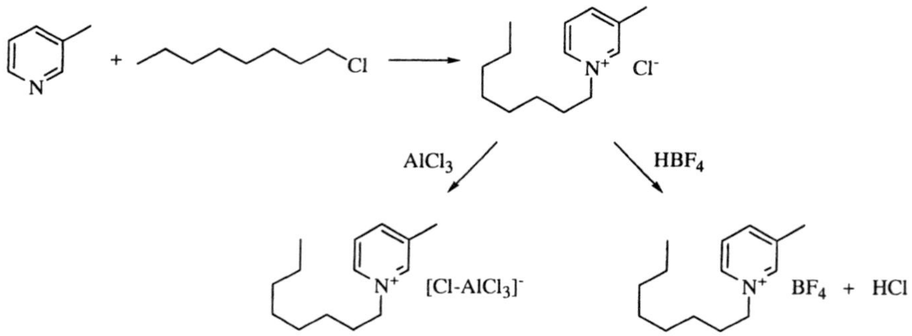

*Figure 5. Route used to prepare N-octylpicolinium-based ionic liquids on a large scale.*

## Outlook

The reactions described herein clearly demonstrate that ionic liquids have a bright future in hydrogenation catalysis, with potential for industrial applications. The approximate number of papers on hydrogenation reactions in alternative solvents are listed in Table I. From the Table it is clear that in comparison to studies in water and even supercritical fluids, hydrogenation reactions in ionic liquids are in their infancy.

**Table I. Results from a Web of Science search on the number of papers with the word hydrogenation and the alternative solvent in the title or abstract.**

| Hydrogenation and search word(s) | Number of papers |
|---|---|
| Water/aqueous | 2210 |
| Supercritical | 345 |
| Ionic liquid/molten salt | 66 |
| Fluorous | 40 |

NOTE: The search results should be interpreted with caution, and does not reflect the true number of papers on each given system, but it does give a good indication of the relative magnitude of each area.

Despite the availability of only a limited number of studies of hydrogenation reactions carried out in ionic liquids, the basic physical data needed to transport a new solvent from an academic curiosity to one that is industrially useful is becoming available at a tremendous rate (*44*). Furthermore, the clear advantages that ionic liquids offer in hydrogenation reactions, their unique properties within the group of alternative solvents and their potential availability on a large scale at a low cost are all good indicators to suggest that it will not be long before an industrial hydrogenation plant is in operation.

# References

1. Seddon, K. R. *J. Chem. Technol. Biotechnol.* **1997**, *68*, 351.
2. Welton, T. *Chem. Rev.* **1999**, *99*, 2071.
3. Wasserscheid, P.; Keim,W. *Angew. Chem. Int. Ed.* **2000**, *39*, 3772.
4. Dupont, J.; Souza, R. F. D.; Suarez, P. A. Z. *Chem. Rev.* **2002**, *102*, 3667.
5. Tzschucke, C. C.; Markert, C.; Bannwarth, W.; Roller, S.; Hebel, A.; Haag, R. *Angew. Chem. Int. Ed.* **2002**, *41*, 3964.
6. Wilkes, J. S. *Green Chem.*, **2002**, *4*, 73.
7. News of the week in *Chem. Eng. News* **2003** March 31, 9.
8. Berger, A.; de Souza, R. F.; Delgado, M. R.; Dupont, J. *Tetrahedron Asymmetry*, **2001**, *12*, 1825.
9. Dyson, P. J.; Laurenczy, G.; Ohlin, C. A.; Vallance, J.; Welton, T. *Chem. Commun.*, **2003**, 2418.
10. Ohlin, C. A.; Dyson, P. J.; Laurenczy, G. *Chem. Commun.*, **2004**, 1070.
11. Dyson, P. J. *Appl. Organometal. Chem.*, **2002**, *16*, 495.
12. Chauvin, Y.; Olivier-Bourbigou, H. *CHEMTECH*, **1995**, *25*, 26

332

13. Suarez, P. A. Z.; Dullius, J. E. L.; Einloft, S.; de Souza, R. F.; Dupont, J. *Polyhedron*, **1996**, *15*, 1217.
14. Mehnert, C. P.; Mozeleski, E. J.; Cook, R. A. *Chem. Commun.*, **2002**, 3010.
15. Huang, J.; Jiang, T.; Han, B.; Gao, H.; Chang, Y.; Yhao, G.; Wu, W. *Chem. Commun.* **2003**, 1654.
16. Zhao, D.; Fei, Z.; Scopelliti, R.; Dyson, P. J. *Inorg. Chem.*, **2004**, *43*, 2197.
17. Dupont, J.; Fonseca, G. S.; Umpierre, A. P.; Fichtner, P. F. P.; Teixeira, S. R. *J. Am. Chem. Soc.*, **2002**, *124*, 4228.
18. Huang, J.; Jiang, T.; Han. B.; Gao, H.; Chang, Y.; Zhao, G.; Wu, W. *Chem. Commun.*, **2003**, 1654.
19. Suarez, P. A. Z.; Dullius, J. E. L.; Einloft, S.; de Souza, R. F.; Dupont, J. *Inorg. Chim. Acta*, **1997**, *255*, 207.
20. Dupont, J.; Suarez, P. A. Z.;. Umpierre, A. P.; de Souza, R. F. *J. Braz. Chem. Soc.*, **2000**, *11*, 293.
21. Zhao, D.; Laurenczy, G.; McIndoe, J. S.; Dyson, P. J. *J. Mol. Catal. A: Chemical*, **2004**, *214*, 19.
22. Dyson, P. J.; Ellis, D. J.; Parker, D. G.; Welton, T. *Chem Commun.*, **1999**, 25.
23. Boxwell, C. J.; Dyson, P. J.; Ellis, D. J.; Welton, T. *J. Am. Chem. Soc.*, **2002**, *124*, 9334.
24. Adams, C. J.; Earle, M. J.; Seddon, K. R. *Chem. Commun.* **1999**, 1043.
25. Brown, R. A.; Pollet, P.; McKoon, E.; Eckert, C. A.; Liotta, C. L.; Jessop, P. G. *J. Am. Chem. Soc.*, **2001**, *123*, 1254.
26. Anderson, K.; Goodrich, P.; Hardacre, C.; Rooney, D. W. *Green Chem.*, **2003**, *5*, 448.
27. Müller, L. A.; Dupont. J.; de Souza, R. F. *Macromol. Rapid Commun.*, **1998**, *19*, 409.
28. Ohlin, C. A.; Laurenczy, G. *High Pressure Res.* **2003**, *23*, 239.
29. Liu, F. C.; Abrams, M. B.; Baker, R.T.; Tumas, W. *Chem Commun.*, **2001**, 433.
30. Dyson, P. J.; Ellis, D. J.; Welton, T. *Canadian J. Chem.*, **2001**, *79*, 705.
31. Dyson, P. J.; McIndoe, J. S.; Zhao, D. *Chem. Commun.*, **2003**, 508.
32. Donohoe, T. J.; Garg, R.; Stevenson, C. A. *Tetrahedron: Assymetry*, **1996**, *7*, 317.
33. Weissermel, K.; Arpe, H.-J. *Industrial Organic Chemistry*, VCH, New York, **1993**.
34. Corma, A.; Martínez, A.; Martínez-Soria, V. *J. Catal.*, **1997**, *169*, 480.
35. Widegren, J. A.; Finke, R. G. *J. Mol. Catal. A: Chemical*, **2003**, *191*, 187.
36. Dyson, P. J. *Dalton Trans.*, **2003**, 2964.
37. Blum, J.; Amer, I.; Zoran, A.; Sasson, Y. *Tetrahedron Lett.*, **1983**, *24*, 4139.
38. Amer, I.; Amer, H.; Ascher, R.; Blum, J. *J. Mol. Catal.*, **1987**, *39*, 185.

39. Blum, J.; Amer, I.; Vollhardt, K. P. C.; Schwarz, H. Höhne, G. *J. Org. Chem.*, **1987**, *52*, 2804.
40. Weddle, K. S.; Aitken III J. D.; Finke, R. G. *J. Am. Chem. Soc.*, **1998**, *120*, 5653.
41. Struijk, J.; d'Angremond, M. Lucas-de Regt, W. J. M.; Scholten, J. J. F. *Appl. Catal. A: General*, **1992**, *83*, 263.
42. Nagahara, H. *Rev. J. Surf. Sci. Technol. Avant-Garde* **1992**, *30*, 951.
43. Asahi Chemical Co. Jpn. Kokai Tokkyo Koho 62-81332, 62-201830, 63-17834, 63-63627.
44. *Ionic Liquids in Synthesis*, Eds. Wasserscheid, P.; Welton, T. Wiley-VCH, **2002**.

Chapter 23

# Supported Ionic Liquid-Phase Catalysis–Heterogenization of Homogeneous Rhodium Phosphine Catalysts

Anders Riisager[1,2], Rasmus Fehrmann[1], Peter Wasserscheid[3], and Roy van Hal[2,3]

[1]Department of Chemistry and Interdisciplinary Research Center for Catalysis, Technical University of Denmark, DK–2800 Lyngby, Denmark
[2]Institut für Technische Chemie und Makromolekulare Chemie, RWTH-Aachen, Worringer Weg 1, D–52074 Aachen, Germany
[3]Lehrstuhl für Chemische Reaktionstechnik, Universität Erlangen-Nürnberg, Egerlandstraße 3, D–91058 Erlangen, Germany

The concept of supported ionic liquid-phase (SILP) catalysis has been demonstrated for gas- and liquid-phase continuous fixed-bed reactions using rhodium phosphine catalyzed hydroformylation of propene and 1-octene as examples. The nature of the support had important influence on both the catalytic performance, i.e. activity and selectivity, as well as stability of the SILP catalysts. Noticeably, a high catalyst ligand content together with presence of ionic liquid solvent are prerequisites for obtaining selective rhodium phosphine SILP catalysts systems.

In homogeneous catalysis with organometallic compounds the development of new and approved strategies to immobilize catalysts has for decades been a research area receiving significant academic and industrial interest(1). In this context, room-temperature ionic liquids have proven to be attractive, alternative solvents for many transition metal catalyzed liquid-liquid two-phase reactions, because the polar, ionic nature of the ionic liquid catalyst phase often induce limited miscibility with reaction products which facilitate catalyst separation(2-4). Also, often ionic liquids offer additionally, unique advantages useful in organometallic catalysis which are not provided by other solvents (including water):

➤ Ionic liquids usually have very low chemical reactivity, i.e. they do not interfere with reactants/products/catalysts during reactions, which makes it possible to perform reactions/use catalysts not applicable with traditional solvents.
➤ Ionic liquids have a large thermal and liquid-phase window, i.e. they remain liquid during operation thereby ensuring complex flexibility, which is often required for homogeneous catalysts to be selective.
➤ Low vapor pressure of ionic liquids prevents solvent evaporation during reactions, i.e. they are well-suited for continuous-flow processes.
➤ Ionic liquids have many adjustable properties (e.g. different gas solubility) and can be functionalized to specific processes, i.e. task-specific ionic liquids (TSILs)(5).

A two-phase catalytic application where ionic liquid solvents have drawn particular attention is in rhodium phosphine catalyzed olefin hydroformylation. Here, reactions involving charged phosphines have shown to induce good solubility and high relative catalyst affinity for the ionic phase compared to the organic product phase. This have resulted in good to excellent conversion of several olefins to the corresponding aldehydes, including 1-pentene(6,7), 1-hexene(8), 1-octene and longer-chained olefins(9-14). However, only in a few of the reaction systems, a combination of good catalyst activity and selectivity with quantitative retention of catalyst components in the ionic liquid-phase after reaction, has been obtained. This includes, e.g. catalyst systems with the biphosphine ligand 1(14) (sulfoxantphos, Figure 1) and related charged ligands with the xantphos structure. For these systems selectivities corresponding to ≥ 95 % of the preferred linear aldehyd (n/iso ratios of 20 or higher) are commonly achieved due to the large P-rhodium-P bite angles formed by the xanthene backbone structure(15). In contrast, low selectivities (usually n/iso ratios ≤ 3) for the linear products are obtained with most other systems, including the common rhodium-tris(m-sulfonatophenyl)phosphine (rhodium-TPPTS) catalyst system(16).

In essentially all of the present hydroformylation applications halogen-containing ionic liquids have been used as solvents, e.g. [bmim][PF₆]. However, such compounds may not be an ideal choice due to safety concerns and environmental impact, toxicology, undesired reactions and other technical aspects potentially associated with ionic liquids reactions involving halogen-containing anions like, e.g. [PF₆]. For example decomposes [bmim][PF₆] with formation of phosphates and HF in reaction mixtures of transition metal complexes and hydrogen in the presence of water(17). These problems can, however, be overcome by using ionic liquids with halogen-free anions, e.g. R-OSO₃⁻ (alkylsulfates), HSO₄⁻ or NO₃⁻. Especially the alkylsulfates are highly interesting anions since most of these anions are widely available as their alkali salts and they have well investigated toxicology, and halogen-free, low-melting [bmim][n-C₈H₁₇OSO₃] ionic liquid has been introduced as a new "greener" solvent for hydroformylation(13).

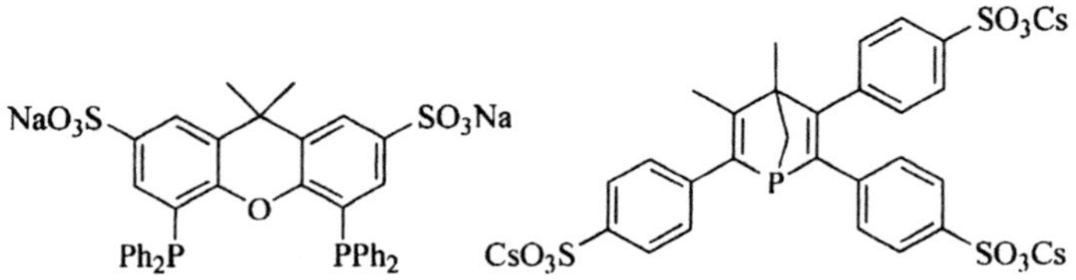

*Figure 1. Phosphine ligands. 1: disodium 2,7-bissulfonate-4,5-bis(diphenyl-phosphino)-9,9-dimethylxanthene (sulfoxantphos), 2: tricesium 3,4-dimethyl-2,5,6-tris(p-sulfonatophenyl)-1-phosphanorbornadiene (NORBOS).*

In liquid-liquid processes where chemical kinetics is fast, e.g. hydroformylation, the reaction may take place at the surface or in the diffusion layer of the ionic liquid rather than in the bulk solvent. In these cases, the ordinary two-phase approach may therefore not be the ideal way to realize a catalytic reaction with an ionic catalyst layer, since only a minor part of the relatively expensive ionic liquid and of the catalyst dissolved therein is effectively used in the reaction. In addition, chemical industry strongly prefers heterogeneous catalyst systems where catalyst separation is easy and where it is possible to use fixed-bed reactors leading to, e.g. continuous operation mode, avoidance of large amount of solvent, and applicability to gas-phase reactions. Therefore, a solid catalyst is highly advantageous instead of a catalyst composed of a liquid solution.

In line with this, an improved type of heterogenized homogeneous catalysts, refered to as a supported ionic liquid-phase (SILP) catalysts, has recently been introduced and examined by us as well as other research groups for hydroformylation*(16, 18, 19)* and hydrogenation*(20, 21)* of olefins. Generally, the SILP catalysts are comprised of a small amount of ionic liquid containing dissolved rhodium-metal complexes, which are heterogenized on a high-area solid support by either physisorption or by anchoring of ionic liquid fragments. Hence, although the catalyst is a solid it performs as a homogeneous catalyst, i.e. the reaction between reactants, substrate and the dissolved catalytically active complexes takes place in the highly dispersed ionic liquid, as illustrated in Figure 2.

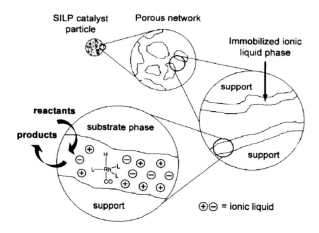

*Figure 2. Schematic representation of the SILP catalyst concept.*

The rhodium phosphine SILP catalysts used by Mehnert *et al.* for 1-hexene hydroformylation were composed of monoarylphosphines PPh$_3$ and TPPTS with [bmim][X] (X = PF$_6$ and BF$_4$) ionic liquids using a modified silylimidazolium silica support (silica with covalent anchored fragments of 1-*n*-butyl-3-[3-(triethoxysilanyl)propyl]-imidazolium). In all cases low selectivities were obtained (*n/iso* ratios of 0.4-2.4). In addition, a significant amount of rhodium was leached to the product phase at high conversion due to depletion of the ionic liquid phase from the support. Importantly, a pronounced catalyst deactivation was also observed at lower conversion during recycling independently of the pre-silylation of the support*(22)*, which shortened the lifetime and limited the applicability of the catalysts significantly. Also noteworthy, in all cases the processes were carried out as batch reactions and not as the technically more attractive continuous flow fixed-bed processes, which we report here.

The principle of the SILP catalytic concept is inspired by prior work on supported liquid-phase (SLP) catalysts including the aqueous version, supported aqueous-phase (SAP)*(23)* catalysts (typically involving rhodium-TPPTS catalysts), which is well known to afford good phase separation of catalysts from products in both gas- and liquid-phase hydroformylation reactions*(24)*. It is unclear, though, whether the supported catalyst remain liquid at reaction conditions in the SAP systems*(25-27)*, which seems to be important for achieving good selectivity. In contrast to the SAP processes the SILP catalysis concept makes use of ionic liquid solvents which have a negligible vapor pressure and remain liquid in a large thermal range. These characteristics ensure that the solvent is retained on the support in its fluid state at reaction conditions, also in continuous processes. Also, compared to conventional two-phase liquid-liquid catalysis in organic/ionic liquid mixtures, the SILP catalyst systems offer superior advantages of very efficient ionic liquid use and relatively short diffusion distances of the compounds involved during the catalytic processes due to the high dispersion of the ionic liquid catalyst solution.

In the present work we have examined supported ionic liquid-phase (SILP) rhodium phosphine catalysts for fixed-bed continuous gas- and liquid-phase hydroformylation of propene and 1-octene. The applied catalysts are modified with rhodium-complexes of sulfonated phosphine ligands 1 (sulfoxantphos) and 2 (NORBOS) (Figure 1) dissolved in neutral, weakly coordinating halogen-containing and halogen-free [bmim] ionic liquids, [bmim][$PF_6$] and [bmim][$n$-$C_8H_{17}OSO_3$], on porous, high-area oxides $SiO_2$, $TiO_2$, $Al_2O_3$ and $ZrO_2$. The NORBOS ligand has previously shown to exhibit high activity in propene hydroformylation in both two-phase aqueous systems*(28)* and in derived SAP systems*(27)* using rhodium-2 catalysts.

# Experimental methods

## Preparation of SILP rhodium phosphine catalysts

The supported ionic liquid-phase rhodium catalyst systems were prepared by impregnation of supports with dry, degassed methanol solutions or mixed methylenchloride/water solutions of Rh(acac)(CO)$_2$ (98 %, Aldrich), phosphine ligands and ionic liquids under inert atmosphere, as previously described*(18,19)*. The ligands 1*(29)* and 2*(27)* (modified method to the one used by Herrmann *et al.(28)*) were prepared according to literature procedures, while the ionic liquids, [bmim][$PF_6$] and [bmim][$n$-$C_8H_{17}OSO_3$]*(13)* were either purchased from Solvent Innovation GmbH, Cologne, Germany (>98 %) or prepared by reported methods. The applied supports were obtained from commercial suppliers and used after drying in vacuum (0.1 mbar, 110 °C, 24 h) unless otherwise mentioned. General characteristics of the used supports are shown in Table I.

**Table I. Characteristics of supports**

|  | $SiO_2$ | $Al_2O_3$ | $TiO_2$ | $ZrO_2{}^a$ |
|---|---|---|---|---|
| Supplier | Merck | Merck | Millennium Co. | MEL Chemicals |
| Trade name | Silica gel 100 | Alumina 90 neutral | TIONA-G5 anatase | MELCAT XZO-882 |
| BET area (m$^2$/g) | 298 | 103 | 309 | 100 |
| Pore volume (cm$^3$/g)$^b$ | 1.02 | 0.25 | 0.37 | 0.13 |
| Mean pore diameter (Å) | 137 | 98 | 47 | 50 |
| Particle sizes ($\mu$m)$^c$ | 63-200 | 63-200 | n/a | 15$^d$ |

$^a$ Recieved as Zr(OH)$_4 \cdot x$H$_2$O. Decomposed to ZrO$_2$ by calcination (500 °C, 2 h).

$^b$ For mesopores having D$_p$ within 18-500 Å.

$^c$ Data from supplier.

$^d$ Particle size of 15 μm corresponds to the 50 % fractile.

## Continuous-flow hydroformylation

The continuous hydroformylation reactions were studied at differential conversion (≤10 %) at isobaric and isothermal reaction conditions using a microcatalytic flow system with the SILP catalysts forming a fixed-bed in a stainless steel tubular reactor. Reaction gasses; propene (99.4 %), CO (99.997 %), H$_2$ (99.9997 %) and (50:50) vol. % CO/H$_2$ (>99.8 %) (Air Liquide, Denmark) were used as received (**CAUTION!** CO gas is an odorless, highly toxic gas, which should be used with precaution only in presence of a CO gas detector in a ventilated hood). 1-octene (98 %, Aldrich) and decane (>99 %, internal reference) were distilled prior to use. Analyses of products from the reactions were done periodically using FID-GC by comparison with samples of the aldehyde products; n-butanal (99 %), iso-butanal (>98 %) and n-nonanal (97 %) (Aldrich), 2-methyloctanal(30) prepared by literature procedure. The observed reaction rates corresponded to the hydroformylation activities since no by-products were observed. The selectivity of the reactions corresponded to the ratio of formation for linear and branced aldehydes, i.e. n/iso ratio. Additional information about the microcatalytic reactor systems and procedures for the gas-*(19)* and liquid-phase*(18)* reactions, respectively, have been reported elsewhere.

## Gas-phase hydroformylation

### Propene hydroformylation with SiO₂ SILP rhodium-1 catalysts

The catalytic performance of $SiO_2$ SILP rhodium-1 catalysts with various ligand-to-rhodium (L/Rh) ratios and different amounts of [bmim] ionic liquids [bmim][$PF_6$] or [bmim][n-$C_8H_{17}OSO_3$] (ionic liquid loading, $\alpha$, reported as volume ionic liquid/support pore volume) were measured for continuous gas-phase hydroformylation of propene at 10 bar ($C_3H_6$:CO:$H_2$ = 1:1:1) and 100 °C. Representative activities and selectivities measured for catalysts during 4-5 hours on stream are shown in Figure 3 and 4, respectively.

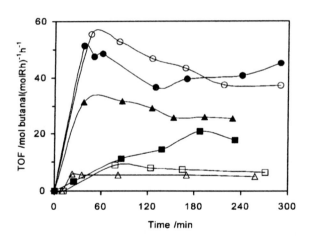

*Figure 3. Propene hydroformylation activity of $SiO_2$ SILP rhodium-1 catalyst with L/Rh ratios of 2.5 (open symbols) and 10 (closed symbols) containing no ionic liquid (circles, $\alpha$ = 0), [bmim][$PF_6$] (triangles, $\alpha$ = 0.5) or [bmim][n-$C_8H_{17}OSO_3$] (squares, $\alpha$ = 0.5).*

The performances of the catalysts were strongly influenced by the ligand content, i.e. L/Rh ratios, and the ionic liquid content, but influenced less by the choice of ionic liquid. However, the catalyst pre-reaction of the rhodium-precursor complex and the ligand **1** giving the catalytically active components, appeared to be slower in [bmim][$n$-$C_8H_{17}OSO_3$] than for the [$PF_6$] ionic liquid, suggesting differences in the solubility of rhodium-precursor and/or **1** in the fluids. In accordance with this, no difference in pre-reaction was apparently observed in the reactions using catalysts having no ionic liquid solvent.

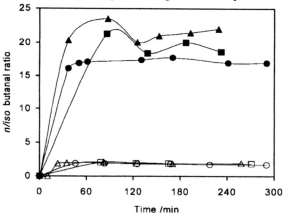

*Figure 4. Propene hydroformylation selectivity of $SiO_2$ SILP rhodium-1 catalyst with L/Rh ratios of 2.5 (open symbols) and 10 (closed symbols) containing no ionic liquid (circles, $\alpha = 0$), [bmim][$PF_6$] (triangles, $\alpha = 0.5$) or [bmim][$n$-$C_8H_{17}OSO_3$] (squares, $\alpha = 0.5$).*

The most active of the examined catalysts after reaction times of 4-5 hours proved to be those without ionic liquid giving activities corresponding to TOFs of 40-45 h$^{-1}$. For the analogous catalysts containing ionic liquid noticeable lower activities were obtained, suggesting that the introduction of ionic liquid in these systems induce phase mass transfer limitation for reactants. Importantly, high selectivity giving *n/iso* ratios $\geq$ 20 was only obtained for reactions using catalyst systems containing ionic liquid as well as high L/Rh ratio, i.e. L/Rh of 10. In contrast, selectivities about 2 were obtained for the analogous catalysts with L/Rh ratios of 2.5. This is different from what are observed for analogous reactions performed in ionic liquid solution alone(*31*), and can therefore be attributed to an effect of the support. Also, for the ionic liquid-free catalyst systems slightly lower selectivities were obtained for catalysts with L/Rh ratio of 10 than for the analogous catalysts having ionic liquid present. Combined, this suggests that the active catalytic components, operating under ionic liquid-free conditions and low ligand content, are mostly ligand-free components possibly immobilized directly on the support. In contrast, the active components operating in the catalysts containing ionic liquid and L/Rh ratio of 10 are probably dissolved [HRh(**1**)(CO)$_2$] complexes, in accordance with previously reported work in ionic liquids*(32)* and other systems*(33)* using the **1** ligand. Hence, the results clearly indicates, that a high ligand content together with the presence of ionic liquid solvents are prerequisites for obtaining selective SiO$_2$ rhodium phosphine SILP catalysts systems.

Importantly, it should be mentioned that despite the fact that steady-state performance of the SiO$_2$ catalysts apparently were obtained within the reported 4-5 hour reaction period, prolonged use resulted in a decrease in catalytic activity and selectivity independent of the type of ionic liquid, ionic liquid content, and applied L/Rh ratio. Thus, it seems that the choice of support might have an unexpected, pronounced influence on the catalyst performance and stability.

## Propene hydroformylation with SILP rhodium-1 catalysts based on alternative supports

In order to examine the effect of the support on the catalytic performance and stability of SILP rhodium-**1** catalysts, we prepared a series of other SILP rhodium-**1**/[bmim][$n$-C$_8$H$_{17}$OSO$_3$] catalysts based on alternative, commercially available high-area oxide supports (Table I). The measured pore distributions for the used supports are shown in Figure 5, while the results obtained from continuous gas-phase hydroformylation of propene using the catalysts systems are shown in Figure 6 (10 bar, C$_3$H$_6$:CO:H$_2$ = 1:1:1, and 100 °C).

The SILP rhodium-**1**/[bmim][$n$-C$_8$H$_{17}$OSO$_3$] catalysts based on the alternative supports generally proved to be less active and selective than the analogous SiO$_2$-based (compared to the initial performance), but more stabile at prolonged reaction. Especially the TiO$_2$-based catalyst remained remarkably stable during the 55 hours of reaction indicating no sign of deactivation. After 55 hours reaction the alternative catalysts gave apparently activities corresponding to TOFs of 3-8 h$^{-1}$ and $n/iso$ ratios of 9-15. For comparison, the analogous SiO$_2$-based catalyst deactivated almost completely within 24 hours of reaction (after 4-5 h of reaction: TOF = 21 h$^{-1}$, $n/iso$ = 20) resulting in a dramatic decrease in activity and selectivity (after 24 h: TOF < 2 h$^{-1}$, $n/iso$ ratio < 10). It this context it should be noted, that the individually catalytic performance of the catalysts are influenced by the difference in ionic liquid content in the catalysts. Thus, it can be expected that presence of a relative large amount of ionic liquid as in the case of the less porous ZrO$_2$ catalyst

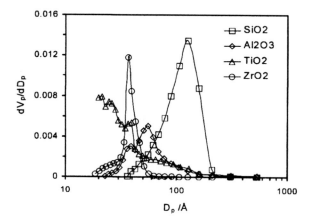

*Figure 5. Pore size distribution of supports.*

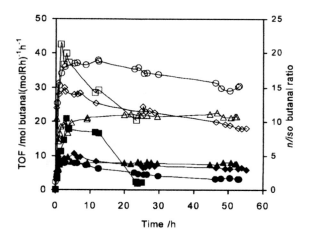

*Figure 6. Propene hydroformylation activities (closed symbols) and selectivities (open symbols) of SILP rhodium-1/[bmim][n-C$_8$H$_{17}$OSO$_3$] catalysts (L/Rh = 10) using different supports: SiO$_2$ (squares, α = 0.5), Al$_2$O$_3$ (diamonds, α = 0.34), TiO$_2$ (triangles, α = 0.27) and ZrO$_2$ (circles, α = 0.67).*

($\alpha = 0.67$) has a negative effect on the activity compared to catalysts containing less ionic liquid.

FT-IR spectra recorded of the catalysts based on the different supports in the period where they were active revealed the presence of CO stretching bands around $v(CO) = 1998$ cm$^{-1}$ and 1940 cm$^{-1}$ corresponding to the presence of catalytically active [HRh(1)(CO)$_2$] complexes (examples are shown in Figure 7 and 8 for TiO$_2$- and SiO$_2$-catalyst, respectively). After catalyst deactivation the CO bands disappeared. Hence, in FT-IR spectra recorded of the TiO$_2$ catalyst prior to and after 55 hours on stream together with pure TiO$_2$ support (Figure 7), the bands are only seen in the spectra, at $v(CO) = 1994$ cm$^{-1}$ and 1936 cm$^{-1}$, after treatment with the reactant gas mixture. In contrast, the CO bands ($v(CO) = 1998$ cm$^{-1}$ and 1940 cm$^{-1}$) are for the SiO$_2$ catalyst only observed on the 4 h spectra (Figure 8), but disappeared after catalyst deactivation (55 h).

No additional bands in the 1700-2100 cm$^{-1}$ region was observed in either of the catalysts excluding, e.g. formation of complexes with bridging CO. Hence, it must be concluded that the deactivation is caused by a support effect. Additional characterization and investigations of the catalysts are in progress in order to understand and explain the observed support effects in more detail.

## Liquid-phase hydroformylation

### 1-octene hydroformylation using SiO$_2$ SILP rhodium-2 catalysts

The applicability of continuous-flow supported ionic liquid-phase hydroformylation was extended from gas-phase reactions to include two-phase, liquid-liquid hydroformylation of 1-octene. The 1-octene substrate was pre-saturation with reactant gas (CO/H$_2$) prior to reaction. This confined the reactions to two phases instead of three, thereby simplifying the process design and avoiding problems that otherwise could be anticipated by operating tri-phasic reactions in continuous mode. However, the pre-saturation unfortunately also limited the reaction efficiency, since the solubility of H$_2$ and CO gasses in 1-octene are low (about 45 mM*(34)* at 14 bar and 115 °C). Anyhow, proff of the concept is demonstrated by the results obtained from continuous-flow two-phase hydroformylation of 1-octene using a SiO$_2$ rhodium-2/[bmim][PF$_6$] catalyst (Figure 9) (14 bar, CO:H$_2$ = 1:1, and 115 °C).

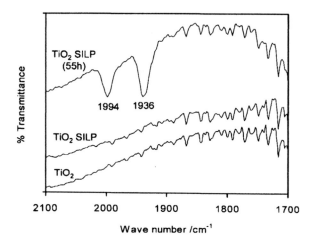

*Figure 7. FT-IR spectra of TiO₂ and TiO₂ SILP rhodium-1/[bmim][n-C₈H₁₇OSO₃] catalyst.*

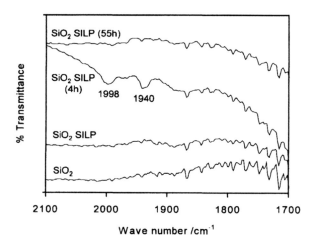

*Figure 8. FT-IR spectra of SiO₂ and SiO₂ SILP rhodium-1/[bmim][n-C₈H₁₇OSO₃] catalyst.*

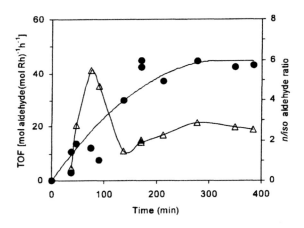

*Figure 9. 1-octene hydroformylation activity (closed circles) and selectivity (open triangles) of SiO₂ SILP rhodium-2/[bmim][PF₆] catalyst (L/Rh = 11.3, α = 0.05). (Reproduced with permission from ref. 18. Copyright 2003 Plenum)*

As expected from the experiments with the catalytically active complexes formed *in situ*, a gradually increase in activity of the catalyst was observed. Since the phase mass transfer of reactant gasses are more pronounced in the liquid system than for analogous gas-phase systems, it is also expected that the pre-reaction is slower in this case than for the gas-phase reactions. After 3-4 hours of reaction an apparent steady-state activity (TOF) of 44 h$^{-1}$ was reached, where after the activity remained unchanged for at least additional 3 hours. During the reaction no indication of deactivation was observed, and analysis (ICP-AES) of outlet samples taken at steady-state conversion demonstrated leaching of rhodium metal to be negligible and below the detection limit. At steady-state conditions a *n/iso* ratio of 2.6 was obtained, although fluctuations were observed initially in the reaction period most likely due to the pre-reaction. This *n/iso* ratio is in the expected range for rhodium catalysts containing monophosphine ligands.

## Summary and conclusion

SILP rhodium catalysts based on different high-area, mesoporous oxide supports containing [bmim] ionic liquids, [bmim][PF$_6$] or [bmim][$n$-C$_8$H$_{17}$OSO$_3$], and dissolved rhodium-1 biphosphine complexes were prepared and examined for fixed-bed, continuous flow propene hydroformylation. Catalytic performances of the catalysts were strongly influenced by the content of ligand and the ionic liquid. Moderate activities were obtained, while a combination of high ligand content and ionic liquid in the catalysts resulted in selective catalyst systems giving *n/iso* ratios > 20. Only minor influence was observed in catalytic performance of catalysts containing the different [bmim] ionic liquids. This indicated that the ionic liquid functioned only as solvent, and demonstrated an catalytic application where a "greener" ionic liquid such as [bmim][$n$-C$_8$H$_{17}$OSO$_3$] can substitute common halogen-containing liquids like, e.g. [bmim][PF$_6$].

In addition, SILP rhodium-1/[bmim][$n$-C$_8$H$_{17}$OSO$_3$] catalysts based on alternative supports, e.g. TiO$_2$, proved to be less active and selective than analogous SiO$_2$-catalysts, but much more stable at prolonged reaction. Thus, it was possible to obtain a TiO$_2$ catalyst system which remained stable for at least 55 hours (end of reaction). FT-IR spectra recorded of the catalysts based on the different supports in the period where they were active, revealed the presence of CO stretching bands corresponding to the presence of catalytic active [HRh(1)(CO)$_2$] complexes. Analogous CO bands were observed with all the supports, indicating that the complex is formed independently of the support. In summary, it can be concluded that the choice of catalyst support has an unexpected large influence on the performance and the stability of the catalyst.

Finally, the concept of continuous-flow two-phase, liquid-liquid supported ionic liquid-phase catalysis was demonstrated for 1-octene hydroformylation using a monophosphine SiO$_2$ rhodium-2/[bmim][PF$_6$] catalyst. However, the applicability of the concept was not fully revealed using the applied process design, since conversion of the olefin was limited by the low solubility of gas reactants in the substrate phase.

## Acknowledgement

The work was supported by the Danish Research Council (Interdisciplinary Research Center for Catalysis) and by the German Federal Ministry for Education and Research via the ConNeCat Lighthouse-Project, *Multiphase Catalysis*. The contributions made by K. Michael Eriksen and Ulla M. Hansen (DTU, Lyngby) in relation to the work are appreciated.

## References

1. Hamilton, D. J. C. *Science* **2003**, *299*, 1702.
2. Zhao, D.; Wu, M.; Kou, Y.; Min, E. *Catal. Today* **2002**, *74*, 157.
3. Olivier-Bourbigou, H.; Magna, L. *J. Mol. Catal. A: Chem.* **2002**, *182-183*, 419.
4. Dupont, J.; de Souza, R. F.; Suarez, P. A. Z. *Chem. Rev.* **2002**, *102*, 3667.
5. *Ionic Liquids in Synthesis;* Editor, Wasserscheid, P.; Welton, VCH-Wiley: New York, NY, 2003.
6. Chauvin, Y.; Mussmann, L.; Olivier, H. *Angew. Chem. Int. Ed. Eng.* **1995**, *34*, 2698.
7. Chauvin, Y.; Olivier, H.; Mussmann, L. EP776880A1, 1997.
8. Favre, F.; Olivier-Bourbigou, H.; Commereuc, D.; Saussine, L. *Chem. Commun.* **2001**, 1360.
9. Brasse, C. C.; Englert, U.; Salzer, A.; Waffenschmidt, H.; Wasserscheid, P. *Organometallics* **2000**, *19*, 3818.
10. Wasserscheid, P.; Waffenschmidt, H.; Machnitzki, P.; Kottsieper, K. W.; Stelzer, O. *Chem. Commun.* **2001**, 451.
11. Brauer, D. J.; Kottsieper, K. W.; Liek, C.; Stelzer, O.; Waffenschmidt, H.; Wasserscheid, P. *J. Organomet. Chem.* **2001**, *630*, 177.
12. Kottsieper, K. W.; Stelzer, O.; Wasserscheid, P. *J. Mol. Catal. A: Chem.* **2001**, *175*, 285.

13. Wasserscheid, P.; van Hal, R.; Bösmann, A. *Green Chem.* **2002**, *4,* 400.
14. Dupont, J.; Silva, S. M.; de Souza, R. F. *Cata. Lett.* **2001**, *77,* 131.
15. Freixa, Z.; van Leeuwen, P. W. N. M. *Dalton Trans.* **2003**, 1890.
16. Mehnert, C. P.; Cook, R. A.; Dispenziere, N. C.; Afeworki, M. *J. Am. Chem. Soc.* **2002**, *124,* 12932.
17. Dupont, J.; Fonseca, G. S.; Umpierre, A. P.; Fichtner, P. F. P.; Teixeira, S. R. *J. Am. Chem. Soc.* **2003**, *124,* 4228.
18. Riisager, A.; Eriksen, K. M.; Wasserscheid, P.; Fehrmann, R. *Catal. Lett.* **2003**, *90,* 149.
19. Riisager, A.; Wasserscheid, P.; van Hal, R.; Fehrmann, R. *J. Catal.* **2003**, *219,* 252.
20. Mehnert, C. P.; Mozeleski, E. J.; Cook, R. A. *Chem. Commun.* **2002**, 3010.
21. Wolfson, A.; Vankelecom, I. F. J.; Jacobs, P. A. *Tetrahedron Lett.* **2003**, *44,* 1195.
22. *Supported ionic liquid catalysis for hydroformylation and hydrogenation reactions;* Mehnert, C. P.; Cook, R. A.; Mozeleski, E. J.; Dispenziere, N. C.; Afeworki, M., 226th ACS National Meeting, New York, 2003.
23. Arhancet, J. P.; Davis, M. E.; Merola, J. S.; Hanson, B. E. Nature **1989**, *339,* 454.
24. Anson, M. S.; Leese, M. P.; Tonks, L.; Williams, J. M. J. *J. Chem. Soc., Dalton Trans.* **1998**, 3529.
25. Bunn, B. B.; Bartik, T.; Bartik, B.; Bebout, W. R.; Glass, T. E.; Hanson, B. E. *J. Mol. Catalysis* **1994**, *94,* 157.
26. Horváth, I. T. *Catal. Lett.* **1990**, *6,* 43.
27. Riisager, A; Eriksen, K. M.; Hjortkjær, J.; Fehrmann, R. *J. Mol. Catal. A: Chem.* **2003**, *193,* 259.
28. Herrmann, W. A.; Kohlpaintner, C. W.; Manetsberger, R. B.; Bahrmann, H.; Kottmann, H. *J. Mol. Catal. A: Chem.* **1995**, *97,* 65.
29. Goedheijt, M. S.; Kamer, P. C. J.; van Leeuwen, P. W. N. M. *J. Mol. Chem. A: Chem.* **1998**, *134,* 243.
30. Nikishin, G. I.; Vinogradov, M. G.; Il'ina, G. P. *J. Org. Chem. USSR* **1972**, *8,* 1422.
31. van Hal, R.; Wasserscheid, P. RWTH-Aachen, Germany, *unpublished.*
32. Silva, S. M.; Bronger, R. P. J.; Freixa, Z.; Dupont, J.; van Leeuwen, P. W. N. M. *New J. Chem.* **2003**, *27* 1294.
33. van Leeuwen, P. W. N. M.; Sandee, A. J.; Reek, J. N. H.; Kamer, P. C. J. *J. Mol. Chem. A: Chem.* **2002**, *182-183,* 107.
34. *Hydrogen and Deuterium;* Editor, Young, C. L.; IUPAC Solubility Data Series; Pergamon Press: New York, 981; Vol. 5/6 and *Carbon Monooxide*; Editor, Cargill, R. W.; IUPAC Solubility Data Series; Pergamon Press: New York, 1990; Vol. 43.

Chapter 24

# Recent Applications of Chloroaluminate Ionic Liquids in Promoting Organic Reactions

## Anil Kumar and Diganta Sarma

### Physical Chemistry Division, National Chemical Laboratory, Pune 411 008, India

Chloroaluminate ionic liquids composed of $AlCl_3$ and organic cations can promote a variety of organic reactions. The composition of these chloroaluminate ionic liquids governs the acidity and basicity of solvent media, in which the reactions are carried out. Useful physico-chemical properties of these ionic liquids and the recent work on their role in accelerating organic reactions are described in the present article.

## Introduction

Considering the urgent need of replacing volatile organic solvents by environmental benign reaction media in the wake of Green Chemistry, the room temperature ionic liquids are emerging as useful media for carrying out several organic reactions. The preference of room temperature ionic liquids over conventional organic solvents in organic synthesis and other areas is established and desribed in several reports.[1-3] The definition and general structural aspects of room temperature ionic liquids and the reasons to prefer them over conventional organic solvents have been discussed in recent literature.[4-6]

In general, ionic liquids are either organic salts or mixtures containing at least one organic component. The most common cations used in ionic liquids are alkylammonium, N-alkylpyridinium, alkylphosphonium and N,N'-

dialkylimidazolium. The nature of the pertinent anion in ionic liquid plays important role in governing its melting point, handling properties, stability and acidic or basic properties. The *N,N'*-dialkylimidazolium ionic liquids have attracted the worldwide attention of researchers, as these ionic liquids can be easily synthesized and remain in the liquid state over a wide range of temperatures.

The enhanced research activities in diversified area have resulted into numerous publications. To cover all these developments in one article is a cumbersome task. We therefore focus on important aspects of an interesting ionic liquid called chloroaluminates, in particular the developments made after 1999. We shall concentrate on the developments pertaining to solution chemistry and the rate enhancement of organic reactions in chloroaluminate ionic liquids. The chapter is organized as follows: The general information on chloroaluminate ionic liquids is given, followed by their physico-chemical properties. Discussion on the synthesis, general and spectroscopic properties of chloroaluminate ionic liquids and their applications to some reactions is avoided here in view of availability of the compiled literature in the form of an excellent review by Tom Welton.[2] Later follows a discussion on recent applications of chloroaluminate ionic liquids to several organic and other pertinent reactions.

## Chloroaluminate Ionic Liquids

Out of several ionic liquids that have been synthesized and subsequently used to promote organic reactions, chloroaluminate ionic liquids are of great significance. Wilkes in a recent article outlined the historical development in the area of ionic liquids.[7] According to the report, the research on chloroaluminate ionic liquids was originated by Lowell A. King in U.S. Air Force Academy in Colorado Springs. These activities have later been supervised by John Wilkes and Richard Carlin. Other significant contributors to the research on chloroaluminate ionic liquids are Robert Osteryoung, Gleb Mamantov, Charles Hussey and others.[8] The details in greater depth are available in Wilkes' article. In our article, we address chloroaluminate ionic liquids, which are constituted of organic species. These chloroaluminates melt below $100^0$ C in contrast to the high melting inorganic chloroaluminates. Chloroaluminate ionic liquids involve mixtures of substituted organic cations with $AlCl_3$ in definite proportions. In general, the substituted organic species of interest are 1-ethyl-3-methylimidazolium chloride [EMIM]Cl (I), 1-methyl-3-ethylimidazolium chloride [MEIM]Cl (II), 1-butyl-3-methylimidazolium chloride [BMIM]Cl (III), 1-butylpyridinium chloride [BP]Cl (IV) and 1-butyl-4-methylpyridinium chloride [BMP]Cl (V). These substituted organic species are shown in Figure 1.

These chloroaluminate ionic liquids are polar in nature. Several organic and inorganic species are easily soluble in these ionic liquids. The chemistry observed in chloroaluminate ionic liquids is strongly affected by their compositions.[9] The chloroaluminate ionic liquids exhibit variable Lewis acidity based upon their

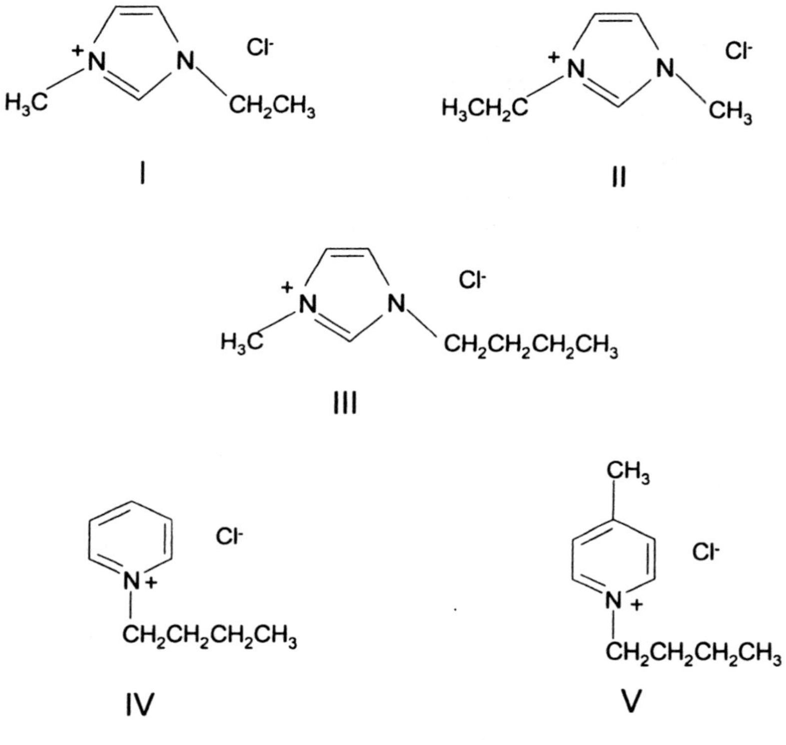

*Figure 1: The substituted organic species in chloroaluminate ionic liquids*

compositions.[10] For example, if the organic species is in excess the ionic liquid offers basic composition. On the other hand, the excess of $AlCl_3$ in the mixture, gives rise to acidic composition. The structural and speciation aspects of chloroaluminates have been aptly reviewed elsewhere.[2] Also described are the spectroscopic evidences in support of the species present in chloroaluminate ionic liquids. The Lewis acidity of chloroaluminate ionic liquids is a special characteristic and hence can be used to alter the rates of organic reactions. The composition of chloroaluminate ionic liquids is expressed in terms of either mole fraction, X or mole % of the component. The chloroaluminate ionic liquids will be acidic if $X_{AlCl3} > 0.5$, basic when $X_{AlCl3} < 0.5$ and neutral at $X_{AlCl3} = 0.5$. The acidic chloroaluminate ionic liquid contains $AlCl_4^-$ and the coordinately unsaturated species, $Al_2Cl_7^-$ species, while the basic ones consist of $AlCl_4^-$ and $Cl^-$ species. Further, Smith et al. have demonstrated that HCl when dissolved in chloroaluminate ionic liquid, [EMIM]Cl-$AlCl_3$ is a Brønsted superacids with Hammett acidity function ranging from - 12.6 (for example, corresponding to [EMIM] : $AlCl_3$ as 0.96 : 1 mol ratio) to -18 (for example, corresponding to [EMIM] : $AlCl_3$ as 0.5 : 1 mol ratio).[11] Later, this fundamental work has been very useful in carrying out rather difficult reactions.[2]

## Physico-chemical Properties

The research group of Osteryoung was the first to synthesize chloroaluminate ionic liquids composed of [BP]Cl - $AlCl_3$.[12,13] Several important studies followed after the work of Osteryoung.[14-18] A useful discussion on the properties of chloroaluminates is given elsewhere.[2, 19-21] The replacement of $Cl^-$ by $Br^-$ in chloroaluminate ionic liquids, though possible has not been followed by further work.[22]

Chloroaluminate ionic liquids have been primarily employed as solvent media for carrying out several processes. Knowledge of the solvent properties in this regard is therefore much desirable. The Gutmann solvent parameters describe the basicity and acidity parameters in terms of donor numbers (DN) and acceptor numbers (AN), respectively. The DN and AN values of the chloroaluminates, composed of $AlCl_3$ with [BMP]Cl and [MEIM]Cl have been determined using the Eu (III) reduction potential and [31]P chemical shift of triethylphosphine oxide. The [BMP]Cl with $X_{AlCl3} > 0.5$ are very poor donor and strong acceptor media. No such information is available in the literature regarding AN properties in the chloroaluminates with $X_{AlCl3} < 0.5$.[23]

The density of chloroaluminate ionic liquids is linearly correlated with the length of the $N'$-alkyl chain on the imidazolium cation. The viscosities of these ionic liquids with $X_{AlCl3} < 0.5$ are reported to increase by a factor of 10 suggesting the formation of hydrogen bonds in between the imidazolium cation and the basic chloride ion.[24]

In order to observe the effect of solvents on solution properties of chloroaluminate ionic liquids, the densities, viscosities and specific conductances of the mixtures of some organic solvents with the [MEIM]Cl - AlCl$_3$ were measured at 298.2 K. These measurements have been extended from basic to acidic environment. In the case of the basic melt ( $X_{AlCl3}$ = 0.44 and 0.49 ), addition of acetonitrile, benzene and dichloromethane caused a significant decrease in viscosity but a large increase in the conductivity of the ionic liquids. However, in the case of the acidic melt ($X_{AlCl3}$ = 0.51, 0.60 and 0.6667), the decrease in the viscosity and increase in the conductivity upon addition of small amount of benzene was relatively small as compared to the basic melt. The effect of the solvents was noted to be in the order of acetonitrile > dichloromethane > benzene.[25, 26] The studies suggest that organic solvents "solvate" the constituent ion of ionic liquids resulting in a decrease in the aggregation of these ions.

The viscosity, density, conductivity, potentiometric titration and cyclic voltammetry have also been reported for [MEIM]Cl-AlCl$_3$.[18, 27]

# Organic Reactions Promoted by Chloroaluminate Ionic Liquids

Several organic reactions have been accelerated in the presence of chloroaluminate ionic liquids. These ionic liquids have played the roles of both of solvent as well as of catalyst. Only recent developments in this regard are described below. The reactions that are already discussed in other reports are ignored until their inclusion in the present write-up was necessary for tracing the origin of such developments. The reactions with protons in the acidic chloroaluminate ionic liquids are excluded from this section, as a comprehensive report is available elsewhere.[2]

### Addition Reactions

*Diels-Alder Reactions*

Diels-Alder reaction **(Scheme 1)** is the most widely employed synthetic method for the production of polycyclic ring systems with excellent control over stereoselectivities. Because of remarkable importance of Diels-Alder reactions in the synthesis of natural products and physiologically active molecules, last two decades have witnessed an upsurge in the research activities aimed at developing new methods to improve yields and stereoselectivities of the desired products.[28] The rates, stereoselectivities and yields of Diels-Alder reactions can be enhanced in water, organic solvents and their salt solutions.[29-31] It was later noted that Diels-Alder reactions were accelerated by Lewis acid catalyst.[32] The first use of ionic liquid in carrying out Diels-Alder reaction was demonstrated by Jaeger and Tucker, who employed a very low melting (m.p. 12°C) ionic liquid called ethylammonium nitrate (EAN) for accelerating the reaction of cyclopentadiene with methyl acrylate.[33] The *endo:exo* ratios for the reaction varied between 5:1 to

9:1 in EAN comparable to in many other organic solvents. In a detailed study later, Lee carried out the same reaction in chloroaluminate ionic liquids and measured the reaction rates and *endo:exo* ratios.[34] Since the composition of chloroaluminate ionic liquids can be easily varied in order to impart Lewis acidity due to the presence of $AlCl_3$, the reaction was carried out with increasing content of $AlCl_3$. The chloroaluminate ionic liquids used in the reaction were constituted of [BP]Cl or [EMIM]Cl and $AlCl_3$ with compositions $X_{AlCl_3}$ varying from 0.48 to 0.51. A dramatic enhancement in *endo:exo* ratios (19:1) in the mixture with $X_{AlCl_3}$ = 0.51 was noted as compared to 4.88:1 obtained in the chloroaluminate mixture with $X_{AlCl_3}$ = 0.48. This enhancement is due to the increase in Lewis acidity of the medium. Similarly, the reaction rate increased by about 24 times in the chloroaluminate with $X_{AlCl_3}$ = 0.51 when compared to in $X_{AlCl_3}$ = 0.48. The rate of reaction in the acidic melt was noted to be 10 times faster than in water and 175 times faster than that observed in EAN. Further, it was noted that rate of the reaction is faster in [EMIM]Cl than in [BP]Cl at the same composition of $AlCl_3$. The reaction reached completion in about 45 min. in $X_{AlCl_3}$ = 0.60 with the [EMIM]Cl chloroaluminate as compared to 4h in organic solvents. Efforts were also made to carry out the reaction of cyclopentadiene with dimethyl maleate with quite similar results.[34]

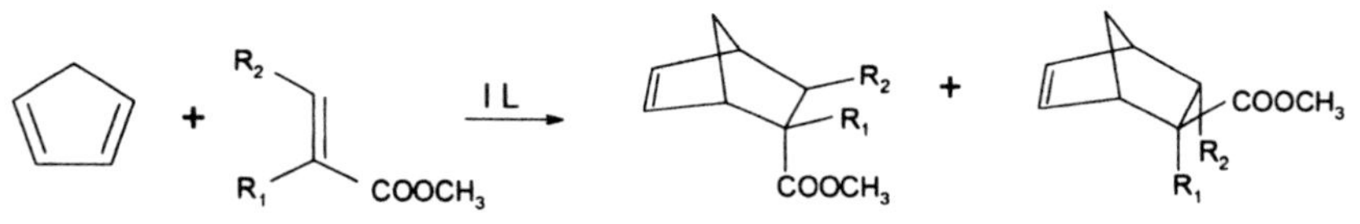

*Scheme 1: Diels-Alder reaction. (Reproduced from reference 36. Copyright 2004 American Chemical Society.)*

The reaction of cyclopentadiene with methyl methacrylate offered interesting results, when carried out in chloroaluminate ionic liquids. This reaction gives higher *exo*-product (lower *endo*-product) in organic solvents.[35] The use of chloroaluminate ionic liquids afforded to convert this *exo*-selective reaction to *endo*-selective one.[36] The *endo:exo* ratio of 0.35:1 for this reaction observed in organic solvents was converted to 3:1 in chloroaluminate ionic liquids with its effective reuse in Diels-Alder reactions. This reversal of the stereoselectivity in Diels-Alder reactions was controlled by the Lewis acid effect caused by the presence of $AlCl_3$. With increasing concentration of $AlCl_3$ the amount of *exo*-

product decreased and higher *endo*-product was obtained. The effect of Lewis acid catalyst on Diels-Alder reaction was further supported by the observation that the basic conditions in the chloroaluminates with $X_{AlCl3}$ = 0.45, 0.47 and 0.49 did not influence the stereoselectivity of the reaction. If the reaction was carried out in the acidic chloroaluminates ($X_{AlCl3}$ > 0.5), the yields increased up to 58% compared to ~10% achieved in organic solvents. A 6-fold increase in the yields for the reaction was observed in the acidic chloroaluminate ionic liquids.

This reaction was completed in about 90 and 75 min. in the [BP]Cl and [EMIM] chloroaluminates, respectively with $X_{AlCl3}$ = 0.60 as compared to in about 10h in organic solvents.

The reaction of cyclopentadiene with methyl-*trans*-crotonate, which offered higher *endo*- product in organic solvents showed a regular increase in *endo*-product with increasing $X_{AlCl3}$ = 0.51 to 0.60. No noticeable difference was observed in *endo*- product obtained in [BP]Cl and [EMIM]Cl with $X_{AlCl3}$ = 0.45 to 0.49. An enhancement in *endo*- product was seen up to 86 and 92% in [BP]Cl and [EMIM]Cl, respectively with $X_{AlCl3}$ = 0.60. It is important to note here that the *endo:exo* ratio reaches to 6:1 in [BP]Cl with $X_{AlCl3}$ = 0.60 as compared to 0.92:1 in [BP]Cl with $X_{AlCl3}$ = 0.45. One can achieve *endo:exo* ratio of 11.5:1 in [EMIM]Cl with $X_{AlCl3}$ = 0.60 in comparison to 0.96:1 in [EMIM]Cl with $X_{AlCl3}$ = 0.45. Similarly, a 7-fold increase in the yields is seen in the chloroaluminate with $X_{AlCl3}$ = 0.60 for the reaction of cyclopentadiene with methyl-*trans*-crotonate when compared to that obtained in 2,2,4-trimethylpentane.

These chloroaluminates for all the above reactions were used six times after their recycle offering nearly the same yields.

Chloroaluminate ionic liquids were found to be ineffective for the reaction of furan with methyl acrylate.[37] The Lewis acidic ionic liquids, like chloroaluminates cannot be employed in the presence of furan and other heteroaromatic diene due to their lack of stability in the presence of common Lewis acids. However, the reversal of an *exo*-selective reaction to an *endo*-selective reaction in the Diels-Alder reaction of furan with methyl acrylate was reported in [BMIM]BF$_4$ and [BMIM]PF$_6$. Several Diels-Alder reactions involving cyclopentadiene and isoprene as dienes and dimethyl maleate, ethyl acrylate, acrylonitrile, etc. as dienophiles have been carried out in neutral ionic liquids such as [BMIM] –trifluoromethanesulfonate, -hexafluorophosphate, - tetrafluoroborate and –lectate.[38] Use of these ionic liquids offered similar *endo:exo* ratios for these Diels-Alder reactions as performed in LiClO$_4$-diethyl ether.[30]

*Dimerization of 1,3 cyclopentadiene*

1,3-cyclopentadiene is frequently used both for the synthetic work and deciphering the mechanistic aspects of cycloaddition. It has a strong tendency to dimerize in order to give dicyclopentadiene during the course of reaction. This dimerization in conventional organic solvents has been studied extensively and understood with regard to its dependence on solubility parameters, viscostities, etc. of solvents.[39] Due to dimerization, less concentration of reactive 1,3-cyclopentadiene becomes available for its reaction with a dienophile.

The dimerization of 1,3-cyclopentadiene was studied in chloroaluminate ionic liquids.[40] The dimerization of 1,3-cyclopentadiene was carried out both in the [BP]Cl and [EMIM]Cl chloroluminates with varying compositions of $AlCl_3$. Higher rates of dimerization of 1,3-cyclopentadiene were noted in the [EMIM]Cl than in [BP]Cl chloroaluminates.

The rate of dimerization of cyclopentadiene increased with the increase in the $AlCl_3$ content. The increase in the rates of dimerization with increasing $AlCl_3$ composition can be ascribed to the Lewis-acid catalysis by $AlCl_3$. The chloroaluminates with $X_{AlCl3} = 0.45$ are basic in nature and hence result into lower $k_2$ values. With increasing $AlCl_3$, the dimerization process is promoted by the Lewis acid catalysis.

The dimerization of 1,3-cyclopentadiene in chloroaluminates was favored by increase in temperature. The energy of activation for the dimerization process in [BP]Cl and [EMIM]Cl are in the range of 7 to 11 kJ mol$^{-1}$.

The kinetics of dimerization of 1,3-cyclopentadiene was also studied for the reactions of 1,3-cyclopentadiene with methyl acrylate, methyl methacrylate and methyl- *trans*- crotonate. No noticeable dimerization was observed for the reaction of 1,3-cyclopentadiene with methyl acrylate. However, the formation of dicyclopentadiene increased for the reactions of 1,3-cyclopentadiene with methyl methacrylate and methyl-*trans*-crotonate with the increase in $AlCl_3$. The dimerization process in [BP]Cl or [EMIM]Cl with $X_{AlCl3} = 0.60$ increased by about 200% as compared to that with $X_{AlCl3} = 0.45$ again showing the effect of Lewis acid catalysis.

## Baylis-Hillman Reactions

Very recently, the Baylis-Hillman reaction (**Scheme 2**) an important carbon-carbon bond-forming process has received remarkable attention of synthetic organic chemists. This reaction is very sluggish in nature and often takes several days to reach its completion with poor yield.[41] In the past, several attempts have been made to accelerate this reaction by using water, salt solutions, different combinations of standard amine catalyst, 1,4-diazabicyclo[2,2,2]octane (DABCO), high pressure, ultrasound method and microwave irradiation.[42]

Very recently, several Baylis-Hillman reactions have been efficiently carried out in chloroaluminates.[43] The reaction of benzaldehyde with methyl acrylate in the presence of DABCO was carried out in chloroaluminates. The reaction in DABCO alone (without chloroaluminates) offered 65% product in 19h. The reaction in [BP]Cl with $X_{AlCl3} = 0.45$ offered 67 % product in 17h, while the same composition with [EMIM]Cl gave a product of 69% in 15h as compared to a yield of 35% in 24 h obtained in $CH_3CN$. The product increased with increasing amounts of $AlCl_3$ in [BP]Cl to reach 76% (in 11.5h; $X_{AlCl3} = 0.57$). A 20-fold rate enhancement was achieved in [BP]Cl with $X_{AlCl3} = 0.60$. A high yield of 80% in 8h is noted in the mixture of [EMIM]Cl with $X_{AlCl3} = 0.60$ with a 23-fold increase in the rates. The [EMIM]Cl - $AlCl_3$ was noted to be more efficient ionic liquid than [BP]Cl - $AlCl_3$, as it offered higher yields in comparatively shorter time.

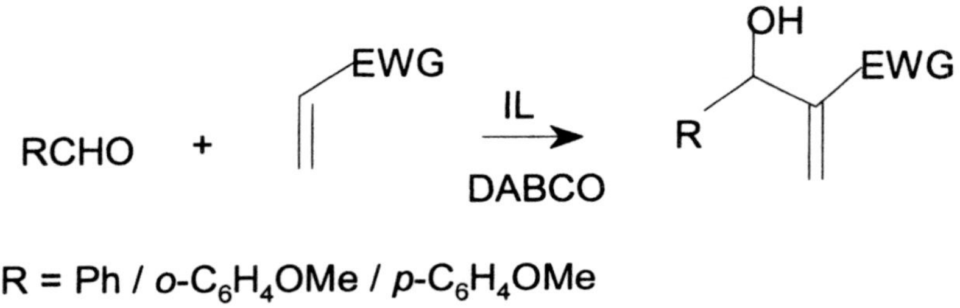

RCHO + (EWG) $\xrightarrow[\text{DABCO}]{\text{IL}}$ R (OH) (EWG)

$$R = Ph \:/\: o\text{-}C_6H_4OMe \:/\: p\text{-}C_6H_4OMe$$

*Scheme 2: Baylis-Hillman reaction. (Reproduced from reference 43 . Copyright 2004 Elsevier B.V.)*

The reaction of benzaldehyde with acrylonitrile, which gave 82% yield in 9h in DABCO alone, became faster by 9-10 times both in the [BP]Cl and [EMIM]Cl with $X_{AlCl3} = 0.60$. Similarly, the reactions of $p$-anisaldehyde and of isobutyraldehyde with ethyl acrylate became 9 and 12 times faster with improved yields in the presence of [BP]Cl and [EMIM]Cl, respectively with $X_{AlCl3} = 0.60$ as compared to in DABCO alone. The reaction of isobutyraldehyde with ethyl acrylate is sluggish giving 17% product in 48h when carried in DABCO. This reaction became faster with increasing $AlCl_3$ to give up to 26 (35h) and 30% (34h) in [BP]Cl and [EMIM]Cl, respectively with $X_{AlCl3} = 0.60$. The rates were enhanced by 8 to 10 times in the [BP]Cl and [EMIM]Cl-chloroaluminates. For the reaction of $o$-anisaldehyde with *tert*-butyl acrylate, a 16% increase in the yield and about 10-fold rate enhancement was achieved in [BP]Cl and [EMIM]Cl with $X_{AlCl3} = 0.60$ as compared to the results obtained in DABCO alone.

The recovered ionic liquid was used six times for the reaction of benzaldehyde with methyl acrylate with 70 to 78% .

The first report on the DABCO-catalysed Baylis-Hillman reaction in an ionic liquid, [BMIM][$PF_6$] was published by Rosa et al.[44] However, it was later shown that these imidazolium salts were deprotonated under mild basic conditions in order to offer reactive nucleophiles. The low yields reported in these ionic liquids resulted from the reaction between aldehyde and the reactive nucleophiles under mild basic conditions.[45] Kim et al. have employed various [BMIM]-based ionic liquids to accelerate Baylis-Hillman reactions.[46] They achieved the maximum rate enhancement for the reaction of benzaldehyde and methyl acrylate with DABCO in [BMIM][PF$_6$]. A moderate acceleration in the reaction rates was observed if the above ionic liquid was used in combination with Lewis acid or H-bond additives. More importantly, the rate-enhancing effects of the ionic liquid and Lewis acid or H-bond donors were not additive. A two-fold rate increase was achieved in the mixtures of [BMIM][PF$_6$], Lanthanum triflate (La (OTf)$_3$) and 2,2',2"-nitrilotris[ethanol]. Kabalka et al. have also employed [BMIM]Br, [BMIM]BF$_4$ and [BMIM]PF$_6$ ionic liquids for the transformation of acetates of Baylis-Hillman adducts into trisubstituted alkenes.[47]

Baylis-Hillman reactions are favored in basic environment. The rate enhancement of Baylis-Hillman reactions due to Lewis acid effect is a puzzling observation consistent with the experimental findings described elsewhere.[42e, 46] No conclusion on the role of Lewis acid in chloroaluminate ionic liquids on the progress of Baylis-Hillman reactions can be drawn based on the available data.

## Prins Cyclization

Tetrahydropyrans are structural features of a variety of biologically active products such as polyether antibiotics, marine toxins and pheromones. Prins cyclization is a powerful method to construct the six-membered tetrahydrpyran derivatives. In general, these derivatives are prepared under acidic environment with long reaction times and with the possibility of formation of side products. The Prins cyclization involves the reaction of homoallyl alcohol with an aldehyde to give tetrahydropyran. Very recently, chloroaluminate ionic liquids have been employed to accelerate Prins cyclization. Symmetric 2,6-disubstituted 4-halotetrahydropyrans were obtained as a result of coupling reaction of aromatic aldehydes with corresponding homoallylic alcohols in chloroaluminate ionic liquids.[48] Similarly, cross coupling reaction between aromatic homoallyl alcohols and aliphatic aldehydes or between aliphatic homoallyl alcohols and aromatic aldehydes offered the corresponding unsymmetrical chloropyrans.

Higher rates (reaction time 5–12 min) and yields (87-95%) by Prins cyclization have been reported in [BMIM]Cl with AlCl$_3$ as compared to in volatile chlorinated hydrocarbons such as chloroform or dichloromethane. An increase in the molar fraction of AlCl$_3$ in chloroaluminates led the rate enhancement indicating the role of Lewis acidity on Prins cyclization. The reaction in chloroaluminate ionic liquid provides easy access to the synthesis of functionalized tetrahydropyrans with diverse chemical structure.

360

*Cyclization of 1-dodecene to cyclododecane*

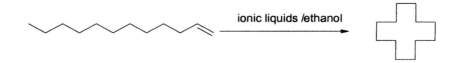

*Scheme 3: Cyclization of 1-dodecene to cyclodecane. (Reproduced from reference 49. Copyright 2003 Elsevier B.V.)*

The reaction of simple alpha olefin to prepare less readily available products is important in chemical industry. The cyclization of 1-dodecene to cyclododecane is one such reaction (**Scheme 3**). However, direct selective cyclization of alpha olefin to cycloalkane was first achieved by Qiao and Deng.[49] The reaction of cyclization of 1-dodecene to cyclododecane was carried out in the ethanol buffered chloroaluminate ionic liquid consisting of [BMIM]Cl with AlCl$_3$ to offer high selectivity under moderate pressure. The combinations of chloroaluminates with either toluene or heptane have been earlier used for this reaction.[50] The rate of conversion increased with the acidity of the chloroaluminates. Since the product is immiscible in the chloroaluminates, the separation of the product was quite easy. The chloroaluminates however, did not promote the cyclization reaction of alkenes with even number carbon atom, like octene, 1-tetradecene, etc.

**Substitution Reactions**

*Friedel-Craft Sulfonylation*

The Friedel-Craft sulfonylation reaction of benzene and substituted benzenes with 4-methylbenzenesulfonyl chloride (**Scheme 4**) has been successfully reported in the [BMIM]Cl based chloroaluminate ionic liquids. [51] In the past the sulfonylation has been carried out by reacting alkyl/arylsulfonylhalides and sulfonic acids with aromatic hydrocarbons in the presence of different acid catalysts.[52] The reaction rates were noted to increase with the increasing AlCl$_3$ content in the chloroaluminates suggesting effective role of Lewis acid catalyst during the reaction. The chloroaluminate-promoted sulfonylation proceeded with enhanced reactivity with quantitative yields of diaryl sulfones under ambient conditions.

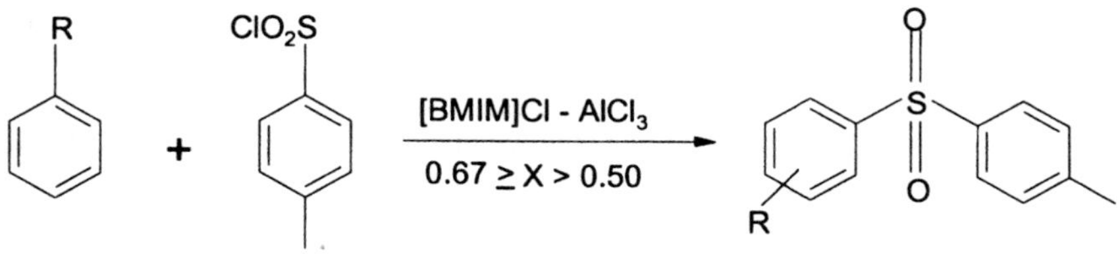

*Scheme 4: Friedel Crafts sulfonylation in chloroaluminates.*
*(Reproduced from reference 51. Copyright 2001 American Chemical Society.)*

The reaction of arenes and isothiocynates in the acidic chloroaluminate ionic liquids offered an effective method to synthesize N-substituted thioamines. The products showed variations with the Lewis acidity of the ionic liquids.

## Friedel- Crafts Acylation

The chloroaluminate ionic liquids have been employed to promote Friedel-Crafts acylation of indoles at normal temperature. The acylation in the [EMIM]Cl-AlCl$_3$ proceeds with high yields to offer 3-substituted indoles. This methodology appears to be much general for less electron-rich indole ring systems.[53]

The reaction of arenes and isothiocyanates in the acidic chloroaluminate ionic liquids offered an effective method to synthesize N-substituted thioamines. The products showed variations with the Lewis acidity and stoichiometry of ionic liquids. A distinct *para* selectivity for the incoming thioamido group on activated arenes was noted under observed conditions.[54]

## Alkylation of Benzene

Alkylation of benzene is an important industrial process, which is achieved by the liquid acid catalyzed alkylation of aromatics. Many conventional processes suffer from disadvantages like formation of aluminate wastage, cumbersome product recovery and non re-use old catalyst. Longer reaction times and low yields are also some of the main discouraging points during conventional alkylation processes.[55]

Very recently, the alkylation of benzene with dodecene and/or chloromethanes have been achieved in chloroaluminate ionic liquids. The chloroaluminate ionic liquids used in the process were [EMIM]Cl, [BP]Cl, [BMIM]Cl and trimethylamine hydrochloride (TMHC). The reactions were noted to be quite fast completing within 5 min. to 8h depending upon the alkylation reagent chosen for the process.[56] These reactions in chloroaluminates offered yields ranging from 38-91%. The alkylation proceeded with more favorable distribution of products and enhanced catalytic activity. The superacidity of the ionic liquids induced by HCl can be attributed to this effect. The reactions could be carried out for five times in chloroaluminate ionic liquids. Another advantage of this work was the easy separation of product with high purity.

## Diaryl Sulfoxides

The sulfoxides and sulfones have been synthesized by organic chemists worldwide because of their varied reactivity as a functional group for transformation into a variety of organo sulfur compounds. These reactions are significant in the synthesis of drugs and sulfur-substituted natural products. Friedel-Crafts sulfonylation of arenes using a catalyst such as $AlCl_3$ or trifluoromethane sulfonic acid is an important method to synthesize diaryl sulfoxides.[57] However, the diaryl sulfoxides have now been synthesized in [BMIM]Cl chloroaluminate with $X_{AlCl_3} = 0.67$ to offer good yields (>90%) in short time (5 min). When the reactions were carried out in the chloroaluminates with varying amounts of $AlCl_3$, the acidic ionic liquid offered higher yields as compared to those in with basic ionic liquid.[58] The studies suggested that the extent of conversion was favored by the Lewis acidity imparted by chloroaluminate ionic liquids. Though, the proposed methodology is effective for synthesizing diaryl sulfoxides, no studies have been done by the authors to check the recyclability of the ionic liquids

## Organometallic Reactions

The acylation of ferrocene with acetic anhydride in the [EMIM]I-AlCl$_3$ chloroaluminate ionic liquid was achieved with monoacetylferrocene as the sole product during the reaction.[59] A large number of arene(cyclopentadienyl) iron(II) complexes have been synthesized in the acidic chloroaluminate ionic liquid.[60] A brief discussion on the use of chloroaluminates in organomettalic chemistry is given by Welton.[2]

## Elimination Reactions

*Ether Cleavage*

Chloroaluminate ionic liquids have also been noted to cleave the aromatic methyl, allyl and benzyl ethers under mild conditions. The chloroaluminate ionic liquids used in the study were trimethylammonium chloride [TMAH][$Al_2Cl_7$], [BMIM][$Al_2Cl_7$] and [EMIM][$Al_2Cl_6I$]. Though comparable results were obtained in all the three ionic liquids, the use of [TMAH][$Al_2Cl_7$] was preferred as this ionic liquid could be prepared in one step with readily available and inexpensive starting materials. The ether cleavage in the presence of [TMAH][$Al_2Cl_7$] using several heterocyclic aromatic compounds bearing a variety of functional groups.[61]

*Synthesis of Coumarin Derivatives*

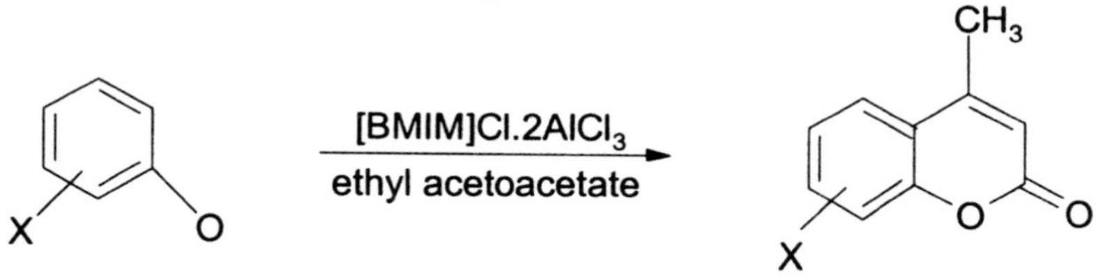

Scheme 5: Coumarin synthesis via Pechhmann condensation in chloroaluminates. (Reproduced from reference 63. Copyright 2002 Elsevier B.V.)

Coumarin derivatives are important compounds in natural and synthetic organic chemistry. The coumarin derivatives were obtained by the Pechmann condensation of phenols with ethyl acetoacetate in chloroaluminate ionic liquids (**Scheme 5**). The reactions reached completion in a very short time and gave high yields. Chloroaluminate ionic liquids played roles of solvent and catalyst in these reactions. [62] The coumarin derivatives were also obtained in good yields *via* the Knovenagel condensation.[63]

*Synthesis of Electrophilic Alkene*

The Knovenagel reaction involves the synthesis of electrophilic olefins from active methylene and carbonyl compounds.[64] This reaction can be accelerated in both homogeneous and heterogeneous catalytic conditions. The Knovenagel condensations of benzaldehyde and substituted benzaldehydes with diethyl malonate were carried out in chloroaluminate ionic liquids to give benzylidene malonates. The chloroaluminates employed for the condensation were [BMIM]Cl and [BP]Cl. The benzylidene malonates underwent Michael additions with diethyl malonate. The Lewis acidity imparted by the chloroaluminates played an important role in the formation of Michael adducts. It was possible to exercise considerable control over various products in these reaction media by variation of the parameters associated with ionic liquids.[63]

*Catalytic Reactions*

Chloroaluminate ionic liquids can solubilize and stabilize the catalytic species in a heterogeneous catalytic reaction. They do not, however react or interact with active catalytic site. These two factors render chloroaluminates as useful solvent systems for such reactions.

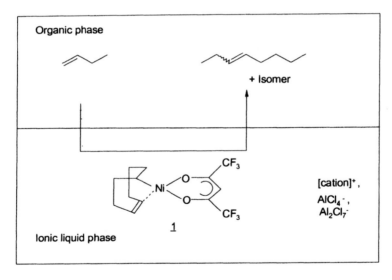

*Figure 2: Biphasic dimerization of 1-butene in chloroaluminates.*
*(Reproduced from reference 65. Copyright 1999 Royal Society of Chemistry.)*

Chauvin et al. were the first to attempt the dimerization of propene to hexenes using nickel (II) complexes as catalyst and chloroaluminates as solvents. The use of 1-ethylaluminium (III) dichloride-based chloroaluminates eliminated the possibility of contamination of product due to a side reaction.[65] The chloroaluminate ionic liquids have been used as catalyst solvent for Ni-complexes for the purpose of biphasic oligomerisation of olefins with significantly enhanced catalytic activity.[66,67] After the reaction in slightly acidic ionic liquid, the ionic catalysts could be separated from the organic products by a simple decantation step allowing complete catalyst recovery. Recently, Wasserscheid and Eichmann employed [EMIM]Cl, [BMIM]Cl and [4-MBP]Cl with AlCl$_3$ with varying acidities using square planar Ni-complexes with $O,O'$-chelating ligands such as (cod)Ni(hfacac) as catalyst (Figure 2).[68] The systems that were prepared by buffering an acidic ionic liquid with weak organic bases were found to be efficient solvents. Pyrrole and $N$-methylpyrrole, when used as bases offered 98% selectivity of dimer. The procedure developed by the authors allowed the reaction to take place in biphasic mode with facile catalyst separation and catalyst recycling. Later, the groups of Wasserscheid and Cavell synthesized nickel(II) heterocyclic carbene complexes of the formula NiI$_2$ (carbene)$_2$ and employed them successfully for the dimerization of 1-propene and 1-butene in choloroaluminate ionic liquids. The catalysts were found to be more active in chloroaluminate ionic liquid as compared to in a conventional organic solvent like toluene. The stabilization of catalysts in chloroaluminate ionic liquids was attributed to be one of the possible reasons for better results obtained in the ionic liquid than in toluene.[69] Wilkes et al. have successfully carried out the regioselective dimerization of propene in chloroaluminate ionic liquids with the cationic π-nickel (II) complexes with phosphine ligands. The nature of ligands was noted to control the regioselectivity of the dimerization reaction.[70] The selective dimerization of ethene has also been achieved in the above ionic liquids with excellent results.[71]

The variations in acidic and basic nature of chloroaluminate ionic liquids offer two interesting results. For example, the polymerization of cyclohexene has been reported in the acidic chloroaluminate ionic liquids by rhodium complexes. On the other hand, if the same reaction is catalysed by rhodium catalyst dissolved in the basic chloroaluminate, cyclohexene undergoes hydrogenation. It is possible to recover the catalyst from the ionic phase.[72] Ethene polymerization has been achieved when AlEtCl$_2$ and TiCl$_4$ are added to the acidic chloroaluminates.[73]

## Miscellaneous Reactions

### Ring Opening Polymerization

Cycilc carbonates of the smallest ring size (five- membered) hardly undergo ring-opening polymerization. The ring-opening polymerization of ethylene carbonate, for example, has been achieved in the presence of metal alkoxides, metal acetylacetonates and metal alkyls in the temperature range of 180–200°C.[74] However, it has been possible to achieve ring-opening polymerization of ethylene carbonate in presence of chloroaluminate ionic liquid [BMIM]Cl-AlCl₃ at about 100°C.[75] The reactions carried out with changing AlCl₃ content did not however favor the rate of polymerization. As the reaction also proceeded smoothly in neutral ionic liquid, it was assumed that the aluminium atom acted as the catalytic site under neutral conditions and the Lewis acidity did not play any role in the ring opening polymerization. The ring-opening polymerization of ethylene carbonate was achieved at lower temperatures in the chloroaluminate than in chlorostannate (with $SnCl_2$).

## Oligomerization of Olefins

Chloroaluminate ionic liquids, [BMIM]Cl have been employed to oligomerize the linear 1-butene, 1-pentene, 1-hexene and 1-octene to offer dimers, trimers, and tetramers. However, addition of titaniumtetrachloride produces branches, atactic polymers with narrow monomodal polydispersities as waxy or oily compounds in high yields.[76]

## Fries Rearrangement

Fries rearrangement of phenyl benzoates is generally achieved in the presence of Lewis acid catalyst. As chloroaluminate ionic liquids impart Lewis acidity, the Fries rearrangement of phenyl benzoates was attempted in the acidic chloroaluminates. The substrates showed significant increase in reactivity with reduced reaction times and substantially improved yields. Fries rearrangement of phenyl benzoates in chloroaluminates followed the first-order kinetics.[77] The reaction procedure in chloroaluminates is simple and avoids cumbersome steps involved during the work-up of the reaction.[78]

Several chloroaluminate ionic liquids have recently been used for protecting and deprotecting of alcohols in the tetrahydropyranylation reactions under microwave conditions.[79]

## Theoretical Development

Though many theoretical investigations have been made to delineate the solvent effects on Diels-Alder reactions, only one study directly addresses the effect of chloroaluminate ionic liquids on their kinetics. In an elegant approach, Acevado and Evanseck have employed the Becke three-parameter density functional theory with the 6-31G(d) basis set in order to compute four-stereospecific Diels-Alder reaction of cyclopentadiene with methyl acrylate in chloroaluminate ionic liquids.[80] The computational model included [EMIM] cation and chloroaluminates e.g. $AlCl_4^-$ and $Al_2Cl_7^-$ in a stacked configuration. Four possible transition structures for the reaction of cyclopentadiene with methyl acrylate were computed. These transition structures were NC (*endo, s-cis* methyl acrylate), XC (*exo, s-cis* methyl acrylate), NT (*endo, s-trans* methyl acrylate) and XT (*exo, s-trans* methyl acrylate) as optimized using the B3LYP/6-31G(d) level of theory. The stationary point determinations were made by using several orientations and possible interactions. The stacked model was noted to approximately explain the experimental data for the reaction of cyclopentadiene with methyl acrylate.

## Acknowledgements

Our research in the area of ionic liquids is sponsored by Department of Science and Technology *via* a grant-in-aid (SR/S1/PC-13/2002). One of us (DS) thanks CSIR, New Delhi for awarding him a Junior Research Fellowship for completing his PhD work. We also thank Suvarna Deshpande for her help in the preparation of this manuscript and to Sanjay Pawar for the laboratory work.

## References

1. (a) Seddon, K. R. *Kinet. Catal.* **1996**, 37, 693. (b) Seddon, K. R. *J. Chem. Tech. Biotechnol.* **1997**, 68, 351. (c) Freemantle, M. *C&E News*, 1998 (March 30) 32.
2. Welton, T. *Chem. Rev.* **1999**, 99, 2071.
3. Wassercheid, P.; Wilhelm, K. *Angew. Chem. Int. Ed.* **2000**, 39, 3772.
4. Sheldon, R. *Chem. Commun.* **2001**, 2399.
5. Gordon, M. *Appl. Catalysis A: General* **2001**, 222, 101.
6. All articles are useful in the monograph: *Green Industrial Applications of Ionic Liquids,* Rogers, R. D.; Seddon, K. R.; Volkov, S. Eds., Kluwer: Dordrecht, The Netherlands, 2003.
7. Wilkes, J. S. *Green Chemistry*, **2002**, 4, 73.
8. The contributions of Bernard Gillbert, Niels Bjerrum, Harald Oye and others are noteworthy in this regard.
9. Osteryoung, R. A. Organic Chloroaluminate Ambient Temperature Molten Salts. In *Molten Salt Chemistry*, Mamantov, G.; Marrasi, R. Eds. Reidel: Dordrecht. The Netherlands, 1987, ASI Series C, Vol. 202, pp. 329.

368

10. (a) Hussey, C. L. *Pure & Appl. Chem.* **1988**, 60, 1763.; (b) Oye, H. A.; Jagtoyen, N.; Oksefjell, T.; Wilkes, J. S. *Mater. Sci. Forum* **1991**, 73 and 183.
11. (a) Smith, G. P.; Dworkin, A. S.; Pagni, R. M.; Zingg, S. P. *J. Am. Chem. Soc.* **1989**, 111, 525; (b) Smith, G. P.; Dworkin, A. S.; Pagni, R. M.; Zingg, S. P. *J. Am. Chem. Soc.* **1989**, 111, 5075.
12. Robinson, J.; Osteryoung, R. A. *J. Am. Chem. Soc.* **1979**, 101, 323.
13. Chum, H. L.; Koch, V. R.; Miller, L. L.; Osteryoung, R. A. *J. Am. Chem. Soc.* **1975**, 97, 3264.
14. Nardi, J. C.; Hussey, C. L.; King, L. A.; US Pat. 4 122 245 1978.
15. Carpio, R. A.; King, L. A.; Lindstrom, R. E.; Nardi, J. C.; Hussey, C. L. *J. Electrochem. Soc.* **1979**, 126, 1644.
16. Hussey, C. L.; King, L. A.; Carpio, R. A. *J. Electrochem. Soc.* **1979**, 126, 1029.
17. Hussey, C. L.; King, L. A.; Wilkes, J. S. *J. Electrochem. Soc. Interfacial Electrochem.* **1979**, 102, 321.
18. Wilkes, J. S.; Levisky, J. A.; Wilson, R. A.; Hussey, C. L. *Inorg. Chem.* **1982**, 21, 1263.
19. Pagni, R. M. Ionic Liquids as Alternatives to Traditional Organic and Inorganic Solvents in *Green Industrial Applications of Ionic Liquids*, Rogers, R. D. ; Seddon, K. R.; Volkov, S. Eds. Kluwer: Dordrecht, The Netherlands, 2003, page 105-127).
20. Carlin, R.T.; Wilkes, J. S. Chemistry and speciation in Room Temperature Chloroaluminate Molten Salts. In *Chemistry of Non-aqueous Solutions: Current Progress*, Mamantov, G. M.; Popov, A. I. Eds. VCH, New York, 1994, pp. 277.
21. (a) Hussey, C. L. *Pure & Appl. Chem.* **1988**, 60, 1763 and refs. cited therein. (b) Hussey, C. L. The Electrochemistry of Room- Temperature Haloaluminate Molten Salts. In *Chemistry of Non-aqueous Solutions*: *Current progress*, Mamantov, G. M.; Popov, A. I. Eds. VCH, New York, 1994, pp. 227-275.
22. Sanders, J. R.; Ward, E. H.; Hussey, C. L. *J. Electrochem. Soc.* **1986**, 133, 325.
23. Zawodzinski, T. A.; Osteryoung, R. A. *Inorg. Chem.* **1989**, 28, 1710.
24. Fennin, A. A.; Floreani, D. A.; King, L. A.; Landers, J. S.; Piersma, B. J.; Stech, D. J.; Vaughn, R. L.; Wilkes, J. S. Williams, L. *J. Phys. Chem.* **1984**, 88, 2614.
25. Perry, R. L.; Jones, K. M.; Scott, W. D.; Liao, Q.; Hussey, C. L. *J. Chem. Eng. Data* **1995**, 40, 615.
26. Liao, Q.; Hussey, C. L. *J. Chem. Eng. Data* **1996**, 41, 1126.
27. Fung, Y. S.; Chau, S. M. *J. Appl. Electrochem.* **1993**, 23, 346.
28. For example see: (a) Lindstrom, U. M. *Chem. Rev.* **2002**, 102, 2751.; (b) Kumar, A. *Chem. Rev.* **2001**, 101, 1.; (c) *Organic Synthesis in Water*, Grieco,

P. A. Ed. Blackie: Glassgow, 1998.; (d) Pindur, U.; Lutz, G.; Otto. C. *Chem. Rev.* **1993**, 93, 741.

29. Rideout, D. C.; Breslow, R. *J. Am. Chem. Soc.* **1980**, 102, 7816.
30. Grieco, P. A.; Nunes, J. J.; Gaul, M. D. *J. Am. Chem. Soc.* **1990**, 112, 4595.
31. (a) Pawar, S. S.; Phalgune, U. D.; Kumar, A. *J. Org. Chem.* **1999**, 64, 7055.; (b) Kumar, A.; Pawar, S. S. *J. Org. Chem.* **2001**, 66, 7646.
32. Forman, M. A.; Dailey, W. P. *J. Am. Chem. Soc.* **1991**, 113, 2761.
33. Jaeger, D. A.; Tucker, C. E. *Tetrahedron Lett.* **1989**, 30, 1785.
34. Lee, C. W. *Tetrahedron Lett.* **1999**, 40, 2461.
35. Berson, J. A.; Hamlet, Z.; Mueller, W. A. *J. Am. Chem. Soc.* **1962**, 84, 297.
36. Kumar, A.; Pawar, S. S. *J. Org. Chem.* **2004**, 69, 1419.
37. Hemeon, I.; DeAmicis, C.; Jenkins, H.; Scammells, P.; Singer, R. D. *Synlett.* **2002**, 1815.
38. Earle, M. J.; McCormac, P. B.; Seddon, K. R. *Green Chem.* **1999**, 1, 23.
39. for example see: Swiss, K. A.; Firestone, R. A. *J. Phys. Chem. A.* **1999**, 103, 5369.
40. Kumar, A.; Pawar, S. S. *J. Mol. Cat. A.* **2004**, 208, 33.
41. (a) Baylis, A. B. and Hillman, M. E. D. German Patent 2155113 (1972); *Chem. Abstr.* **1972**, 77, 34174q; (b) for excellent review, Basavaiah, D.; Rao, A. J.; Satyanarayana, T. *Chem. Rev.* **2003**, 103, 811.
42. for example see: (a) Ameer, F.; Drewes, S. E.; Freese, S.; Kaye, P. T. *Synth. Commun.* **1988**, 18, 495.; (b) Roos, G. H. P.; Ramprasadh, P. *Synth. Commun.* **1993**, 21, 1261.; (c) Auge, J.; Lubin, N.; and Lubineau, A. *Tetrahedron Lett.* **1994**, 35, 7947.; (d) Aggarwal, V. K.; Mereu, A. *Chem. Commun.* **1999**, 2311.; (e) Kawamura, M.; Kobayashi, S. *Tetrahedron Lett.* **1999**, 40, 1539.; (f) Aggarwal, V. K.; Dean, D. K.; Mereu, A.; Williams, R. *J. Org. Chem.* **2002**, 67, 510.; (g) Kumar, A.; Pawar, S. S. *Tetrahedron* **2003**, 59, 5019.
43. Kumar, A.; Pawar, S. S. *J. Mol. Cat. A Chem.* **2004**, 211, 43.
44. Rosa, J. N.; Afonso, C. A. M.; Santos, A. G.; *Tetrahedron* **2000**, 57, 4189.
45. Aggarwal, V. K.; Emme, I.; Mereu, A. *Chem. Commun.* **2002**, 1621.
46. Kim, J.; Ko, S. Y.; Song, C. E. *Helv. Chim. Acta* **2003**, 86, 894.
47. Kabalka, G. W.; Venkataiah, B.; Dong, G. *Tetrahedron Lett.* **2003**, 44, 4473.
48. Yadav, J. S.; Reddy, B. S.; Reddy, M. S.; Niranjan, N.; Prasad, A. R. *Eur. J. Org. Chem.* **2003**, 1779.
49. Qiao, K.; Deng, Y. *Tetrahedron Lett.* **2003**, 44, 2191.
50. Ellis, B.; Keim, W.; Wasserscheid, P. *Chem. Commun.* **1999**, 337.
51. Nara, S. J.; Harjani, J. R.; Salunkhe, M. M. *J. Org. Chem.* **2001**, 66, 8616.
52. for example: Olah, G. A. *Friedel-Craft and Related Reactions*; Interscience: New York, 1964: Vol. III. pp. 1319 and refs. therein.
53. Yeung, K. S.; Farkas, M. E.; Que, Z.; Yang, Z. *Tetrahedron Lett.* **2002**, 43, 5793.
54. Naik, P. U.; Nara, S. J.; Harjani, J. R.; Salunkhe, M. M. *Can. J. Chem.* **2003**, 81, 1057.

370

55. Stevens, R.V. *Organic Synthesis*, Vol. 55, Wiley, New York, 1976, page 7.
56. Qiao, K.; Deng, Y. *J. Mol. Cat. A Chem.* **2001**, 171, 81.
57. Olah, G. A.; Nishimura, J. *J. Org. Chem.* **1974**, 39, 1203.
58. Mohite, S. S.; Potdar, M. K.; Salunkhe, M. M. *Tetrahedron Lett.* **2003**, 44, 1255.
59. Surette, J. K. D.; Green, L.; Singer, R. D. *J. Chem. Soc. Chem. Commun.* **1996**, 2753.
60. Dyson, P. J.; Grossel, M. C.; Srinivasan, N.; Vine, T.; Welton, T.; Williams, D. J.; White, A. J. P.; Zigras, T. *J. Chem. Soc. Dalton Trans.* **1997**, 3465.
61. Kemperman, G. J.; Roeters, T. A.; Hilberink, P. W. *Eur. J. Org. Chem.* **2003**, 1681.
62. Potdar, M. K.; Mohile, S. S.; Salunkhe, M. M. *Tetrahedron Lett.* **2001**, 42, 9285.
63. Harjani, J. R., Nara, S. J., Salunkhe, M. M. *Tetrahedron Lett.* **2002**, 43, 1127.
64. Jones, G. *Organic Reactions*; Wiley: New York, 1967, vol. XV, p.204.
65. Chauvin, Y.; Gilbert, I.; Guibard. *J. Chem. Soc. Chem. Commun.* **1990**, 1715.
66. Chauvin, Y.; Einloft, H.; Olivier, H. *Ind. Eng. Chem. Res.* **1995**, 34, 1149.
67. Chauvin, Y.; Olivier, H.; Wyrvalski, C.; Simon, L.; de Souza, E. *J. Cata.* **1997**, 165, 275.
68. Wasserscheid, P.; Eichmann, M. *Catalysis Today*, **2001**, 309.
69. McGuinness, D. S.; Mueller, W.; Wasserscheid, P.; Cavell, K. J.; Skelton, B. W.; White, A. H.; Englert, U. *Organometallics*, **2002**, 21, 175.
70. Wilke, G.; Bogdanovic, B.; Hardt, P.; Heimbach, O.; Kroner, W.; Oberkirck, W.; Tanaka, K.; Steinrucke, E.; Walter, D.; Aimmerman, H. *Angew. Chem. Int. Ed. Engl.* **1966**, 5, 151
71. Einloft, S.; Dietrich, F. K.; de Souza, R. F.; Dupont, J. *Polyhedron*, **1996**, 15, 3257.
72. Suarez, P. A. Z.; Dullius, J. E. L.; Einloft, S.; de Souza, R. F. Dupont, J. *Polyhedron*, **1996**, 15, 1217.
73. Carlin, R. T.; Osteryoung, R. A.; Wilkes, J. S.; Rovang, J. *Inorganic Chem.* **1990**, 29, 3003.
74. Rokicki, G. *Prog. Polym. Sci.* **2000**, 25, 259.
75. Kadokawa, J.; Iwasaki, Y.; Tagaya, H. *Macromol. Rapid Commun.* **2002**, 23, 757.
76. Stenzel, O.; Brull, R.; Wahner, U. M.; Sanderson, R. D.; Raubenheimer, H. G. *J. Mol. Cat. A Chem.* **2003**, 192, 217.
77. Harjani, J. R.; Nara, S. J.; Salunkhe, M. M. *Tetrahedron Lett.* **2001**, 42, 1979.
78. Baltzky, R.; Ide, W. S.; Phillips, A. P. *J. Am. Chem. Soc.* **1955**, 77, 2522.
79. Namboodiri, V.; Varma, R. S. *Chem. Commun.* **2002**, 4342.
80. Acevedo, O.; Evanseck, J. D. Transition Structure Models of Organic Reactions in Chloroaluminate Ionic Liquids, in *Ionic Liquids as Green Solvents*, Rogers, R. D. ed.; ACS Symposium Series, Vol. 856, 2003, 174.

# Indices

# Author Index

# Subject Index

## A

Acetonitrile
  enantioselective resolution of *N*-acetyl amino acids, 121, 122*t*
  *See also* Amino acids
Acid-base systems. *See* Brønsted acid-base systems
Acid scavenging
  BASIL™, 129–131
  methylimidazole, 129, 131
  suspensions, 127–128
Actinide separation, ionic liquids as alternative solvents, 36–37
Activation enthalpy and entropy. *See* Nucleophilic substitution reactions
Addition reactions
  Baylis–Hillman, 357–359
  chloroaluminate ionic liquids, 354–360
  cyclization of 1-dodecene to cyclododecane, 360
  Diels–Alder, 354–356
  dimerization of 1,3-cyclopentadiene, 356–357
  Prins cyclization, 359
Adsorption technology, deep desulfurization, 86
Alkanolamines
  carbon dioxide scrubbing, 50*f*
  removal of acid gases, 50
Alkenes, hydrogenation in ionic liquids, 324–327
Alkoxyphenylphosphines
  BASIL™, 129, 131
  synthesis, 128
Alkyl chain length, strontium extraction using crown ether, 5, 7, 8*f*

1-(*N*-Alkylpyridin-3-yl)butane-1,3-dione hexafluorophosphates, ionic compounds, 29, 30*f*
1-Alkyl substituted-4-amino-1,2,4-triazolium salts
  cation-anion contacts of 1-ethyl-4-amino-1,2,4-triazolium bromide (II), 277, 282*f*, 283
  cation-anion contacts of 1-isopropyl-4-amino-1,2,4-triazolium bromide (IV) synthesis, 285, 287*f*
  cation-anion contacts of 1-isopropyl-4-amino-1,2,4-triazolium nitrate (XVIII), 292, 294*f*
  cation-anion contacts of 1-methylcyclopropyl-4-amino-1,2,4-triazolium nitrate (XX), 295, 297*f*, 298
  cation-anion contacts of 1-*n*-heptyl-4-amino-1,2,4-triazolium bromide (IX), 290, 292, 293*f*
  cation-anion contacts of 1-*n*-hexyl-4-amino-1,2,4-triazolium bromide (VIII), 285, 289*f*, 290
  cation-anion contacts of 1-*n*-propyl-4-amino-1,2,4-triazolium bromide (III), 283, 286*f*
  crystal data and details of structure determinations, 278–281
  crystal structure of 1-ethyl-4-amino-1,2,4-triazolium bromide (II), 275, 277, 282*f*
  crystal structure of 1-isopropyl-4-amino-1,2,4-triazolium bromide (IV), 285, 286*f*
  crystal structure of 1-isopropyl-4-amino-1,2,4-triazolium nitrate (XVIII), 292, 293*f*